Core-Shell Nanomaterials

Core-Shell Nanomaterials

From Fundamentals to Applications

Shreya Sharma, Peeyush Phogat, Jahanvi Thakur, and N. L. Singh

WILEY VCH

Authors

Shreya Sharma
CSIR-National Institute of Science Communication and Policy Research
H.No-696, Sector-23, HUDA
Bhiwani, Haryana 127021, India

Peeyush Phogat
CSIR-National Institute of Science Communication and Policy Research
H.No-2176, Sector-13, HUDA
Bhiwani, Haryana 127021, India

Jahanvi Thakur
Netaji Subhas University of Technology
Dwarka, Sector-3, New Delhi 110078, India

N. L. Singh
Delhi Skill Entrepreneurship University
Dwarka, Sector-9, New Delhi 110077, India

Cover Design: Wiley

Cover Images: © Peeyush Phogat, © MZinchenko/Shutterstock, © theasis/Getty Images

Typesetting: Lumina Datamatics, Madurai, India
Printing: CPI Group (UK) Ltd, Croydon CR0 4YY

Library of Congress Card No.: applied for

British Library Cataloguing-in-Publication Data
A catalogue record for this book is available from the British Library.

Bibliographic information published by the Deutsche Nationalbibliothek The Deutsche Nationalbibliothek lists this publication in the Deutsche Nationalbibliografie; detailed bibliographic data are available on the Internet at <http://dnb.d-nb.de>.

Hardback ISBN 9783527356324
ePDF ISBN 9783527855759
ePub ISBN 9783527855742
oBook ISBN 9783527855766

C9783527356324_111225

ॐ नमः पार्वती पतये, हर-हर महादेवः

कर्मण्येवाधिकारस्ते मा फलेषु कदाचन।

मा कर्मफलहेतुर्भूर्मा ते सङ्गोऽस्त्वकर्मणि॥

To our beloved parents and loved ones—Thank you for your unwavering support, encouragement, and the sacrifices you made that allowed us to pursue our passions. This work would not have been possible without your constant belief in us.

We also dedicate this book to all present and future seekers of knowledge, whose curiosity, perseverance, and spirit of discovery continue to shape a better world.

> "The strong should aid and protect the weak. Then, the weak will become strong, and they in turn will aid and protect those weaker than them. That is the law of nature."
>
> *—Tanjiro Kamado, Demon Slayer*

May this work inspire you to seek boldly, serve selflessly, and always—set your heart ablaze.

Contents

About the Authors

Dr. Shreya Sharma is a physicist who earned her PhD in Physics from Netaji Subhas University of Technology, where her research was centered on the exploration of exceptional properties exhibited by nanomaterials and their potential contributions to advancing renewable energy technologies. Her primary focus lies in investigating the unique properties of 2D materials and their electrochemical study, specifically transition metal dichalcogenides, with the aim of applying these findings to enhance solar cells, supercapacitors, photodetectors, hydrogen production, and electrochemical sensing. Her dedication to leveraging advanced materials for sustainable energy solutions reflects her vision of contributing to societal progress through innovative technology. Currently, she is working to identify sustainable energy resources, with a specific focus on blue energy foresight. She is also the author and co-editor of 6 scientific books and more than 40 research papers in esteemed journals. Driven by a deep commitment to environmental sustainability, Dr. Sharma's work embodies her passion for scientific advancement in service to future energy needs.

Dr. Peeyush Phogat completed his PhD in the Department of Physics at Netaji Subhas University of Technology. His research emphasizes the synthesis and characterization of advanced materials for energy applications, focusing on innovations in energy storage, photodetection, and sustainability. A prolific researcher, Peeyush has authored 36 research papers in reputed journals, contributing to advancements in electrochemical applications and renewable energy materials. He has also published many review articles providing in-depth insights into emerging trends in material science and energy technologies. In addition, Peeyush has contributed 18 book chapters across diverse domains in material science and co-authored three books: *Next-Generation Electrochemical Devices: From Principles to Applications, The Future of Energy: How Innovative Devices Shape Our Environment*, and *AI-Driven Innovations in Electrochemical Technologies for Sustainable Energy Solutions*. Driven by a commitment to scientific progress, Peeyush aims to advance sustainable energy solutions through material science innovations and interdisciplinary research.

Ms. Jahanvi Thakur, a Physics graduate from University of Delhi has completed her MSc in Physics at Netaji Subhas University of Technology where she did research on nanomaterials for renewable energy, particularly 2D transition metal dichalcogenides. Currently, she is pursuing PhD in Physics. Her focus lies on the research for highly efficient solar cells, supercapacitors, and photodetectors, driven by a commitment to sustainable energy solutions. She has co-authored several publications, including, journal papers, conference proceedings and book chapter.

Dr. N. L. Singh is the current Dean of Sciences at Delhi Skill and Entrepreneurship University (DSEU). He has an extensive academic and research background, specializing in nuclear research and material science, particularly in the field of energy devices. With a distinguished career in teaching and academic administration, Dr. Singh has contributed significantly to curriculum development, faculty mentorship, and pioneering research advancements in his domain. He has authored numerous research papers in reputed international journals, actively participated in national and international conferences, and led several funded research projects. His leadership at DSEU is directed toward promoting academic excellence, strengthening industry-academia collaborations, and advancing skill-based, application-oriented education in scientific disciplines.

Preface

Nanotechnology has revolutionized the landscape of material science, particularly through the development of tailored nanostructures with enhanced physicochemical properties. Among these, **core-shell nanomaterials** have emerged as a versatile and powerful class of materials, offering multifunctionality, controlled morphology, and tunable interfaces. These unique attributes have unlocked unprecedented opportunities across domains such as energy storage, catalysis, environmental remediation, biomedicine, and optoelectronics.

This book, *Core-Shell Nanomaterials: From Fundamentals to Applications*, represents a concerted effort to provide a comprehensive and forward-looking examination of the synthesis strategies, structural diversity, characterization methodologies, and cutting-edge applications of core-shell systems. Our aim is to bridge the gap between fundamental design principles and advanced technological applications by integrating insights from experimental studies, theoretical modeling, and recent innovations.

The volume is structured to guide readers from the foundational understanding of core-shell architectures to their deployment in advanced devices and sustainable technologies. Key chapters cover the synthesis of metal and metal oxide-based core-shell nanostructures, carbonaceous and polymeric shells, heterojunction designs, electrochemical and photocatalytic functionalities, and their role in next-generation energy and environmental solutions. A critical emphasis is placed on the synergy between core and shell, which dictates performance, stability, and interfacial dynamics.

We believe that this book will serve as a valuable reference for graduate students, researchers, and professionals in materials science, nanotechnology, chemistry, and engineering disciplines. By drawing from both academic research and emerging industrial trends, the book offers readers a multidimensional perspective on the future trajectory of core-shell nanomaterials.

We extend our sincere gratitude to our academic mentors, research collaborators, and institutional support systems for their continuous encouragement. We also acknowledge the vibrant global community of researchers whose pioneering contributions continue to shape this exciting field. It is our hope that this volume will inspire further innovation and cross-disciplinary collaborations aimed at addressing pressing global challenges through material design at the nanoscale.

Dr. Shreya Sharma
Dr. Peeyush Phogat
Ms. Jahanvi Thakur
Prof. N.L. Singh

Acknowledgements

We extend our heartfelt gratitude to **Prof. Ranjana Jha**, Vice Chancellor, IGDTUW, for her inspiring leadership, guidance, and continued encouragement throughout our academic journey. We are deeply thankful to **Prof. Sukhvir Singh** of DSEU for their invaluable insights, mentorship, and unwavering support at every step of this endeavor.

Our sincere appreciation goes to the **Wiley publishing team** for recognizing the value of this work and providing us with the opportunity to share our vision with the global research community. Your professionalism and support have made this project a reality.

We also thank our **research team**, whose dedication, resilience, and collaborative spirit have been a constant source of strength and motivation. Your contributions, both seen and unseen, have helped shape the foundation of this book.

Lastly, to our families and loved ones—thank you for your endless patience, sacrifices, and unconditional love. This work is a reflection of all the support you've given us.

— Dr. Shreya Sharma, Dr. Peeyush Phogat,
Ms. Jahanvi Thakur and Prof. N. L. Singh

Disclosure of AI Use

In accordance with publishing guidelines and ethical authorship practices, the authors of this book affirm the following regarding the use of Artificial Intelligence (AI) tools in the preparation of this manuscript:

1. AI tools (e.g., OpenAI's ChatGPT) were utilized under human supervision to assist in:
   - Structuring content outlines and chapter flow
   - Drafting preliminary versions of technical explanations
   - Language polishing and consistency checking
   - Generating illustrative figure captions and table summaries All AI-generated content was **thoroughly reviewed, edited, and verified** by the authors to ensure **accuracy, originality, and contextual relevance**.
2. All final content decisions, interpretations of scientific data, and conceptual frameworks presented in this book were **formulated, validated, and approved by the human authors**. AI was **not used for ideation, conclusion formulation, or original research interpretation**. The authors take **full responsibility for the integrity and reliability** of the content.
3. All figures, tables, and illustrations included in this book are either:
   - Created by the authors using professional scientific software
   - Derived from publicly available and cited data
   - Commissioned via professional graphic design based on author-provided input No standalone generative AI image tools were used without human modification or verification.
4. No personal data, confidential research material, or copyrighted content was used in AI prompts or inputs. All data complies with **academic integrity standards, open-access policies, and third-party rights**.
5. This declaration is made to ensure transparency for readers, editors, and peer reviewers. The authors affirm that the use of AI did not replace critical thinking, domain expertise, or the ethical responsibilities of authorship.

1

Introduction to Core-shell Nanomaterials

Core-shell nanomaterials have emerged as a transformative class of materials, offering tailored functionalities by combining two or more distinct components into a single nanoscale structure. These materials consist of a core, which serves as the central framework, and a shell that encapsulates it, modifying and enhancing its physical, chemical, and functional properties. The precise engineering of core-shell structures allows for unprecedented control over stability, reactivity, and selectivity, making them indispensable in a wide range of applications spanning energy, environment, and medicine. The concept of core-shell architectures has its roots in classical colloidal chemistry, where early studies focused on coating particles with thin layers to alter surface properties [1]. Over time, advances in synthesis techniques and nanofabrication methods have enabled the design of highly sophisticated core-shell structures with tailored compositions, morphologies, and functionalities. The ability to integrate multiple materials into a single entity has given rise to novel phenomena, including synergistic electronic interactions, enhanced mechanical strength, and controlled surface chemistry, leading to superior performance compared to their single-component counterparts.

In modern research and industrial applications, core-shell nanomaterials play a pivotal role in addressing global challenges. In the energy sector, they are revolutionizing battery technology, fuel cells, and photocatalytic energy conversion by enhancing charge transport and stability. In environmental science, they contribute to pollution remediation, water purification, and catalytic processes aimed at sustainable development. In medicine, core-shell structures are redefining drug delivery, biosensing, and imaging by providing controlled release mechanisms and biocompatibility. Their versatility extends to emerging fields such as flexible electronics, wearable sensors, and next-generation computing, where their unique properties enable groundbreaking innovations. The continuous evolution of core-shell nanomaterials reflects the dynamic progress in materials science and nanotechnology. With ongoing advancements in synthesis techniques, characterization tools, and computational modeling, the potential for developing new architectures with unprecedented functionalities continues to expand. Understanding the fundamental

Core-Shell Nanomaterials: From Fundamentals to Applications, First Edition. Shreya Sharma et al.

principles governing these structures is essential for unlocking their full potential and harnessing their benefits across diverse scientific and technological domains.

1.1 Basic Concept of Core-shell Structures

Core-shell nanomaterials are defined as composite structures where one material (the core) is encapsulated by another (the shell), forming a hierarchical architecture with distinct physicochemical properties. The core can be metallic, ceramic, polymeric, or carbon-based, while the shell can be composed of materials such as oxides, polymers, metals, or even organic molecules. This structural arrangement allows for precise control over surface chemistry, reactivity, and stability, leading to tailored properties that cannot be achieved with homogeneous nanomaterials. Core-shell nanomaterials represent a fundamental advancement in the field of nanoscience, providing an innovative means of tailoring material properties at the nanoscale. The ability to selectively engineer the core and shell with different compositions, morphologies, and interfacial properties has led to a broad spectrum of applications across scientific and industrial domains. The importance of core-shell nanomaterials stems from their ability to combine the advantages of both core and shell materials while mitigating their individual limitations. The shell can act as a protective barrier, preventing core degradation, oxidation, or unwanted interactions with the external environment [2]. Conversely, it can also enhance the surface properties of the core, improving its solubility, biocompatibility, or catalytic efficiency. Additionally, core-shell structures enable controlled release mechanisms in drug delivery, selective adsorption in environmental remediation, and enhanced charge transfer in energy conversion technologies. Their ability to be fine-tuned for specific functions makes them indispensable in scientific research and technological advancements.

Compared to single-component nanomaterials, core-shell structures offer several critical advantages that make them superior in functionality and applicability. The most significant benefits include enhanced stability, tunable properties, multifunctionality, and improved performance across various applications [3]. One of the primary challenges with conventional nanomaterials is their inherent instability, particularly in reactive environments. Metal nanoparticles, for instance, often suffer from oxidation, aggregation, or dissolution, which severely limits their practical use. The shell in core-shell nanomaterials serves as a protective layer, shielding the core from environmental degradation and improving its long-term stability. This feature is particularly important in catalysis, where nanoparticles need to maintain their structural integrity under harsh reaction conditions. Another crucial advantage is the tunability of physical and chemical properties. By modifying the shell composition, thickness, or structure, researchers can control surface interactions, optical absorption, and electronic conductivity [4]. This tunability is highly beneficial in optical and electronic applications, where the core-shell architecture enables fine control over plasmonic resonance, bandgap engineering, and charge transport mechanisms. For example, in quantum dot–based solar cells, core-shell designs improve light absorption efficiency while minimizing energy loss through recombination.

Core-shell nanomaterials also offer multifunctionality, allowing for the integration of complementary properties within a single material. This synergy is particularly evident in biomedical applications, where a magnetic core can be combined with a biocompatible shell for targeted drug delivery and imaging. Similarly, catalytic nanoparticles can be encapsulated within porous shells to enhance selectivity and reaction rates while minimizing unwanted side reactions. This multifunctional capability extends their utility in diverse fields, ranging from biosensing to environmental remediation. Additionally, core-shell structures enable enhanced dispersion and processability compared to single-component nanomaterials. Many nanoparticles tend to aggregate due to high surface energy, limiting their uniformity and application potential [5]. The presence of a stabilizing shell improves colloidal stability, ensuring better dispersion in solvents, polymers, and other matrices. This property is particularly advantageous in nanocomposite manufacturing and coating technologies, where uniform distribution of nanomaterials is essential for performance optimization.

1.2 Historical Development of Core-shell Materials

The concept of core-shell materials has evolved significantly over time, from early observations in colloidal chemistry to sophisticated nanoscale engineering in modern materials science. The development of core-shell structures has been driven by the need to manipulate material properties at the microscopic and nanoscopic levels, enabling advancements in catalysis, optics, energy storage, and biomedicine. The history of core-shell materials is intertwined with progress in surface chemistry, thin-film deposition, and nanofabrication techniques, each contributing to the precision and functionality of these hybrid architectures. The continuous refinement of synthesis methods, along with the growing understanding of interfacial interactions, has led to the emergence of highly customized core-shell systems designed for specific applications [2].

The foundation of core-shell materials can be traced back to the early studies in colloidal chemistry during the late 19th and early 20th centuries. Researchers investigating colloidal suspensions observed the formation of layered structures where one material surrounded another, often due to differences in surface energy and chemical affinity. These early discoveries laid the groundwork for the controlled synthesis of heterogeneous particles, where distinct compositions could be maintained in a single entity. One of the earliest practical demonstrations of core-shell structures was in the form of gold-coated colloidal particles, which were explored for their optical properties in stained glass and early colorimetric assays. The phenomenon of surface plasmon resonance, observed in these materials, highlighted the impact of shell layers on the optical behavior of metallic cores [6]. These findings spurred further investigations into how core-shell structures could be systematically engineered to achieve desirable optical, electronic, and chemical properties.

By the mid-20th century, advances in coating and thin-film technologies provided new pathways for constructing core-shell materials with greater precision. Techniques such as electrochemical deposition, sol–gel processing, and chemical vapor deposition (CVD) enabled researchers to deposit uniform layers of materials onto various substrates [7]. These methods were initially applied to macroscopic materials but were later adapted for nanoscale applications as interest in nanotechnology grew. The development of metal-oxide coatings marked a significant step in core-shell material research, as oxide shells provided stability and passivation to reactive metallic cores. This was particularly important for applications in catalysis and corrosion resistance, where the controlled exposure of active sites could be achieved through selective shell design. Additionally, polymer coatings were introduced to enhance dispersibility and biocompatibility, paving the way for biomedical applications of core-shell nanoparticles [8]. As deposition techniques improved, multilayer core-shell structures emerged, where multiple shell layers could be stacked to achieve enhanced functionality. This advancement was particularly useful in optical coatings, where interference effects could be manipulated to control reflectivity and absorption across different wavelengths. The evolution of these technologies set the stage for modern nanofabrication approaches, enabling the production of tailored core-shell architectures for a wide range of applications.

The rise of nanotechnology in the late 20th and early 21st centuries revolutionized the design and synthesis of core-shell materials. The ability to manipulate materials at the atomic and molecular levels allowed for unprecedented control over core and shell properties, leading to materials with precisely engineered functionalities. Breakthroughs in chemical synthesis, such as the Stöber method for silica coating and seed-mediated growth for metal nanoparticles, provided reliable approaches for constructing uniform core-shell structures [9]. These methods enabled the fine-tuning of shell thickness, composition, and crystallinity, ensuring optimal performance in specific applications. Additionally, self-assembly techniques facilitated the spontaneous organization of nanoscale components into core-shell architectures, leveraging molecular interactions to achieve controlled morphologies. At the same time, advancements in physical deposition techniques, including atomic layer deposition (ALD) and pulsed laser deposition, allowed for conformal coating of nanomaterials with atomic precision [10]. These techniques were particularly valuable in electronic and catalytic applications, where nanoscale uniformity is crucial for performance optimization. The integration of computational modeling further accelerated the development of core-shell materials, enabling the prediction of interfacial behavior and guiding experimental design.

As synthesis methods became more sophisticated, the complexity of core-shell architectures evolved beyond simple core-shell designs into multi-shell and hybrid structures. These advanced architectures were developed to address specific functional requirements in emerging technologies. One notable advancement was the introduction of yolk-shell structures, where a hollow space exists between the core and shell. These materials provided enhanced reactivity and transport properties, making them particularly valuable in energy storage and catalysis. Similarly, asymmetric core-shell structures were designed to achieve directional interactions, enabling applications in targeted drug delivery and directional light emission [11]. Another key development was the emergence of multifunctional core-shell systems, where different material properties could be integrated within a single nanostructure. For example, magnetic-plasmonic core-shell nanoparticles combined

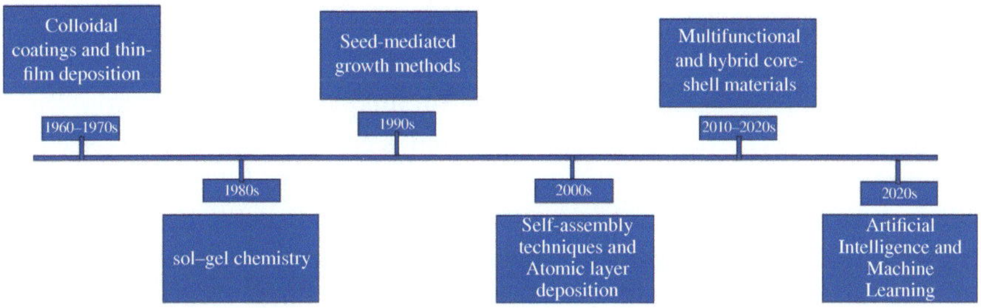

Figure 1.1 This timeline illustrates key milestones in the evolution of core-shell materials, from early colloidal coatings and sol–gel chemistry to advanced self-assembly techniques and AI-driven design.

optical and magnetic functionalities, allowing for dual-mode imaging and therapeutic applications. Similarly, semiconductor-metal core-shell systems were designed for enhanced photocatalytic activity, leveraging synergistic effects at the core-shell interface [12]. The continued evolution of core-shell architectures has been driven by the need for materials with tunable, application-specific properties. By leveraging interfacial engineering and hybrid material design, researchers have expanded the capabilities of core-shell structures, enabling their use in increasingly diverse and complex technological applications.

The development of core-shell materials has been marked by several key milestones, each representing a significant leap in scientific understanding and technological application. Figure 1.1 shows the historical development of core-shell materials from the early methods to ongoing researches.

1960s–1970s: Early studies on colloidal coatings and thin-film deposition laid the foundation for controlled core-shell synthesis.

1980s: The development of sol–gel chemistry enabled the formation of silica-coated nanoparticles, a breakthrough for biomedical and catalytic applications.

1990s: Seed-mediated growth methods were introduced for metal nanostructures, leading to precise control over size and morphology in plasmonic materials.

2000s: Advances in self-assembly techniques and ALD allowed for the fabrication of complex multilayer and hollow-core architectures.

2010s: The rise of multifunctional and hybrid core-shell materials led to breakthroughs in energy storage, targeted drug delivery, and high-performance coatings.

2020s and Beyond: Ongoing research focuses on the integration of artificial intelligence and machine learning to design next-generation core-shell materials with optimized properties for emerging applications.

The historical progression of core-shell materials demonstrates the convergence of chemistry, physics, and materials science in developing highly engineered nanostructures. As new fabrication techniques and theoretical insights continue to emerge, core-shell architectures are expected to play a crucial role in the advancement of nanotechnology, enabling innovations in fields ranging from renewable energy to personalized medicine.

1.3 Unique Properties of Core-shell Structures

Core-shell nanomaterials exhibit a wide range of unique properties that arise from their distinct architecture, where a core material is encapsulated within a shell of a different composition. This structural arrangement allows for the combination of diverse material properties, leading to enhanced functionality beyond what is achievable with single-component nanomaterials. By carefully selecting the core and shell materials, researchers can design nanostructures with tailored optical, electronic, magnetic, mechanical, and catalytic characteristics, making them highly versatile for applications in energy storage, environmental remediation, and biomedicine. The presence of a shell provides additional stability to the core, preventing degradation, oxidation, or unwanted interactions with the surrounding environment. This is particularly crucial for materials that are prone to chemical instability, such as metal nanoparticles, which require protective shells to maintain their reactivity and longevity [13]. Furthermore, the interface between the core and shell plays a significant role in determining the overall behavior of the nanomaterial, as it influences charge transfer, surface interactions, and structural integrity. Core-shell structures also enable precise control over surface chemistry, allowing for selective functionalization with biomolecules, catalysts, or other active species. This ability to manipulate surface properties without altering the core's intrinsic characteristics is valuable in applications such as targeted drug delivery, where surface modifications can enhance biocompatibility while preserving the therapeutic properties of the core. Additionally, the tunability of optical, electronic, and magnetic properties through core-shell engineering has opened new possibilities in fields such as plasmonics, spintronics, and quantum dot solar cells [14]. By integrating multiple functionalities within a single nanostructure, core-shell materials provide a platform for designing advanced materials with superior performance and multifunctionality. Their adaptability and tunable nature have made them a cornerstone of modern nanotechnology, driving innovations in a wide range of scientific and industrial fields. The unique properties of core-shell structures shown in Figure 1.2,

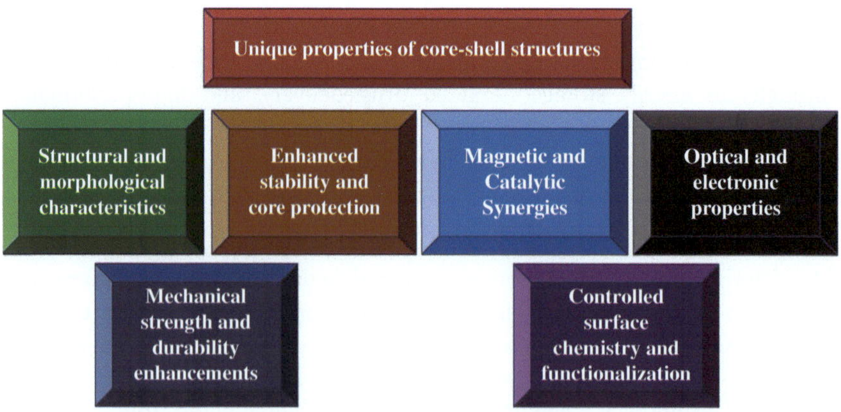

Figure 1.2 Unique properties of core-shell structures including structural stability, enhanced protection, magnetic and catalytic synergies, optical and electronic properties, mechanical strength, and controlled surface chemistry.

such as enhanced stability, catalytic synergies, and controlled surface functionalization, play a crucial role in various advanced applications.

1.3.1 Structural and Morphological Characteristics

The structural and morphological characteristics of core-shell nanomaterials are fundamental to their unique properties and functional advantages. The interplay between core and shell materials governs the overall physical and chemical behavior, influencing stability, reactivity, and application-specific performance. Core-shell nanomaterials exhibit well-defined geometries, which can be precisely engineered through controlled synthesis methods such as sol–gel processing, templated growth, and ALD [15]. Their morphology varies from spherical and cubic to rod-like, hollow, or hierarchical structures, depending on the synthesis conditions and the material system employed. The structural and morphological characteristics of core-shell nanomaterials critically influence their physicochemical properties, determining their stability, reactivity, electronic behavior, and mechanical performance. The architecture of a core-shell nanostructure is defined by the size, shape, and composition of both the core and the shell, as well as the interfacial interactions between them. The ability to manipulate these parameters enables precise control over the material's functionality, making core-shell structures highly versatile for diverse applications [16]. One of the primary structural advantages of core-shell nanomaterials is their ability to exhibit synergistic properties that arise from the combination of core and shell materials. For instance, in plasmonic nanoparticles, the shell can modify the localized surface plasmon resonance (LSPR) of the core by altering the dielectric environment, leading to tunable optical absorption and scattering properties. This principle is extensively utilized in surface-enhanced Raman spectroscopy (SERS) and biosensing applications. Similarly, in quantum dots (QDs), the shell passivates surface defects in the core, reducing non-radiative recombination and enhancing photoluminescence efficiency, which is crucial for applications in optoelectronic devices and bioimaging.

Morphology plays a crucial role in determining charge transport and carrier dynamics in electronic and energy applications. For example, in core-shell nanowires used in photovoltaic devices, a well-ordered shell with a uniform thickness can facilitate efficient charge separation while minimizing recombination losses at the core-shell interface. The structural integrity of the shell is particularly important in lithium-ion battery anodes, where it acts as a buffer layer, accommodating volume expansion and preventing mechanical degradation of high-capacity anode materials such as silicon [17]. The interfacial structure between the core and shell also significantly affects catalytic activity. In core-shell catalysts, lattice strain at the interface modifies the electronic structure of active sites, often leading to enhanced catalytic efficiency. For example, in Pt-based core-shell electrocatalysts for fuel cells, the shell can alter the d-band center of the core, optimizing the adsorption energy of reactants and improving catalytic turnover rates. Additionally, the porous morphology of certain core-shell structures enhances mass transport of reactants and products, further boosting catalytic performance [18].

The interfacial compatibility between the core and shell is a crucial factor, as lattice mismatch, strain effects, and interfacial bonding dictate the overall stability and mechanical integrity of the structure. In many cases, the core-shell architecture enables the confinement of charge carriers, phonons, or spin interactions within a specific region, leading to enhanced electronic, optical, and magnetic behaviors. The shell material can also introduce strain-induced effects that modify the energy band structure of the core, thereby tuning electronic states and optical responses. Furthermore, precise control over shell thickness at the nanometer scale allows for gradual modification of physical properties, such as light absorption in plasmonic nanostructures or exciton confinement in semiconductor QDs [19]. The morphological features of core-shell nanomaterials also play a crucial role in catalytic applications, where high surface area and porous shell structures facilitate reactant diffusion while protecting the active core from deactivation. Similarly, in biomedical applications, the shell morphology dictates cellular uptake, biocompatibility, and controlled release kinetics of drug-loaded nanocarriers. By tailoring the core-shell interface and overall morphology, scientists can fine-tune the properties of these nanostructures for specific technological applications, making them indispensable in fields ranging from energy storage and environmental sensing to nanomedicine and optoelectronics. Thus, the structural and morphological characteristics of core-shell nanomaterials are not merely descriptive features but are deeply connected to their functional attributes. By precisely engineering the size, shape, and interface of core-shell structures, researchers can tailor their optical, electronic, catalytic, and mechanical properties to meet the specific demands of advanced applications in nanotechnology.

1.3.2 Enhanced Stability and Core Protection

The core-shell architecture is fundamentally designed to enhance the stability and protect the core material from degradation, oxidation, dissolution, or unwanted interactions with the surrounding environment. The shell serves as a physical and chemical barrier, significantly extending the functional lifespan of the core material while maintaining or even improving its intrinsic properties. The degree of stability enhancement depends on factors such as shell composition, thickness, crystallinity, and interfacial bonding with the core [20]. One of the most significant advantages of core-shell nanomaterials is their role in preventing oxidation and corrosion, particularly in metallic and semiconductor nanoparticles. For example, noble metal cores such as silver and copper are prone to oxidation in ambient conditions, leading to loss of conductivity and plasmonic properties. Encapsulating these cores with inert shells like silica (SiO_2), alumina (Al_2O_3), or carbon effectively prevents oxidative degradation by forming a diffusion barrier that inhibits oxygen and moisture penetration [21]. Similarly, in semiconductor QDs, passivating the core with a higher-bandgap shell such as ZnS over CdSe QDs significantly improves photostability by preventing surface oxidation and non-radiative recombination at defect sites. The core-shell structure also enhances thermal stability by providing structural rigidity and reducing thermal-induced sintering or coalescence. This is particularly relevant in catalytic applications where metal nanoparticles experience high-temperature conditions. For instance, in Pt-based catalysts for fuel cells, the presence of a protective shell such as TiO_2 or ZrO_2 prevents nanoparticle aggregation and deactivation under thermal stress. The shell can also act as a thermal buffer, redistributing heat more efficiently and reducing localized overheating effects that could destabilize the core.

In addition to chemical and thermal stability, core-shell materials improve colloidal stability in liquid-phase applications. Bare nanoparticles tend to agglomerate due to van der Waals attractions or uncontrolled surface interactions, which reduces their dispersibility and functional performance. Coating these nanoparticles with stabilizing shells such as polyethylene glycol (PEG), silica, or polymers introduces electrostatic or steric repulsion, thereby preventing aggregation and maintaining a stable dispersion over extended periods [22]. This is particularly important in biomedical and catalytic applications, where maintaining uniform particle distribution is crucial for consistent performance. Core-shell structures also play a pivotal role in electrochemical stability, particularly in energy storage and conversion devices. In lithium-ion batteries, transition metal oxide nanoparticles used as cathodes suffer from structural degradation due to continuous lithium intercalation and deintercalation. Coating these cathodes with protective layers such as Li_3PO_4 or Al_2O_3 prevents unwanted side reactions with the electrolyte and mitigates phase transitions that lead to capacity fading [23]. Similarly, in electrocatalysts for water splitting, core-shell architectures reduce leaching of active metal sites, thereby extending the durability of the catalyst under harsh electrochemical conditions.

The core-shell interface is another critical factor in stability enhancement. Strong interfacial adhesion ensures mechanical robustness, preventing shell delamination or collapse under operational conditions. The degree of lattice mismatch between core and shell materials influences the overall stability, with well-matched interfaces leading to improved stress distribution and reduced defect formation. For instance, epitaxially grown shells in semiconductor heterostructures provide better structural coherence, minimizing strain-induced defects that could otherwise compromise electronic and optical performance [24]. Thus, enhanced stability and core protection in core-shell nanomaterials are achieved through a combination of chemical passivation, physical shielding, electrostatic stabilization, and interfacial engineering. These attributes make core-shell structures indispensable in advanced applications where long-term reliability, resistance to environmental factors, and preservation of functional properties are paramount.

1.3.3 Controlled Surface Chemistry and Functionalization

The surface chemistry of core-shell nanomaterials plays a crucial role in determining their interfacial interactions, stability, and functional performance across various applications. Controlled surface chemistry and functionalization allow precise tuning of properties such as solubility, catalytic activity, biocompatibility, and selective binding, making core-shell structures highly versatile for targeted applications in catalysis, sensing, drug delivery, and energy storage [25]. The ability to engineer surface characteristics through the composition and modification of the shell provides a fundamental advantage over conventional single-component nanomaterials.

One of the key aspects of surface chemistry control in core-shell nanostructures is the selective functionalization of the shell with chemical groups that impart desired surface properties. For instance, in biomedical applications, the outer shell of nanoparticles is often modified with hydrophilic polymers, such as PEG, to enhance biocompatibility and prevent nonspecific protein adsorption [26]. Similarly, the functionalization of silica-coated nanoparticles with amino (—NH_2), carboxyl (—COOH), or thiol (—SH) groups enables selective conjugation with biomolecules, facilitating applications in targeted drug delivery, bioimaging, and biosensing [27]. The ability to modify the surface charge and hydrophobicity of core-shell nanomaterials is also critical for dispersion stability and controlled interactions with surrounding environments. For example, in colloidal systems, electrostatic or steric stabilization through functionalization prevents nanoparticle aggregation. The use of surfactants, polyelectrolytes, or zwitterionic coatings can regulate the zeta potential of the shell, ensuring stable dispersions in aqueous or organic media. This approach is particularly beneficial in catalytic applications where maintaining high surface area and accessibility to reactants is crucial.

In catalysis, controlled surface chemistry of core-shell nanomaterials directly influences reaction kinetics and selectivity. The shell can be engineered to create specific active sites or enhance mass transport by incorporating functional groups that regulate adsorption–desorption dynamics. For instance, in core-shell structured catalysts with noble metal cores (e.g., Pt, Pd) and transition metal oxide shells (e.g., TiO_2, CeO_2), the shell can introduce oxygen vacancies that promote catalytic activity through charge transfer effects. Similarly, in enzyme-mimicking nanozymes, surface functionalization of the shell with ligands or co-catalysts can enhance catalytic efficiency by optimizing substrate affinity and reaction pathways [28].

In electrochemical applications, surface functionalization of core-shell materials is employed to improve ion transport, electronic conductivity, and chemical stability. For lithium-ion batteries, carbon-coated transition metal oxides enhance electrical conductivity while preventing electrolyte decomposition at the electrode surface. In supercapacitors, functionalized shells with nitrogen- or sulfur-doped carbon frameworks increase pseudocapacitive behavior, leading to higher energy storage capacity [29]. In fuel cells, surface modification of core-shell electrocatalysts with organic ligands or heteroatoms optimizes charge transfer kinetics, improving catalytic efficiency for oxygen reduction or hydrogen evolution reactions (HER) [30].

Surface chemistry control is also critical in environmental applications, particularly in pollutant removal and water purification. Functionalized core-shell nanoparticles can be designed with selective adsorption sites for heavy metal ions, organic pollutants, or toxic gases. For example, silica-coated magnetic nanoparticles functionalized with thiol (—SH) groups exhibit high affinity for mercury removal from water, while core-shell metal-organic frameworks (MOFs) with functionalized ligands provide tunable adsorption capacities for volatile organic compounds (VOCs) [31].

The engineering of surface chemistry extends beyond simple functionalization to include dynamic and stimuli-responsive coatings. In smart drug delivery systems, pH-sensitive or thermo-responsive polymeric shells can modulate drug release behavior, ensuring site-specific and controlled therapeutic action. Similarly, in adaptive optical materials, surface-modified core-shell nanoparticles respond to external stimuli such as light, electric fields, or magnetic fields, enabling tunable optical and electronic properties [32]. Thus, controlled surface chemistry and functionalization of core-shell nanomaterials provide a powerful tool for tailoring material properties to specific applications. By precisely engineering the shell composition, functional groups, and interfacial interactions, researchers can optimize performance in catalysis, energy storage, biomedical applications, and environmental remediation, demonstrating the profound impact of surface chemistry in nanotechnology.

1.3.4 Tunable Optical and Electronic Properties

Core-shell nanomaterials exhibit highly tunable optical and electronic properties due to their unique structural configurations, which allow precise control over quantum confinement effects, charge transport dynamics, and light–matter interactions. These properties arise from the interplay between the core and shell materials, their relative band structures, and interfacial effects that influence charge separation, recombination, and energy transfer mechanisms. By engineering the composition, thickness, and morphology of the shell, researchers can modulate these properties to achieve tailored functionalities for applications in optoelectronics, photovoltaics, photocatalysis, and quantum dot–based technologies.

One of the most significant optical features of core-shell nanomaterials is the ability to tune their absorption and emission spectra through quantum size effects. In semiconductor core-shell QDs, the bandgap energy can be precisely controlled by varying the core size or introducing a shell with a higher or lower bandgap [14]. For example, CdSe/ZnS QDs exhibit enhanced photoluminescence efficiency and stability due to the ZnS shell, which passivates surface trap states and reduces non-radiative recombination [33]. This property makes them highly useful for applications in bioimaging, light-emitting diodes (LEDs), and display technologies. Beyond quantum confinement, the optical properties of core-shell materials are strongly influenced by plasmonic interactions, especially in metal-core/dielectric-shell structures. In noble metal-based core-shell systems, such as Au@SiO_2 or Ag@TiO_2, LSPR effects can be tuned by modifying the core size, shell thickness, and dielectric environment. The presence of a shell alters the electromagnetic field distribution around the metal core, enabling precise spectral tuning of plasmonic absorption for applications in surface-enhanced Raman scattering (SERS), optical sensors, and photothermal therapy [34]. These tunable plasmonic features also enhance light absorption in solar cells and photocatalysts, improving their efficiency by facilitating hot-electron transfer processes. The electronic properties of core-shell nanomaterials are governed by band alignment at the core-shell interface, which dictates charge separation and carrier transport behavior. In heterostructured semiconductor core-shell materials, such as TiO_2@CdS or ZnO@CdSe, type I, type II, or quasi-type II band alignments can be engineered to optimize charge carrier dynamics. Type II heterojunctions, where the conduction and valence band edges of the core and shell are staggered, promote efficient spatial separation of electrons and holes, reducing recombination losses and enhancing photocatalytic or photovoltaic performance. This design principle is widely applied in artificial photosynthesis, perovskite solar cells, and photodetectors.

Another crucial factor in tuning electronic properties is the modulation of carrier mobility and conductivity through shell engineering. In core-shell nanowires and nanotubes, a conductive shell material can serve as an efficient charge transport layer, reducing interfacial resistance and improving electrical performance [35]. For example, Si@C core-shell nanowires exhibit enhanced lithium-ion transport in battery anodes due to the conductive carbon shell, which mitigates volume expansion and prevents capacity degradation [36]. Similarly, in transparent conducting electrodes, metal-oxide core-shell structures like Ag@ITO or Cu@ZnO provide high electrical conductivity while maintaining optical transparency, making them ideal for next-generation flexible electronics and displays. Core-shell nanomaterials also enable tunable nonlinear optical (NLO) responses and photonic applications by leveraging interfacial effects. In metal–semiconductor hybrid structures, such as Au@ TiO_2 or Pt@CdS, strong plasmon–exciton coupling can modify the optical response, leading to enhanced two-photon absorption, second-harmonic generation, and other nonlinear effects. These properties are highly valuable for applications in laser technologies, optical limiters, and ultrafast photonic devices.

Furthermore, core-shell architectures can be designed to exhibit dynamic, stimuli-responsive optical and electronic behaviors. Thermochromic, electrochromic, and mechanochromic materials utilize core-shell structures to achieve reversible color changes under external stimuli [37]. For instance, core-shell nanocrystals with phase-change materials in the shell (e.g., GeTe@SiO_2) show tunable infrared reflectivity, which can be applied in smart windows and adaptive thermal coatings. Similarly, ferroelectric core-shell nanoparticles exhibit field-tunable dielectric properties, enabling applications in tunable capacitors and high-frequency electronic components [38]. Thus, the ability to precisely manipulate optical and electronic properties through core-shell nanostructuring has opened new avenues in energy conversion, sensing, optoelectronics, and quantum technologies. By leveraging band engineering, interfacial control, and plasmonic enhancements, core-shell materials provide a highly adaptable platform for next-generation functional materials with superior performance and versatility.

1.3.5 Magnetic and Catalytic Synergies

Core-shell nanomaterials exhibit remarkable magnetic and catalytic properties due to the synergistic effects between the core and shell components. By integrating different functional materials in a hierarchical structure, these nanomaterials achieve enhanced performance compared to their single-component counterparts. The interplay between magnetism and catalysis is particularly valuable in applications such as targeted drug delivery, magnetic separation, environmental remediation, and heterogeneous catalysis. The ability to tune these properties by altering the composition, shell thickness, and interfacial interactions makes core-shell architectures highly versatile for advanced nanotechnology applications. The magnetic properties of core-shell nanomaterials are strongly influenced by factors such as core size, shell composition, interfacial strain, and spin interactions. Magnetic core-shell structures typically consist of a ferromagnetic or superparamagnetic core (e.g., Fe, Fe_3O_4, Co, Ni) encapsulated by a protective shell made of oxides, noble metals, or polymers [39].

The shell serves multiple roles, such as preventing oxidation, reducing magnetic aggregation, and providing functionalization sites for biomedical or catalytic applications. For example, Fe_3O_4@SiO_2 nanoparticles retain their superparamagnetic behavior while gaining chemical stability and surface modifiability, making them ideal for magnetically guided drug delivery and bioseparation [40].

Exchange coupling effects at the core-shell interface can also lead to enhanced magnetic anisotropy and coercivity, particularly in structures such as Fe@Fe_3O_4 or Co@Pt. These effects enable precise control over magnetic relaxation dynamics, which is crucial for magnetic hyperthermia treatments in cancer therapy. In magnetic resonance imaging (MRI), core-shell nanostructures such as Fe_3O_4@Au or Fe_3O_4@Dopamine exhibit improved relaxivity, leading to superior contrast enhancement compared to conventional MRI contrast agents. In the field of catalysis, core-shell nanomaterials enhance reaction efficiency through multiple mechanisms, including electronic modulation, active site protection, and surface area optimization [13]. Core-shell catalysts often feature a catalytically active core surrounded by a shell that fine-tunes reactivity while improving stability and selectivity. In metal@metal oxide systems (e.g., Pt@CeO_2, Au@TiO_2), the shell regulates charge transfer at the interface, boosting catalytic activity in oxidation and hydrogenation reactions. These effects are particularly beneficial in energy applications such as fuel cells, HER, and CO_2 reduction.

One of the most significant advantages of core-shell nanocatalysts is their ability to prevent catalyst deactivation due to sintering, poisoning, or leaching. Noble metal catalysts such as Pt or Pd are prone to aggregation under high-temperature conditions, leading to performance degradation. Encapsulating them with a thermally stable oxide shell, as in Pt@ZrO_2 or Pd@Al_2O_3, prevents nanoparticle migration and maintains catalytic efficiency. Similarly, carbon-shell coatings in Fe@C or Ni@C catalysts enhance durability while preserving active site accessibility. Magnetic core-shell catalysts offer additional benefits in heterogeneous catalysis, as they enable facile recovery and recyclability using an external magnetic field. This approach is widely used in wastewater treatment, biomass conversion, and industrial chemical synthesis. For example, Fe_3O_4@SiO_2-Pd nanoparticles function as effective catalysts for Suzuki coupling reactions and can be magnetically separated from reaction mixtures, minimizing catalyst loss and reducing purification costs. Bifunctional core-shell nanostructures that combine magnetic and catalytic properties further expand their applicability. In Fe_3O_4@Au or Co@Pd systems, the magnetic core enables external control over the catalyst's position, while the noble metal shell provides high catalytic activity for redox reactions. This synergy is particularly useful in enzymatic mimicry, where magnetically responsive peroxidase-like catalysts (e.g., Fe_3O_4@Pt) facilitate reactive oxygen species (ROS) generation for antibacterial and environmental remediation applications. The integration of magnetic and catalytic functionalities in core-shell architectures also plays a vital role in emerging technologies such as artificial photosynthesis, spintronics, and magnetically switchable catalysis [41]. In magneto-plasmonic nanostructures (e.g., Fe_3O_4@Ag, Fe@Au), the interplay between LSPR and magnetic effects enables light-driven catalytic processes with tunable reactivity. These hybrid materials are finding applications in sustainable energy conversion, photocatalytic water splitting, and pollutant degradation.

By leveraging interfacial effects, strain engineering, and composition tuning, core-shell nanomaterials provide an advanced platform for designing multifunctional materials with superior magnetic and catalytic performance. The ability to engineer these properties with precision opens new possibilities for nanotechnology-driven applications in medicine, energy, and environmental sustainability.

1.3.6 Mechanical Strength and Durability Enhancements

Core-shell nanomaterials offer significant improvements in mechanical strength and durability, making them ideal for a wide range of applications where structural integrity and long-term stability are critical. The combination of a strong core material with a protective or functionalized shell provides synergistic effects that enhance the mechanical properties of the composite [42]. By tailoring the composition, thickness, and interface between the core and shell, researchers can achieve materials with superior hardness, toughness, and resistance to wear, corrosion, and fatigue compared to single-component nanomaterials. This enhancement of mechanical properties is particularly useful in industries such as aerospace, automotive, biomedical, and construction, where high-performance materials are essential. The core, typically composed of a material with inherent strength (e.g., metals, ceramics, or carbon-based materials), provides structural support and ensures the material can withstand external forces without failure [43]. The shell, often made of a more flexible or durable material, serves to protect the core from mechanical stress, preventing direct exposure to harsh environmental conditions. In composites such as $SiC@TiO_2$ or $Al_2O_3@Ni$, the hard ceramic or metal core is shielded by a tough oxide or polymer shell, which helps absorb and dissipate mechanical energy, reducing the likelihood of fracture under stress.

One of the most significant advantages of core-shell nanomaterials is their ability to improve toughness without compromising strength. The shell can act as a buffer, preventing the propagation of cracks and defects that would typically weaken the material. In systems such as $Cu@SiO_2$ or Ti@C, the tough shell material imparts energy-dissipating properties, leading to enhanced resistance to crack growth and failure under impact. This is particularly important in high-stress applications, such as structural coatings, armor materials, and lightweight composites used in aerospace engineering. Furthermore, the shell's ability to undergo plastic deformation before fracture can prevent catastrophic failure, a phenomenon that is particularly valuable in load-bearing components [44]. For instance, in core-shell composites like Fe@C or carbon nanotube (CNT) core-shell materials, the shell undergoes deformation under load, which helps to redistribute stresses and prevent brittle fracture [45]. This mechanism can be harnessed in the design of materials for high-impact environments, including automotive crash protection systems and sporting goods, where durability and safety are paramount. In addition to improving mechanical strength, core-shell nanomaterials exhibit enhanced wear and abrasion resistance, a critical factor in industrial applications. Materials such as $TiO_2@Al_2O_3$ and $SiC@SiO_2$ show remarkable resistance to surface wear due to the hardness of the shell material, which protects the core from erosion and abrasion. This property is highly beneficial in the manufacture of cutting tools, bearings, and coatings for machinery that must endure harsh operating conditions. Durability is also enhanced in core-shell nanomaterials through their ability to resist corrosion and environmental degradation. The shell acts as a protective barrier, shielding the core from exposure to corrosive agents, moisture, and high temperatures [46]. For example, in core-shell materials like Zn@Cu or Au@Ag, the shell material prevents oxidation and rust formation on the core, ensuring long-term stability and extending the lifespan of the material. This property is particularly valuable in applications such as marine coatings, medical implants, and electronic devices, where prolonged exposure to aggressive environments can cause material degradation. The ability of core-shell nanomaterials to enhance fatigue resistance is another crucial benefit. In cyclic loading conditions, the core-shell design helps to reduce the initiation and propagation of cracks that would otherwise lead to material failure. This is especially important in applications requiring long-term performance under repeated stress, such as in turbine blades, engine components, and structural elements in bridges and buildings [47]. The shell's ability to absorb and redistribute stress allows the material to withstand many cycles of loading without significant degradation, providing reliability and longevity in critical applications.

Finally, the versatility of core-shell nanomaterials allows for further tuning of mechanical properties through functionalization and surface modification. By introducing functional groups or coating the shell with additional layers of material, the mechanical properties of core-shell nanomaterials can be further enhanced to meet the specific requirements of various applications. This tunability makes core-shell nanomaterials highly adaptable for emerging technologies, such as flexible electronics, wearable devices, and energy storage systems, where both strength and durability are crucial.

1.4 Role in Advancing Nanotechnology

Core-shell nanomaterials are at the forefront of advancing nanotechnology due to their exceptional versatility and unique properties. Their ability to combine the strengths of multiple materials within a single structure enables the tailoring of properties that are essential for a wide range of technological applications. From energy storage to environmental sustainability, biomedicine, and electronics, the incorporation of core-shell structures has led to significant breakthroughs in these fields, pushing the boundaries of what is possible with traditional materials. The functionalization of core-shell nanomaterials allows for precise control over their properties, leading to innovations in various domains that were previously unachievable [48]. The role of core-shell nanomaterials in advancing nanotechnology is closely linked to their adaptability, which makes them indispensable for both current and emerging applications. Their ability to exhibit tunable properties, such as optical, electronic, and mechanical characteristics, is critical for developing next-generation technologies that address challenges across energy, health, and environmental sectors [2]. These materials are also central to the development of sustainable solutions, offering enhanced efficiency, longevity, and performance in a variety of contexts. As research continues to uncover new ways to exploit the

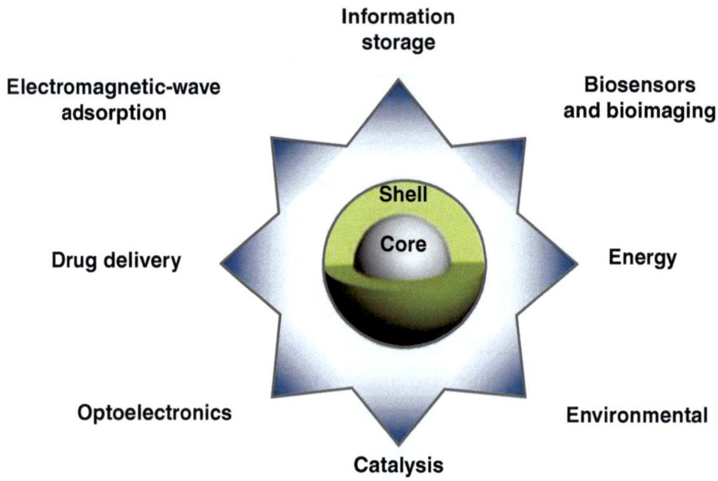

Figure 1.3 Multifunctional roles of core-shell structures in advanced nanotechnology including catalysis, optoelectronics, drug delivery, electromagnetic-wave adsorption, biosensing, energy storage, environmental remediation, and information storage. [13]/with permission of Royal Society of Chemistry.

potential of core-shell nanomaterials, their impact on technological advancement is becoming more profound. From improving the efficiency of energy systems to enabling targeted drug delivery and driving innovations in smart technologies, core-shell materials are playing a pivotal role in shaping the future of nanotechnology. The following subsections explore some of the most important areas where core-shell nanomaterials are contributing to technological advancement. Core-shell structures have gained significant attention due to their versatile applications in advanced nanotechnology, as shown in Figure 1.3, including catalysis, optoelectronics, drug delivery, and energy storage.

1.4.1 Energy Storage and Conversion

Core-shell nanomaterials are revolutionizing energy storage and conversion technologies by offering enhanced performance, stability, and efficiency over traditional materials. These materials are particularly valuable in energy systems that require high energy density, fast charge/discharge cycles, and long-term durability, such as batteries, supercapacitors, and fuel cells. The unique properties of core-shell structures, including their ability to combine dissimilar materials with tailored interfaces, enable the optimization of electrochemical behavior, leading to significant improvements in energy storage and conversion devices. In battery technologies, core-shell nanomaterials are increasingly being used to enhance the performance of anodes and cathodes. For example, in lithium-ion batteries, core-shell particles, such as silicon-based cores coated with carbon or metal oxide shells, help address one of the key challenges in battery design—volume expansion during cycling [49]. Silicon, with its high theoretical capacity for lithium-ion storage, undergoes significant expansion and contraction during charge/discharge cycles, which can lead to material degradation and capacity loss [50]. By encapsulating the silicon core with a thin, flexible carbon or metal oxide shell, this issue is mitigated, as the shell provides structural integrity, preventing excessive expansion while allowing for high-capacity charge storage. The result is a significant increase in the lifespan and cycling stability of the battery, alongside improved energy density. Similarly, in sodium-ion batteries and other alternative battery technologies, core-shell nanomaterials provide enhanced performance by improving conductivity, reducing internal resistance, and ensuring more efficient ion transfer between the core and shell [51]. The shell material can be engineered to provide a stable and conductive environment for the core material, resulting in a higher rate of charge/discharge and better overall performance. The design of core-shell nanomaterials for energy storage is not limited to batteries; they are also widely used in supercapacitors. In supercapacitors, where high power density and fast charging/discharging are essential, core-shell structures can improve the electrode surface area and increase the charge storage capacity, thus enhancing the overall energy storage efficiency [52].

For energy conversion, core-shell nanomaterials also play a crucial role in optimizing the efficiency of solar cells, fuel cells, and thermoelectric devices. In photovoltaic applications, the use of core-shell nanostructures in materials like QDs or metal-semiconductor hybrids has shown promising results in improving light absorption, charge separation, and charge carrier mobility [53]. The unique optical properties of core-shell structures, which allow for the tuning of absorption wavelengths and enhanced light trapping, contribute to better energy conversion efficiency in solar energy harvesting. The core-shell interface can also facilitate better electron transport and reduce recombination losses, increasing the overall performance of solar cells [54]. In fuel cells, core-shell nanomaterials are employed to enhance catalytic activity and reduce the use of expensive precious metals such as platinum. By designing core-shell catalysts with a noble metal core and a non-noble metal shell, researchers have been able to improve the electrocatalytic performance of the fuel cell while lowering material costs [55]. The shell's role is to facilitate the adsorption of reactants and promote faster reaction kinetics, while the core provides the necessary stability and conductivity to maintain consistent performance over time. Furthermore, core-shell nanomaterials are being explored in thermoelectric devices, where the conversion of heat into electrical energy is a key concern [56]. In these applications, the unique structure of core-shell materials can help optimize the thermal and electrical conductivity properties, allowing for improved thermoelectric efficiency. The core and shell materials can be selected to have complementary properties that enhance heat management and electron mobility, making core-shell nanomaterials ideal candidates for future thermoelectric energy conversion devices.

Overall, the versatility of core-shell nanomaterials in energy storage and conversion applications lies in their ability to combine the best properties of different materials at the nanoscale. By carefully tailoring the core and shell components, it is possible to create materials that address specific challenges in energy efficiency, stability, and performance. As research in this area continues to grow, core-shell nanomaterials are expected to play an increasingly important role in advancing the next generation of energy storage and conversion technologies, contributing to the development of more efficient, sustainable, and high-performance energy systems.

1.4.2 Environmental Applications and Sustainability

Core-shell nanomaterials are increasingly recognized for their potential in environmental applications, particularly in areas related to sustainability, pollution control, and resource management. Their unique properties enable them to address key environmental challenges, including water purification, air quality management, waste treatment, and energy efficiency. The multifunctional nature of core-shell structures allows for the development of materials that not only improve environmental outcomes but also enhance the overall sustainability of industrial processes and everyday technologies. One of the most significant applications of core-shell nanomaterials in environmental protection is in water treatment and purification. Water scarcity and contamination are critical global issues, and the demand for effective water purification technologies is growing [57]. Core-shell materials, especially those with photocatalytic or adsorptive properties, are highly effective in breaking down harmful contaminants such as heavy metals, organic pollutants, and microbial pathogens. For instance, core-shell nanoparticles made of metal oxide cores (e.g., titanium dioxide) with a shell of functionalized carbon or polymer coatings can serve as highly active photocatalysts under UV light. These nanomaterials can degrade organic pollutants and kill bacteria, providing a sustainable approach to water decontamination [58]. The core-shell structure allows for optimal exposure of the catalytic surface, enhancing the material's reactivity and stability, and ensuring long-lasting performance in purification processes. In addition to water treatment, core-shell nanomaterials are gaining attention for their role in air pollution control. Nanocatalysts, composed of a reactive core material surrounded by a protective shell, are used in the catalytic conversion of harmful gases such as carbon monoxide (CO), nitrogen oxides (NOx), and VOCs [59]. The shell material, often a metal oxide or carbon, acts as a stabilizer, protecting the active core and facilitating the adsorption and transformation of pollutants into less harmful substances. This is particularly beneficial in automotive exhaust systems and industrial applications, where air quality regulations require efficient removal of toxic gases. The enhanced surface area and tunable chemical properties of core-shell nanomaterials make them highly effective in these applications, offering better catalytic efficiency and longer operational lifetimes compared to conventional catalysts.

Another promising application is in the field of environmental remediation and waste treatment. Core-shell nanomaterials can be engineered to capture and degrade hazardous substances, such as heavy metals and pesticides, from contaminated soils or industrial effluents. For example, core-shell nanoparticles with magnetic cores and functionalized shells can be used for the selective removal of toxic metals from wastewater [60]. The magnetic properties of the core allow for easy separation and recycling of the nanoparticles after the cleanup process, ensuring that the materials can be reused, which adds to the sustainability aspect of this technology. Additionally, core-shell materials can be employed to remove oils and hydrocarbons from water and soil, addressing oil spills and other environmental disasters with enhanced efficiency and lower environmental impact [61]. Core-shell nanomaterials are also being explored for energy-efficient environmental solutions, such as the development of energy-efficient coatings for buildings and other infrastructure.

The use of core-shell composites in thermal insulating materials allows for better heat retention and temperature regulation, reducing the need for external energy inputs in heating and cooling systems [62]. These energy-saving applications contribute to the reduction of carbon footprints and energy consumption in both residential and industrial buildings, making them an important component of sustainable construction and urban planning.

In the field of renewable energy, core-shell nanomaterials contribute to environmental sustainability by improving the efficiency and performance of solar energy devices and other renewable technologies. For example, in solar cells, core-shell QDs and nanostructures can enhance light absorption and improve energy conversion efficiency, reducing the overall carbon footprint of energy production [15]. The integration of core-shell materials in solar technologies can also lower the cost of energy production, making renewable sources more competitive with traditional fossil fuels. Beyond direct applications in pollution control and resource management, core-shell nanomaterials contribute to the broader concept of sustainability through their potential in recycling and waste reduction. For instance, in electronic waste (e-waste) recycling, core-shell nanoparticles can be used to selectively extract valuable metals such as gold, silver, and copper from discarded electronic devices [63]. The shell material can be designed to interact with specific metals, ensuring efficient extraction and minimizing the release of hazardous substances. This capability not only supports the recovery of valuable materials but also reduces the environmental burden associated with e-waste disposal. The ability to engineer core-shell nanomaterials with specific functional properties makes them highly suitable for a wide range of environmental applications [2]. By addressing key challenges such as water and air pollution, waste treatment, energy efficiency, and renewable energy development, these materials are paving the way for a more sustainable and environmentally friendly future. As research and development continue, core-shell nanomaterials are expected to play an even more significant role in advancing global sustainability efforts, providing solutions to some of the most pressing environmental issues of our time.

1.4.3 Biomedical Innovations and Drug Delivery

Core-shell nanomaterials are rapidly advancing biomedical applications, particularly in drug delivery, imaging, and therapeutic treatments. The unique structural and chemical properties of core-shell nanoparticles make them ideal candidates for targeted drug delivery, enhanced bioimaging, and the development of new therapeutic modalities [64–66]. These materials can be engineered to offer highly controllable release profiles, biocompatibility, and multifunctionality, positioning them at the forefront of modern medicine. The ability to design core-shell structures with tailored properties enables the development of nanomaterials that can address the complex demands of biomedical applications, improving treatment efficacy and reducing side effects. One of the most significant applications of core-shell nanomaterials in biomedicine is in targeted drug delivery [67]. The core-shell architecture allows for the encapsulation of therapeutic agents within the core, while the shell can be functionalized to provide specific targeting capabilities. For instance, the shell can be decorated with ligands, antibodies, or peptides that bind selectively to receptors on the surface of target cells or tissues, such as cancer cells or inflamed tissue [68]. This selective targeting reduces the systemic distribution of the drug, thereby minimizing off-target effects and enhancing the therapeutic efficacy of the treatment. The core material can be designed to encapsulate a wide range of therapeutic agents, including small molecules, proteins, RNA, and even gene-editing tools, such as CRISPR [69]. The shell can provide stability and protection to these fragile agents, preventing their premature degradation and ensuring their efficient delivery to the target site.

The controlled release of drugs is another major advantage of core-shell nanomaterials in drug delivery. By altering the composition of the core and shell, the release rate of the encapsulated drug can be precisely regulated, enabling sustained or triggered release depending on the needs of the treatment. For example, the core can be made from a biodegradable polymer that breaks down over time, allowing for the gradual release of the drug. Alternatively, the shell can be responsive to external stimuli, such as pH, temperature, or light, enabling the release of the drug in response to the specific conditions at the disease site [70]. This "smart" drug delivery approach allows for more efficient treatments by ensuring that drugs are delivered only when and where they are needed, improving patient outcomes and reducing the frequency of dosing. Core-shell nanomaterials are also playing a pivotal role in the development of nanocarriers for gene therapy [71]. The core-shell structure can encapsulate genetic material, such as DNA or RNA, and protect it from degradation in the bloodstream, while the shell can be modified to enhance cellular uptake and facilitate endosomal escape. This capability is especially important for the development of non-viral gene delivery systems, which offer a safer alternative to viral vectors that can carry the risk of immune reactions and insertional mutagenesis. Core-shell nanoparticles can also improve the stability of genetic material, providing an effective vehicle for gene-editing tools and therapies [72]. As gene therapies continue to advance, core-shell nanomaterials will play an increasingly vital role in their translation from research to clinical application.

Beyond drug delivery, core-shell nanomaterials are making significant contributions to biomedical imaging and diagnostic applications. The core-shell structure allows for the incorporation of various imaging agents, such as fluorescent dyes, magnetic nanoparticles, or radioisotopes, within the core, while the shell can be designed to enhance the stability and biocompatibility of the particles. These nanoparticles can be used in a variety of imaging modalities, including MRI, computed tomography (CT), positron emission tomography (PET), and fluorescence imaging. By integrating both therapeutic and diagnostic functions into a single nanoparticle (a strategy known as "theranostics"), core-shell nanomaterials enable more precise diagnosis, real-time monitoring of treatment efficacy, and the delivery of targeted therapies. Core-shell nanomaterials are also being explored for their potential in regenerative medicine [73]. The core can be made from biocompatible materials that promote tissue regeneration, while the shell can be functionalized with growth factors, peptides, or biomolecules that stimulate cell proliferation and tissue repair. For instance, core-shell nanoparticles have been used in bone tissue engineering to deliver osteogenic growth factors to sites of bone injury or degeneration.

By combining the regenerative properties of the core with the bioactive molecules on the shell, core-shell nanomaterials can accelerate the healing process and improve the outcomes of tissue repair and regeneration [74]. The development of core-shell nanomaterials for biomedical applications requires careful consideration of factors such as biocompatibility, biodegradability, and toxicity. However, advances in material design and surface functionalization have led to the creation of nanoparticles that are not only effective but also safe for use in medical applications. The versatility and multifunctionality of core-shell nanomaterials provide significant opportunities for improving healthcare outcomes, from more effective drug delivery and gene therapy to enhanced diagnostic imaging and regenerative medicine.

1.4.4 Electronics, Photonics, and Wearable Technologies

Core-shell nanomaterials are making significant strides in the fields of electronics, photonics, and wearable technologies due to their ability to provide enhanced properties such as conductivity, optical response, and mechanical flexibility. Their unique architecture, with a functional core and a shell that can be tailored for specific functions, enables the design of materials with a combination of characteristics that are difficult to achieve with conventional single-component nanomaterials. As a result, core-shell structures are increasingly being explored for applications in flexible electronics, optoelectronics, sensors, and next-generation wearable devices [75]. In electronics, core-shell nanomaterials offer exceptional advantages in the development of high-performance conductors and semiconductors. For example, the core can consist of a material with high electrical conductivity, such as gold, copper, or CNTs, while the shell can be made from a semiconducting or insulating material [76]. This combination allows for the design of nanoparticles that exhibit both high conductivity and tailored electronic properties, making them ideal candidates for use in transistors, integrated circuits, and other electronic components. Additionally, the core-shell structure can be used to enhance the stability of nanomaterials in electronic devices, protecting the conductive core from oxidation or degradation while maintaining efficient charge transport properties.

Core-shell nanomaterials also hold great potential in photonics, particularly in applications such as LEDs, solar cells, and photodetectors [77]. The optical properties of core-shell nanoparticles can be finely tuned by adjusting the composition of the core and shell, allowing for the design of materials with specific absorption and emission characteristics. For example, core-shell QDs are widely used in photonics due to their ability to emit light at precisely controlled wavelengths, which is crucial for applications in displays, lighting, and solar energy conversion [78]. By altering the shell material, the surface states of the core can be controlled, leading to improvements in photoluminescence efficiency, stability, and optical properties. In the field of solar energy, core-shell nanomaterials are being explored for use in next-generation solar cells. The core can be made from a highly efficient light-absorbing material, such as cadmium telluride or silicon, while the shell can be engineered to improve charge separation, enhance stability, and reduce recombination losses [15]. The shell can also protect the core material from environmental degradation, such as moisture and oxidation, which is a significant challenge in the long-term performance of solar devices. Core-shell structures have the potential to increase the efficiency and durability of photovoltaic devices, making them a promising solution for renewable energy applications. Wearable technologies represent one of the most exciting areas where core-shell nanomaterials are making an impact. As demand grows for lightweight, flexible, and durable devices that can be worn on the body, core-shell nanomaterials offer a unique solution due to their ability to combine strength, flexibility, and conductivity. For example, core-shell nanoparticles can be incorporated into flexible substrates to create stretchable electronics, such as sensors, displays, and communication devices [79]. The shell material can be designed to provide mechanical flexibility, while the core material can impart conductivity and electronic functionality, allowing for the creation of devices that can withstand deformation without compromising performance. This combination of properties is critical for applications in health monitoring, smart clothing, and other wearable technologies.

Core-shell nanomaterials are also being used in the development of sensors for wearable applications. The core can be engineered to have specific interactions with target molecules or environmental conditions, such as temperature, humidity, or pH, while the shell can be functionalized with recognition elements or catalysts that enable sensitive and selective detection. For instance, core-shell nanomaterials have been used in wearable biosensors that monitor glucose levels, detect environmental pollutants, or track physiological parameters such as heart rate and sweat composition [80]. These sensors are essential for personalized healthcare, providing real-time monitoring and data collection that can inform treatment decisions and improve patient outcomes. In the rapidly growing field of optoelectronics, core-shell nanomaterials are being investigated for their potential in applications such as light-emitting devices, photodetectors, and lasers. By manipulating the core-shell architecture, researchers can control the band gap, optical absorption, and emission characteristics of the material, enabling the design of more efficient devices for use in communication systems, imaging, and sensing technologies. For example, core-shell QDs are being explored for use in LEDs, where they offer improved color purity and energy efficiency compared to traditional semiconductor materials [81]. Similarly, in photodetector applications, core-shell structures can be engineered to enhance light absorption and improve the device's sensitivity, paving the way for advancements in optical communication and sensing technologies. Core-shell nanomaterials are also integral to the development of flexible, stretchable, and highly efficient batteries and energy storage devices for wearable technologies [82]. The core-shell structure allows for improved charge storage capacity and cycling stability, which is essential for portable electronics that require long battery life. By designing core-shell electrodes with tailored ion diffusion paths and optimized structural stability, it is possible to increase the efficiency and longevity of wearable energy storage systems. Additionally, the mechanical flexibility of core-shell materials makes them ideal for integration into wearable devices without sacrificing performance, enabling the development of lightweight, long-lasting energy solutions for the growing demand in wearable electronics.

1.4.5 Emerging Fields and Future Prospects

The unique characteristics of core-shell nanomaterials position them as a key technology for emerging fields and future innovations across various domains. As the understanding of these materials deepens and synthesis techniques advance, their potential applications continue to expand, offering exciting possibilities in areas like quantum computing, environmental remediation, and advanced manufacturing. The adaptability of core-shell structures—where the properties of both the core and the shell can be tailored independently—makes them particularly well-suited to meet the demands of cutting-edge technologies in energy, medicine, and beyond. In the field of quantum computing, core-shell nanomaterials are gaining attention for their ability to facilitate quantum state manipulation and enhance coherence times. QDs, for example, can be designed with specific core-shell architectures that control electronic interactions and allow for more stable and efficient qubit operation. The shell can be engineered to reduce electron scattering, which is crucial for maintaining the quantum state over longer periods. Furthermore, core-shell nanomaterials could lead to the development of optically active quantum devices, where the core can absorb and emit light at precisely controlled wavelengths, facilitating the creation of photonic qubits. This application is still in its early stages but holds great promise for the advancement of quantum technologies that will revolutionize computing power. In environmental remediation, core-shell nanomaterials are being explored for their ability to address some of the most pressing challenges related to pollution, resource recovery, and sustainability. Their ability to selectively interact with and remove harmful pollutants, such as heavy metals, organic contaminants, and radioactive substances, makes them ideal candidates for water and soil purification applications. For instance, core-shell nanoparticles with a reactive core (such as iron oxide) and a protective shell (such as carbon or silica) can be used to capture and degrade pollutants through catalytic processes. The shell can also prevent the release of toxic substances from the core, enhancing both the stability and effectiveness of the materials in environmental applications. Additionally, core-shell structures can be designed to be magnetically responsive, facilitating the easy removal of contaminants from water systems.

As advanced manufacturing techniques continue to evolve, core-shell nanomaterials are likely to play an essential role in the fabrication of next-generation devices and components. These materials can be incorporated into additive manufacturing processes, such as 3D printing, to create highly functionalized, multi-material components with specific properties tailored to their intended applications. The ability to integrate different materials into the core and shell of nanoparticles allows for the development of hybrid materials with unprecedented combinations of mechanical, electrical, and thermal properties. For example, core-shell nanomaterials could be used to produce high-strength composites that retain flexibility, enabling the creation of lightweight, durable structures for use in aerospace, automotive, and construction industries. These materials could also be used in the creation of multifunctional coatings for various applications, such as corrosion resistance, self-healing materials, and antifouling surfaces. In energy harvesting and storage, the exploration of core-shell nanomaterials is opening new possibilities for improving the efficiency and scalability of technologies like solar cells, supercapacitors, and batteries. For solar energy, core-shell materials are being studied for their potential to increase light absorption and improve charge transport efficiency, leading to the development of more efficient solar cells. In energy storage systems, core-shell nanostructures are being developed to enhance the performance of electrodes by optimizing ion diffusion pathways and improving charge/discharge cycles. Additionally, as the demand for flexible and wearable energy devices grows, core-shell materials offer the mechanical flexibility and performance needed for next-generation energy solutions, such as bendable batteries and energy-harvesting textiles.

The medical field is also witnessing breakthroughs in the use of core-shell nanomaterials for personalized medicine and targeted therapies. For example, in drug delivery, core-shell nanoparticles can be engineered to encapsulate drugs in the core, while the shell can be functionalized with targeting ligands to direct the delivery to specific cells or tissues. This approach not only improves the precision and efficacy of treatments but also minimizes side effects by reducing off-target interactions. Furthermore, core-shell nanomaterials can be used in diagnostic applications, such as imaging and biosensing, where the core material provides contrast or detection properties, and the shell can protect the core from degradation or functionalize it for specific interactions with biomarkers. As the field of nanomedicine continues to evolve, core-shell structures will be central to the development of innovative therapies and diagnostic tools that could revolutionize healthcare.

Looking toward the future, core-shell nanomaterials are poised to drive advances in sustainable technologies. As the demand for eco-friendly and renewable technologies increases, the design of core-shell structures with optimized environmental performance will play a significant role in areas such as green energy, sustainable agriculture, and waste management. For instance, core-shell nanoparticles could be designed for efficient light harvesting in bio-photovoltaic systems or to enable sustainable agricultural practices by delivering fertilizers or pesticides in a controlled, environmentally friendly manner. Additionally, the integration of core-shell materials into catalytic processes could pave the way for more efficient industrial applications that reduce energy consumption and waste. The future prospects of core-shell nanomaterials are vast and varied, with their unique properties enabling breakthroughs across a wide range of emerging fields. As research in nanotechnology progresses, the potential for core-shell materials to revolutionize industries—from quantum computing and energy storage to environmental sustainability and personalized medicine—continues to grow. With further advancements in synthesis methods, characterization techniques, and application development, core-shell nanomaterials will undoubtedly continue to play a pivotal role in shaping the next generation of technological innovations.

1.5 Conclusion

Core-shell nanomaterials represent a transformative class of materials that have revolutionized the landscape of nanotechnology. Their unique ability to combine the properties of distinct materials in a single structure offers unprecedented opportunities across diverse fields such as energy, environment, and medicine. By encapsulating a core material within a shell, these structures enable a level of control over physical, chemical, and electronic properties that single-component nanomaterials cannot match. From enhanced stability and protection of the core to tunable optical and electronic behaviors, the versatility of core-shell architectures positions them at the forefront of scientific research and technological development. The historical evolution of core-shell materials, from early colloidal chemistry concepts to modern advancements in nanotechnology and synthesis methods, has paved the way for their widespread application. Their significance in both industrial and research domains is undeniable, offering solutions to complex challenges in energy storage, energy conversion, environmental sustainability, and medical innovations. By enabling precise control over material properties at the nanoscale, core-shell nanomaterials have also facilitated progress in new fields such as quantum computing, advanced manufacturing, and targeted drug delivery, demonstrating their vast potential for future developments. As the research and development of core-shell nanomaterials continue to advance, the possibilities for their application expand further. Their impact on the future of nanotechnology is profound, driving innovations in fields ranging from flexible electronics and renewable energy to personalized healthcare and environmental remediation. Moving forward, the continued exploration of these materials will undoubtedly lead to even more groundbreaking discoveries, solidifying their place as a cornerstone of modern nanoscience and engineering.

References

1 Afolabi RO. A comprehensive review of nanosystems' multifaceted applications in catalysis, energy, and the environment. J Mol Liq. 2024 Mar 1;397:124190.

2 Ghosh Chaudhuri R, Paria S. Core/shell nanoparticles: Classes, properties, synthesis mechanisms, characterization, and applications. Chem Rev [Internet]. 2012 Apr 11 [cited 2025 Feb 5];112(4):2373–433. Available from: https://pubs.acs.org/doi/full/10.1021/cr100449n

3 Wan F, Thakur A, Thakur P. Classification of nanomaterials (carbon, metals, polymers, bio-ceramics). In: Suhag D, Thaku, A, Thakur P, editors. Integrated Nanomaterials and their Applications [Internet]. 2023 Dec 6 [cited 2025 Feb 5];37–55. Available from: https://link.springer.com/chapter/10.1007/978-981-99-6105-4_3

4 Mirzaei A, Kim JH, Kim HW, Kim SS. How shell thickness can affect the gas sensing properties of nanostructured materials: Survey of literature. Sens Actuators B Chem. 2018 Apr 1;258:270–94.

5 Mendes PCD, Song Y, Ma W, Gani TZH, Lim KH, Kawi S, et al. Opportunities in the design of metal@oxide core-shell nanoparticles. Adv Phys X [Internet]. 2023 Dec 31 [cited 2025 Feb 5];8(1):2175623. Available from: https://www.tandfonline.com/doi/abs/10.1080/23746149.2023.2175623

6 Lorenzo R, Arben M, Víctor P. Designing advanced nanomaterials for next generation in vitro diagnostics : Development of optical and electrochemical biosensors for determination of viral and bacterial infections based on hollow

AuAg nanoparticles. 2019 [cited 2025 Feb 5]. Available from: https://ddd.uab.cat/record/207910

7 Savale PA. Comparative study of various chemical deposition methods for synthesis of thin films: A review. Asian J Res Chem. 2018;11(1):195.

8 Rezaei B, Harun A, Wu X, Iyer PR, Mostufa S, Ciannella S, et al. Effect of polymer and cell membrane coatings on theranostic applications of nanoparticles: A review. Adv Healthc Mater [Internet]. 2024 Oct 1 [cited 2025 Feb 5];13(26):2401213. Available from: https://onlinelibrary.wiley.com/doi/full/10.1002/adhm.202401213

9 Jankiewicz BJ, Jamiola D, Choma J, Jaroniec M. Silica–metal core–shell nanostructures. Adv Colloid Interface Sci. 2012 Jan 15;170(1–2):28–47.

10 Escorcia-Díaz D, García-Mora S, Rendón-Castrillón L, Ramírez-Carmona M, Ocampo-López C. Advancements in nanoparticle deposition techniques for diverse substrates: A review. Nanomaterials [Internet]. 2023 Sep 19 [cited 2025 Feb 5];13(18):2586. Available from: https://www.mdpi.com/2079-4991/13/18/2586/htm

11 Liu C, Li F, Lai-Peng M, Cheng HM. Advanced materials for energy storage. Adv Mater [Internet]. 2010 Feb 23 [cited 2025 Feb 6];22(8):E28–62. Available from: https://onlinelibrary.wiley.com/doi/full/10.1002/adma.200903328

12 Levin CS, Hofmann C, Ali TA, Kelly AT, Morosan E, Nordlander P, et al. Magnetic-plasmonic core-shell nanoparticles. ACS Nano [Internet]. 2009 Jun 23 [cited 2025 Feb 6];3(6):1379–88. Available from: https://pubs.acs.org/doi/full/10.1021/nn900118a

13 Gawande MB, Goswami A, Asefa T, Guo H, Biradar AV, Peng DL, et al. Core–shell nanoparticles: Synthesis and applications in catalysis and electrocatalysis. Chem Soc Rev [Internet]. 2015 Oct 19 [cited 2025 Feb 6];44(21):7540–90. Available from: https://pubs.rsc.org/en/content/articlehtml/2015/cs/c5cs00343a

14 Jang Y, Shapiro A, Isarov M, Rubin-Brusilovski A, Safran A, Budniak AK, et al. Interface control of electronic and optical properties in IV–VI and II–VI core/shell colloidal quantum dots: A review. Chem Commun [Internet]. 2017 Jan 17 [cited 2025 Feb 6];53(6):1002–24. Available from: https://pubs.rsc.org/en/content/articlehtml/2017/cc/c6cc08742f

15 Li W, Elzatahry A, Aldhayan D, Zhao D. Core–shell structured titanium dioxide nanomaterials for solar energy utilization. Chem Soc Rev [Internet]. 2018 Nov 12 [cited 2025 Feb 6];47(22):8203–37. Available from: https://pubs.rsc.org/en/content/articlehtml/2018/cs/c8cs00443a

16 Zhang Q, Lee I, Joo JB, Zaera F, Yin Y. Core-shell nanostructured catalysts. Acc Chem Res [Internet]. 2013 Aug 20 [cited 2025 Feb 6];46(8):1816–24. Available from: https://pubs.acs.org/doi/full/10.1021/ar300230s

17 Liu X, Iocozzia J, Wang Y, Cui X, Chen Y, Zhao S, et al. Noble metal–metal oxide nanohybrids with tailored nanostructures for efficient solar energy conversion, photocatalysis and environmental remediation. Energy Environ Sci [Internet]. 2017 Feb 15 [cited 2025 Feb 6];10(2):402–34. Available from: https://pubs.rsc.org/en/content/articlehtml/2017/ee/c6ee02265k

18 Oezaslan M, Hasché F, Strasser P. Pt-based core-shell catalyst architectures for oxygen fuel cell electrodes. J Phys Chem Lett [Internet]. 2013 Oct 3 [cited 2025

Feb 6];4(19):3273–91. Available from: https://pubs.acs.org/doi/full/10.1021/jz4014135

19 Kumar V, Nisika N, Kumar M. Modified absorption and emission properties leading to intriguing applications in plasmonic–excitonic nanostructures. Adv Opt Mater [Internet]. 2021 Jan 1 [cited 2025 Feb 6];9(2):2001150. Available from: https://onlinelibrary.wiley.com/doi/full/10.1002/adom.202001150

20 Wu Z, Cheng HW, Jin C, Yang B, Xu C, Pei K, et al. Dimensional design and core–shell engineering of nanomaterials for electromagnetic wave absorption. Adv Mater [Internet]. 2022 Mar 1 [cited 2025 Feb 6];34(11):2107538. Available from: https://onlinelibrary.wiley.com/doi/full/10.1002/adma.202107538

21 Ajayan PM, Schadler LS, and Braun PV. Nanocomposite Science and Technology. Nanocomposite Science and Technology 2003 Aug 26. Available from: https://onlinelibrary.wiley.com/doi/book/10.1002/3527602127

22 Ramimoghadam D, Bagheri S, Abd Hamid SB. Stable monodisperse nanomagnetic colloidal suspensions: An overview. Colloids Surf B Biointerfaces. 2015 Sep 1;133:388–411.

23 Veeranmaril SK. Designing sinter resistant monometallic and bimetallic nanomaterials for catalysis [Doctoral thesis] [Internet]. 2020 [cited 2025 Feb 6]. Available from: http://hdl.handle.net/10388/12902

24 Hou Z, Cui C, Li Y, Gao Y, Zhu D, Gu Y, et al. Lattice-strain engineering for heterogenous electrocatalytic oxygen evolution reaction. Adv Mater [Internet]. 2023 Sep 1 [cited 2025 Mar 7];35(39):2209876. Available from: https://onlinelibrary.wiley.com/doi/full/10.1002/adma.202209876

25 Comanescu C. Recent advances in surface functionalization of magnetic nanoparticles. Coatings [Internet]. 2023 Oct 15 [cited 2025 Mar 7];13(10):1772. Available from: https://www.mdpi.com/2079-6412/13/10/1772/htm

26 Gamucci O, Bertero A, Gagliardi M, Bardi G. Biomedical nanoparticles: Overview of their surface immune-compatibility. Coatings [Internet]. 2014 Feb 12 [cited 2025 Mar 7];4(1):139–59. Available from: https://www.mdpi.com/2079-6412/4/1/139/htm

27 Sperling RA, aus Wiesbaden. Surface modification and functionalization of colloidal nanoparticles. 2008.

28 Ding L, Wang L, Zhang J, Ren N, Wang A, Liu H, et al. Nanozymes regulated by nitrogen element: Mechanism, design, and application. Adv Powder Mater. 2024 Aug 1;3(4):100191.

29 Shaheen Shah S, Abu Nayem SM, Sultana N, Saleh Ahammad AJ, Abdul Aziz M. Preparation of sulfur-doped carbon for supercapacitor applications: A review. ChemSusChem [Internet]. 2022 Jan 10 [cited 2025 Mar 7];15(1):e202101282. Available from: https://onlinelibrary.wiley.com/doi/full/10.1002/cssc.202101282

30 Huang C, Wang F, Chen X, Li J, Shao M, Wei Z. Innovative strategies for designing and constructing efficient fuel cell electrocatalysts. Chem Commun [Internet]. 2025 Feb 6 [cited 2025 Mar 7];61(13):2658–83. Available from: https://pubs.rsc.org/en/content/articlehtml/2025/cc/d4cc05928j

31 Younis SA, Maitlo HA, Lee J, Kim KH. Nanotechnology-based sorption and membrane technologies for the treatment of petroleum-based pollutants in natural ecosystems and wastewater streams. Adv Colloid Interface Sci. 2020 Jan 1;275:102071.

32 Yeo RJ, Yeo JCC, Yu Tan S, Sng A, Tomczak N, Muiruri JK, et al. Core-shell micro- and nano-structures for the modification of light-surface interactions. Adv Opt Mater [Internet]. 2024 Feb 1 [cited 2025 Mar 7];12(4):2301955. Available from: https://onlinelibrary.wiley.com/doi/full/10.1002/adom.202301955

33 Huang B, Yang H, Zhang L, Yuan Y, Cui Y, Zhang J. Effect of surface/interfacial defects on photo-stability of thick-shell CdZnSeS/ZnS quantum dots. Nanoscale [Internet]. 2018 Oct 4 [cited 2025 Mar 7];10(38):18331–40. Available from: https://pubs.rsc.org/en/content/articlehtml/2018/nr/c8nr04224a

34 Zhang YJ, Radjenovic PM, Zhou XS, Zhang H, Yao JL, Li JF, et al. Plasmonic core–shell nanomaterials and their applications in spectroscopies. Adv Mater [Internet]. 2021 Dec 1 [cited 2025 Mar 7];33(50):2005900. Available from: https://onlinelibrary.wiley.com/doi/full/10.1002/adma.202005900

35 Chen H, Daneshvar F, Tu Q, Sue HJ. Ultrastrong carbon nanotubes-copper core-shell wires with enhanced electrical and thermal conductivities as high-performance power transmission cables. ACS Appl Mater Interfaces [Internet]. 2022 Dec 21 [cited 2025 Mar 7];14(50):56253–67. Available from: https://pubs.acs.org/doi/full/10.1021/acsami.2c13686

36 Mei Y, He Y, Zhu H, Ma Z, Pu Y, Chen Z, et al. Recent advances in the structural design of silicon/carbon anodes for lithium ion batteries: A review. Coatings [Internet]. 2023 Feb 15 [cited 2025 Mar 7];13(2):436. Available from: https://www.mdpi.com/2079-6412/13/2/436/htm

37 Ke Y, Chen J, Lin G, Wang S, Zhou Y, Yin J, et al. Smart windows: electro-, thermo-, mechano-, photochromics, and beyond. Adv Energy Mater [Internet]. 2019 Oct 1 [cited 2025 Mar 7];9(39):1902066. Available from: https://onlinelibrary.wiley.com/doi/full/10.1002/aenm.201902066

38 Liang X, Matyushov A, Hayes P, Schell V, Dong C, Chen H, et al. Roadmap on magnetoelectric materials and devices. IEEE Trans Magn. 2021 Aug 1;57(8):1–57.

39 Al-Saadi A. Preparation and characterisation of encapsulation magnetic metal iron oxide nanoparticles [PhD thesis]. University of Oxford. 2012.

40 Noganta S, Titinchi S, Abbo H. Photocatalytic degradation of organic pollutants using Ag-Fe_3O_4/SiO_2/TiO_2 nanocomposite [Internet]. 2015 [cited 2025 Mar 7]. Available from: https://hdl.handle.net/10566/14550

41 Chopra N. Multifunctional and multicomponent heterostructured one-dimensional nanostructures: Advances in growth, characterisation, and applications. Mater Technol [Internet]. 2010 Sep 1 [cited 2025 Mar 7];25(3–4):212–30. Available from: https://www.tandfonline.com/doi/abs/10.1179/175355510X12723642365124

42 Zhao G, Zhang X, Qiang W, Zhang X. Enhancing the interlaminar toughness and impact resistance of CFRP using MWCNTs and core-shell rubber synergistic strategy. Thin-Walled Struct [Internet]. 2025 Mar 4 [cited 2025 Mar 7];212:113146. Available from: https://linkinghub.elsevier.com/retrieve/pii/S026382312500240X

43 Huang X, Su S, Xu Z, Miao Q, Li W, Wang L. Advanced composite materials for structure strengthening and resilience improvement. Buildings [Internet]. 2023 Sep 22 [cited 2025 Mar 7];13(10):2406. Available from: https://www.mdpi.com/2075-5309/13/10/2406/htm

44 Zhai W, Zhou W, Yu Y, Nai SML. Additive manufacturing of TiB2 particles enabled high-performance 316L with a unique core-shell melt pool structure. J Manuf Process. 2025 Apr 15;139:144–55.
45 Xiao F, Wang YC, Wu ZP, Chen G, Yang F, Zhu S, et al. Recent advances in electrocatalysts for proton exchange membrane fuel cells and alkaline membrane fuel cells. Adv Mater [Internet]. 2021 Dec 1 [cited 2025 Mar 7];33(50):2006292. Available from: https://onlinelibrary.wiley.com/doi/full/10.1002/adma.202006292
46 An S, Lee MW, Yarin AL, Yoon SS. A review on corrosion-protective extrinsic self-healing: Comparison of microcapsule-based systems and those based on core-shell vascular networks. Chem Eng J. 2018 Jul 15;344:206–20.
47 Zhang X and Zheng DF. Mechanical behaviors of individual core-shell microspheres and their polymeric composites under different strain rates. 2017 [cited 2025 Mar 7]. Available from: http://hdl.handle.net/10356/72868
48 Zou Y, Sun Z, Wang Q, Ju Y, Sun N, Yue Q, et al. Core-shell magnetic particles: Tailored synthesis and applications. Chem Rev [Internet]. 2025 Jan 22 [cited 2025 Mar 7];125(2):972–1048. Available from: https://pubs.acs.org/doi/full/10.1021/acs.chemrev.4c00710
49 Qi Y, Wang G, Li S, Liu T, Qiu J, Li H. Recent progress of structural designs of silicon for performance-enhanced lithium-ion batteries. Chem Eng J. 2020 Oct 1;397:125380.
50 Chadha U, Hafiz M, Bhardwaj P, Padmanaban S, Sinha S, Hariharan S, et al. Theoretical progresses in silicon anode substitutes for Lithium-ion batteries. J Energy Storage. 2022 Nov 1;55:105352.
51 Yu C, Wu F, Qian M, Li H, Wang R, Wang J, et al. Core-shell engineering of titanium-based anodes toward enhanced electrochemical lithium/sodium storage performance: A review. Mater Today Energy. 2024 Jul 1;43:101589.
52 Ho KC, Lin LY. A review of electrode materials based on core–shell nanostructures for electrochemical supercapacitors. J Mater Chem A Mater [Internet]. 2019 Feb 19 [cited 2025 Mar 7];7(8):3516–30. Available from: https://pubs.rsc.org/en/content/articlehtml/2019/ta/c8ta11599k
53 Yang T, Lu B, Zuo Y, Huang J. Configuration engineering of plasmonic-metal/semiconductor nanohybrids for solar fuel production†. Chem Mater [Internet]. 2025 [cited 2025 Mar 7];37(5):1685–1715. Available from: https://pubs.acs.org/doi/full/10.1021/acs.chemmater.4c03170
54 Singh Selopal G, Zhao H, Wang ZM, Rosei F, Selopal GS, Wang ZM, et al. Core/shell quantum dots solar cells. Adv Funct Mater [Internet]. 2020 Mar 1 [cited 2025 Mar 7];30(13):1908762. Available from: https://onlinelibrary.wiley.com/doi/full/10.1002/adfm.201908762
55 Wang C, An C, Qin C, Gomaa H, Deng Q, Wu S, et al. Noble metal-based catalysts with core-shell structure for oxygen reduction reaction: Progress and prospective. Nanomaterials [Internet]. 2022 Jul 19 [cited 2025 Mar 7];12(14):2480. Available from: https://www.mdpi.com/2079-4991/12/14/2480/htm
56 Ibáñez M, Zamani R, Gorsse S, Fan J, Ortega S, Cadavid D, et al. Core-shell nanoparticles as building blocks for the bottom-up production of functional nanocomposites: PbTe-PbS thermoelectric properties. ACS Nano [Internet]. 2013

Mar 26 [cited 2025 Mar 7];7(3):2573–86. Available from: https://pubs.acs.org/doi/full/10.1021/nn305971v

57 Mondal K, Sharma A. Recent advances in the synthesis and application of photocatalytic metal–metal oxide core–shell nanoparticles for environmental remediation and their recycling process. RSC Adv [Internet]. 2016 Sep 2 [cited 2025 Mar 7];6(87):83589–612. Available from: https://pubs.rsc.org/en/content/articlehtml/2016/ra/c6ra18102c

58 Shafiee A, Rabiee N, Ahmadi S, Baneshi M, Khatami M, Iravani S, et al. Core–shell nanophotocatalysts: Review of materials and applications. ACS Appl Nano Mater [Internet]. 2022 Jan 28 [cited 2025 Mar 7];5(1):55–86. Available from: https://pubs.acs.org/doi/full/10.1021/acsanm.1c03714

59 Fei X, Wang P, Zhang D, Wang H, Wu Z. Confined catalysts application in environmental catalysis: Current research progress and future prospects. ChemCatChem [Internet]. 2021 May 19 [cited 2025 Mar 7];13(10):2313–36. Available from: https://onlinelibrary.wiley.com/doi/full/10.1002/cctc.202001578

60 Jin S, Park BC, Ham WS, Pan L, Kim YK. Effect of the magnetic core size of amino-functionalized Fe_3O_4-mesoporous SiO_2 core-shell nanoparticles on the removal of heavy metal ions. Colloids Surf A Physicochem Eng Asp. 2017 Oct 20;531:133–40.

61 Iftekhar S, Deb A, Heidari G, Sillanpää M, Lehto VP, Doshi B, et al. A review on the effectiveness of nanocomposites for the treatment and recovery of oil spill. Environ Sci Pollut Res [Internet]. 2023 Jan 7 [cited 2025 Mar 7];30(7):16947–83. Available from: https://link.springer.com/article/10.1007/s11356-022-25102-1

62 Bhattacharjee Y, Chatterjee D, Bose S. Core-multishell heterostructure with excellent heat dissipation for electromagnetic interference shielding. ACS Appl Mater Interfaces [Internet]. 2018 Sep 12 [cited 2025 Mar 7];10(36):30762–73. Available from: https://pubs.acs.org/doi/full/10.1021/acsami.8b10819

63 Bahadoran A, De Lile JR, Masudy-Panah S, Sadeghi B, Li J, Sabzalian MH, et al. Photocatalytic materials obtained from e-waste recycling: Review, techniques, critique, and update. J Manufact Mater Process [Internet]. 2022 Jun 26 [cited 2025 Mar 7];6(4):69. Available from: https://www.mdpi.com/2504-4494/6/4/69/htm

64 Zou H, Luo Z, Yang X, Xie Q, Zhou Y. Toward emerging applications using core–shell nanostructured materials: A review. J Mater Sci. 2022 Jun 1;57(24):10912–42.

65 Gu Q, Zhu J, Weng G-J, Li J-J, Zhao J-W. Core-satellite nanostructures and their biomedical applications. Microchimica Acta. 2022 Dec 1;189(12):470.

66 Ramadhan SA, Ali DS. Innovations in core–shell nanoparticles: advancing drug delivery solutions and precision medicine. OMICS [Internet]. 2025 Mar 1 [cited 2025 Mar 7];29(3):73–86. Available from: https://www.liebertpub.com/doi/10.1089/omi.2024.0182

67 Albinali KE, Zagho MM, Deng Y, Elzatahry AA. A perspective on magnetic core–shell carriers for responsive and targeted drug delivery systems. Int J Nanomedicine [Internet]. 2019 [cited 2025 Mar 7];14:1707–23. Available from: https://www.tandfonline.com/action/journalInformation?journalCode=dijn20

68 Das M, Mohanty C, Sahoo SK. Ligand-based targeted therapy for cancer tissue. Expert Opin Drug Deliv [Internet]. 2009 Mar [cited 2025 Mar 7];

6(3):285–304. Available from: https://www.tandfonline.com/doi/abs/10.1517/17425240902780166

69 Han X, Alu A, Liu H, Shi Y, Wei X, Cai L, et al. Biomaterial-assisted biotherapy: A brief review of biomaterials used in drug delivery, vaccine development, gene therapy, and stem cell therapy. Bioact Mater. 2022 Nov 1;17:29–48.

70 Shi Y, Zhang Y, Zhu L, Miao Y, Zhu Y, Yue B. Tailored drug delivery platforms: Stimulus-responsive core–shell structured nanocarriers. Adv Healthc Mater [Internet]. 2024 Jan 1 [cited 2025 Mar 7];13(1):2301726. Available from: https://onlinelibrary.wiley.com/doi/full/10.1002/adhm.202301726

71 Xu H, Li Z, Si J. Nanocarriers in gene therapy: A review. J Biomed Nanotechnol. 2014 Dec 1;10(12):3483–507.

72 Wei Hu S, Ding T, Tang H, Guo H, Cui W, Shu Y. Nanobiomaterial vectors for improving gene editing and gene therapy. Mater Today. 2023 Jun 1;66:114–36.

73 Khalilov R. A comprehensive review of advanced nano-biomaterials in regenerative medicine and drug delivery. Adv Biol Earth Sci. 2023;8(1):5–18.

74 Du Z, Cao G, Li K, Zhang R, Li X. Nanocomposites for the delivery of bioactive molecules in tissue repair: Vital structural features, application mechanisms, updated progress and future perspectives. J Mater Chem B [Internet]. 2020 Nov 25 [cited 2025 Mar 7];8(45):10271–89. Available from: https://pubs.rsc.org/en/content/articlehtml/2020/tb/d0tb01670e

75 Wang Z, Shen G. Flexible optoelectronic sensors: Status and prospects. Mater Chem Front [Internet]. 2023 Apr 11 [cited 2025 Mar 7];7(8):1496–519. Available from: https://pubs.rsc.org/en/content/articlehtml/2023/qm/d2qm01319c

76 Tehrani M. Advanced electrical conductors: An overview and prospects of metal nanocomposite and nanocarbon based conductors. Physica Status Solidi (A) [Internet]. 2021 Apr 1 [cited 2025 Mar 7];218(8):2000704. Available from: https://onlinelibrary.wiley.com/doi/full/10.1002/pssa.202000704

77 Zhuo MP, Wang XD, Liao LS. Construction and optoelectronic applications of organic core/shell micro/nanostructures. Mater Horiz [Internet]. 2020 Dec 7 [cited 2025 Mar 7];7(12):3161–75. Available from: https://pubs.rsc.org/en/content/articlehtml/2020/mh/d0mh00967a

78 Chen M, Lu L, Yu H, Li C, Zhao N. Integration of colloidal quantum dots with photonic structures for optoelectronic and optical devices. Adv Sci [Internet]. 2021 Sep 1 [cited 2025 Mar 7];8(18):2101560. Available from: https://onlinelibrary.wiley.com/doi/full/10.1002/advs.202101560

79 Koo JH, Kim DC, Shim HJ, Kim TH, Kim DH. Flexible and stretchable smart display: Materials, fabrication, device design, and system integration. Adv Funct Mater [Internet]. 2018 Aug 1 [cited 2025 Mar 7];28(35):1801834. Available from: https://onlinelibrary.wiley.com/doi/full/10.1002/adfm.201801834

80 Khan B, Riaz Z, Ahmad RuS, Khoo BL. Advancements in wearable sensors for cardiovascular disease detection for health monitoring. Mater Sci Eng R: Rep. 2024 Jun 1;159:100804.

81 Chen O, Wei H, Maurice A, Bawendi M, Reiss P. Pure colors from core–shell quantum dots. MRS Bull [Internet]. 2013 Sep [cited 2025 Mar 7];38(9):696–702. Available from: https://www.cambridge.org/core/journals/mrs-bulletin/article/

abs/pure-colors-from-coreshell-quantum-dots/4B8AC6E7E8B1B554420EE8702AC724E5

82 Zhang X, Jiang C, Liang J, Wu W. Electrode materials and device architecture strategies for flexible supercapacitors in wearable energy storage. J Mater Chem A Mater [Internet]. 2021 Apr 6 [cited 2025 Mar 7];9(13):8099–128. Available from: https://pubs.rsc.org/en/content/articlehtml/2021/ta/d0ta12299h

2

Fundamentals of Core-shell Nanomaterials

Core-shell nanomaterials exhibit a remarkable degree of tunability in their properties, which stems from the precise engineering of their architectures. These structures are not limited to a single design; instead, they encompass a broad spectrum of configurations, each tailored for specific functional applications. The fundamental design principles of core-shell architectures revolve around the interplay between core and shell materials, their structural arrangements, and interfacial interactions. Understanding these aspects is essential for optimizing their performance in targeted applications such as catalysis, energy storage, and biomedical technologies. The diversity in core-shell architectures arises from the variations in their structural configurations, material compositions, and the nature of the interfaces between the core and the shell. These variations dictate key properties such as optical absorption, electronic conductivity, mechanical robustness, and chemical reactivity. For example, while solid core-shell structures offer enhanced stability and controlled surface functionalities, hollow and multi-shell architectures provide increased surface area and tunable porosity, making them highly effective for applications like drug delivery and photocatalysis [1]. Janus and asymmetric core-shell systems introduce additional complexity by incorporating two or more distinct functional regions, leading to anisotropic properties beneficial for directed self-assembly and interfacial engineering [2]. Beyond their structural diversity, core-shell materials also exhibit complex interfacial phenomena that influence their behavior at the nanoscale.

The interaction between the core and the shell governs charge transfer, phonon coupling, and lattice strain effects, which can dramatically impact their optical, electronic, and catalytic performance. Theoretical and computational approaches play a critical role in elucidating these interactions, providing insight into the fundamental principles that dictate core-shell behavior. By leveraging advanced modeling techniques, researchers can predict and fine-tune the material properties before synthesis, accelerating the development of next-generation nanomaterials. Another crucial aspect of core-shell nanomaterials is their composition, which spans metals, metal oxides, polymers, carbon-based materials, and hybrid systems. The choice of materials for the core and shell components is dictated by the intended application, with noble metals, transition metals, and semiconductor oxides frequently

Core-Shell Nanomaterials: From Fundamentals to Applications, First Edition. Shreya Sharma et al.

employed for optical and electronic applications, while polymeric and biomimetic core-shell systems find extensive use in biomedical and environmental technologies. The integration of multiple material types in hybrid and composite core-shell architectures further expands the range of achievable properties, enabling multi-functional materials with tailored characteristics [3]. A key challenge in core-shell nanomaterial design is establishing precise structure–property relationships, where factors such as core and shell thickness, interface characteristics, and compositional gradients must be carefully controlled. These factors influence the stability, mechanical strength, electronic behavior, and catalytic efficiency of the materials. The ability to fine-tune these parameters through rational design and controlled synthesis opens new frontiers for the development of highly specialized nanomaterials.

This chapter provides an in-depth exploration of core-shell architectures, beginning with an overview of different structural configurations and their unique attributes. It then delves into the theoretical principles governing core-shell interactions, emphasizing interfacial effects and their implications on material performance. A comprehensive discussion on the diverse material compositions follows, highlighting how different combinations of core and shell components give rise to distinct functionalities. Finally, the chapter examines the intricate structure–property relationships that dictate the overall behavior of core-shell systems, providing a scientific foundation for their design and optimization in advanced applications.

2.1 Core-shell Configurations and Types

The structural design of core-shell nanomaterials plays a pivotal role in determining their physical, chemical, and functional properties. Different configurations arise based on how the core and shell components interact, the geometric arrangement of the shell around the core, and the resulting interfacial properties. These architectures are tailored to meet specific performance criteria in applications such as catalysis, energy storage, drug delivery, and environmental remediation. Core-shell configurations can be broadly categorized based on their structural integrity, material distribution, and functional asymmetry [4]. Some structures feature a dense, uniform shell encasing a core, while others incorporate porous or hollow shells to modulate diffusion properties and surface interactions, as shown in Figure 2.1. Additionally, certain architectures consist of multiple shells layered around a core, introducing complexity that enhances performance in optical, electronic, and catalytic applications. Other variations include asymmetric or Janus-type structures, where the shell is non-uniformly distributed, leading to anisotropic behavior beneficial for directional assembly and selective surface interactions [5]. The interfacial properties between the core and shell are crucial in determining the overall stability, reactivity, and efficiency of these nanomaterials. A well-defined interface can facilitate controlled charge transfer, phonon confinement, and strain engineering, all of which influence key material characteristics. Additionally, the ability to tune the thickness, porosity, and composition of the shell provides a means to manipulate these properties further. Understanding the various core-shell configurations

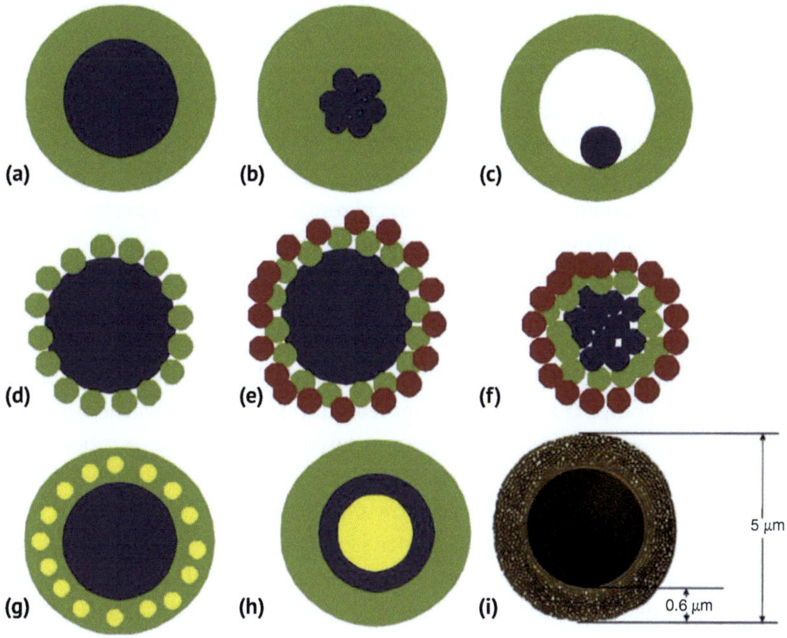

Figure 2.1 Schematic illustration of various core–shell particle types. Reproduced from [6]/Elsevier/CC BY 3.0.

and their structural implications is essential for designing advanced nanomaterials with tailored functionalities. The following subsections explore different types of core-shell architectures, detailing their structural characteristics, advantages, and application-specific benefits.

2.1.1 Solid Core-shell Structures

Solid core-shell nanostructures are among the most fundamental and widely studied architectures in nanotechnology. These structures consist of a dense, nonporous core fully encapsulated by a solid shell of a different material. The combination of distinct core and shell compositions allows for tailored physicochemical properties that can be optimized for specific applications in catalysis, energy storage, biomedical fields, and optical technologies. The solid nature of both the core and shell provides structural integrity, high stability, and controlled surface functionality, making them highly advantageous compared to other core-shell configurations. Unlike hollow or multi-shell architectures, solid core-shell nanostructures maintain a compact and dense structure throughout, without internal voids or multilayered coatings [7]. This difference fundamentally alters their mechanical stability, thermal resistance, and electronic interactions. While hollow core-shell structures excel in applications requiring high surface area and low density, solid core-shell materials are often preferred when durability, mechanical robustness, and controlled surface properties are more critical [8]. In contrast to Janus and asymmetric core-shell structures, which exhibit anisotropic behavior due to their asymmetric composition, solid core-shell systems offer uniformity in chemical and physical characteristics, ensuring consistent performance in bulk applications [9]. Another key distinction is in the way solid core-shell nanomaterials handle interfacial interactions. The well-defined interface between the core and shell facilitates controlled charge and energy transfer, which is particularly beneficial for applications in catalysis, photovoltaics, and electronic devices. The tight integration between core and shell components often enhances electronic coupling, strain effects, and mechanical reinforcement, all of which contribute to superior performance characteristics.

Solid core-shell materials exhibit a range of tunable properties based on the composition, thickness, and morphology of the shell, as well as the nature of the core material. The dense, protective shell prevents oxidation, corrosion, or degradation of the core material, making these structures ideal for long-term applications [10]. The shell composition can be engineered to introduce specific surface functionalities, allowing for selective binding, catalytic activity, or controlled diffusion properties. Core-shell interactions can be exploited to modulate bandgap energies, plasmonic resonances, and electronic conductivity, which is crucial for applications in semiconductors and photonics [11]. The shell can provide additional structural strength to the core, improving resistance to mechanical deformation or thermal expansion. Compared to single-component nanomaterials, solid core-shell structures offer better resistance to harsh chemical environments and high-temperature conditions. The unique properties of solid core-shell structures enable their use in a broad spectrum of technological and industrial applications. Many catalytic reactions require stable and active nanoparticles with high surface area and controlled electronic properties. Core-shell structures, particularly those with noble metal shells (e.g., Au@Pd, Pt@Ru), offer enhanced catalytic efficiency by tuning surface energy and charge transfer dynamics [12]. In lithium-ion batteries, supercapacitors, and fuel cells, solid core-shell materials improve charge transport, reduce degradation, and enhance energy storage capacity. Core-shell metal oxides (e.g., SiO_2@TiO_2) are commonly used for electrode materials due to their electrochemical stability [13]. Core-shell nanoparticles with biocompatible coatings (e.g., Fe_3O_4@SiO_2, Au@PEG) are widely used in drug delivery, bioimaging, and diagnostics. The solid shell protects the core from premature degradation while allowing surface modification for targeted interactions with biological tissues. The tunable optical properties of core-shell materials make them suitable for use in LEDs, sensors, and photonic devices. Gold-and silver-based core-shell nanostructures are particularly useful in plasmonic applications and surface-enhanced Raman spectroscopy (SERS) [14]. Solid core-shell structures play a crucial role in environmental remediation, including pollutant degradation, water purification, and gas sensing. The stability of the shell ensures prolonged activity and selectivity in harsh environments.

What sets solid core-shell nanostructures apart is their ability to integrate distinct functionalities within a compact, stable framework. The precise control over shell thickness, composition, and morphology enables the fine-tuning of chemical reactivity, optical response, and electronic properties, which is often not achievable with single-component nanomaterials. Additionally, their robustness makes them highly desirable for applications that require long-term stability under extreme conditions.

Moreover, advances in synthesis techniques, such as controlled deposition, atomic layer deposition (ALD), and solvothermal methods, have enabled the fabrication of highly uniform and defect-free core-shell nanomaterials [15]. These advancements further expand their applicability in next-generation materials for sustainable energy, nanomedicine, and advanced manufacturing.

2.1.2 Hollow Core-shell Structures

Hollow core-shell structures are a unique class of nanomaterials characterized by a well-defined shell surrounding a central cavity. Unlike solid core-shell materials, which have a dense core, hollow core-shell architectures feature an empty or porous interior, offering a distinctive combination of low density, high surface area, and tunable shell permeability. These characteristics make them highly desirable for applications in drug delivery, catalysis, energy storage, and environmental remediation. By carefully engineering the core-shell interface, material composition, and shell thickness, researchers can optimize their performance for specific functionalities. The key distinction between hollow core-shell nanomaterials and other configurations lies in their interior cavity, which provides advantages in terms of reduced weight, enhanced mass transport, and improved surface reactivity [16]. While solid core-shell structures emphasize stability and mechanical strength, hollow architectures are designed for applications that require high porosity, efficient encapsulation, or rapid molecular exchange [17]. In contrast to multi-shell or layered architectures, which incorporate multiple concentric shells, hollow core-shell nanomaterials typically have a single outer shell, though variations exist with multiple internal layers for additional functionality. Their unique geometry allows for efficient loading and release of active species, such as drugs, catalysts, or ions, which is not possible with dense solid-core materials. Additionally, the hollow structure can accommodate volume changes more effectively, making them highly resilient in applications like lithium-ion batteries, where expansion and contraction during charge–discharge cycles must be managed [18].

The properties of hollow core-shell nanomaterials are primarily governed by their shell composition, shell thickness, porosity, and the presence of functional groups. The presence of an internal cavity increases the effective surface area, improving interactions with external molecules and enhancing catalytic or adsorption capabilities. The hollow structure reduces material weight while maintaining structural integrity, making them ideal for lightweight materials in aerospace, energy storage, and biomedical fields [19]. The hollow core facilitates rapid diffusion of reactants and products, which is particularly beneficial in catalysis and drug delivery applications. By adjusting the shell thickness, porosity, or chemical composition, the permeability of the shell can be controlled, allowing selective diffusion of molecules for controlled release applications. The hollow nature can accommodate stress and deformation more effectively than solid structures, making them useful in impact-resistant coatings and flexible electronics. Hollow core-shell materials can be designed with thermally stable and chemically inert shells, protecting the core from degradation in harsh environments. Hollow core-shell structures offer a wide range of applications due to their unique properties [20]. Hollow nanostructures are widely used for controlled drug release, where the cavity serves as a reservoir for therapeutic agents. Biodegradable polymer or silica-based hollow core-shell nanoparticles (e.g., SiO_2@lipid, PLGA@PEG) can encapsulate drugs, proteins, or genes and release them in response to stimuli such as pH, temperature, or enzymes. The high surface area and accessible interior of hollow core-shell structures enhance catalytic efficiency. Hollow metal or metal oxide shells (e.g., Pd@SiO_2, Au@CeO_2) provide active sites while minimizing material usage, improving reaction kinetics, and enhancing selectivity in catalytic processes. In batteries and supercapacitors, hollow core-shell nanostructures improve charge storage and cycling stability. For example, hollow carbon-coated metal oxides (e.g., Fe_3O_4@C) provide high capacity while reducing volume expansion issues in lithium-ion batteries [21]. In hydrogen storage, hollow materials help trap and release hydrogen efficiently.

The high adsorption capacity of hollow structures enables their use in removing pollutants, heavy metals, and organic contaminants from water and air. Porous hollow metal oxides, such as TiO_2@SiO_2, have been extensively studied for photocatalytic degradation of pollutants. Hollow core-shell materials with tunable optical properties, such as plasmonic nanostructures (e.g., Au@SiO_2), are used in biosensing, SERS [22], and light-harvesting applications. Their ability to trap and guide light enhances sensitivity and detection capabilities. The hollow core allows for encapsulation of functional agents such as corrosion inhibitors, self-healing materials, or flame retardants, which are released upon external stimulus, providing smart coating solutions. The defining feature of hollow core-shell structures is their ability to integrate a functional shell with an internal void, enabling diverse applications that require lightweight materials, high reactivity, or controlled cargo release. Their tunable permeability and surface chemistry provide significant advantages over solid core-shell and other nanostructures. Advances in synthesis techniques, such as template-assisted fabrication, self-assembly, and Kirkendall effect-based methods, have enabled precise control over shell thickness, porosity, and composition, allowing for the design of highly tailored hollow nanomaterials. By leveraging these features, researchers continue to explore new applications in areas such as nanomedicine, energy storage, and catalysis, making hollow core-shell structures a cornerstone of advanced nanotechnology.

2.1.3 Multi-shell (Layered) Architectures

Multi-shell or layered core-shell architectures are an advanced class of nanomaterials characterized by multiple concentric shells surrounding a central core [23]. Unlike conventional single-layer core-shell structures, these materials incorporate additional layers that enhance their functionality, stability, and tunability. By carefully designing the composition, thickness, and sequence of layers, researchers can tailor multi-shell nanomaterials for a variety of applications, including photovoltaics, catalysis, biomedical drug delivery, and energy storage. The hierarchical organization of these structures allows for unique combinations of optical, electronic, and mechanical properties within a single nanosystem. One of the defining features of multi-shell core-shell structures is their ability to integrate multiple functionalities. The additional shell layers contribute to improved charge separation in solar cells, extended drug release in biomedical applications, and enhanced catalytic efficiency [24]. These structures can be classified based on their organization, including concentric multi-shell systems, where sequential layers encapsulate the core; hierarchical multi-shell structures, which have varying degrees of porosity and selective permeability; and gradient multi-shell systems, which feature gradual transitions in composition or doping concentration to optimize interfacial interactions. The key properties of multi-shell architectures stem from their complex structural design. One of the most significant advantages is their enhanced stability and durability, as the outer shells serve as protective barriers against oxidation, heat, and chemical degradation [20]. Additionally, multi-shell systems facilitate efficient charge and energy transfer, which is crucial in optoelectronic and energy storage applications. The layered structure also allows for controlled mass transport, making these materials highly useful in catalysis, drug delivery, and sensing technologies. Moreover, the tunable optical and electronic properties of multi-shell nanostructures arise from the interaction between the core and shell layers, enabling tailored absorption, fluorescence, and plasmonic behavior for advanced optical applications [25].

Due to their versatile nature, multi-shell nanostructures find applications in a wide range of scientific and industrial fields. In photovoltaics, materials such as Si@TiO_2@MoS_2 are used to improve light absorption and charge separation, leading to enhanced solar cell efficiency. In catalysis, multilayered catalysts like Pt@CeO_2@TiO_2 exhibit superior catalytic activity due to improved electron transfer and stability. In biomedical sciences, polymeric and inorganic multi-shell nanoparticles allow for controlled drug release, with each shell layer responding to different environmental stimuli such as pH or enzymatic activity. Furthermore, multi-shell materials enhance lithium-ion battery performance by preventing volume expansion and improving ion transport, making them valuable for energy storage. In environmental remediation, photocatalytic multi-shell materials such as ZnO@TiO_2@SiO_2 are employed to degrade organic pollutants and remove heavy metals from wastewater. The unique ability of multi-shell architectures to combine multiple functionalities within a single material makes them an attractive platform for next-generation nanotechnologies. Advances in fabrication techniques such as layer-by-layer deposition and ALD have enabled precise control over shell thickness and composition, further expanding their potential applications. Future research in this field aims to develop smart multi-shell nanostructures with self-healing properties, adaptive functionalities, and dynamic property modulation. The integration of multi-shell nanomaterials with artificial intelligence (AI)-driven materials science, flexible electronics, and nanomedicine is expected to pave the way for revolutionary advancements in technology and industry.

2.1.4 Janus and Asymmetric Core-shell Systems

Janus and asymmetric core-shell nanostructures represent a unique class of materials characterized by their heterogeneous composition and anisotropic (directionally dependent) properties [26]. Unlike traditional symmetric core-shell nanomaterials, which feature a uniform shell surrounding a core, Janus and asymmetric structures have distinct regions with different chemical compositions, surface functionalities, or physical properties. This spatial asymmetry imparts these materials with highly tunable functionalities, enabling applications in fields such as targeted drug delivery, self-propelling nanomachines, catalysis, and advanced optoelectronics. Janus particles are named after the Roman god Janus, who had two faces, symbolizing the duality of their surface properties [27]. These particles typically have two distinct hemispheres with contrasting characteristics, such as hydrophilic–hydrophobic, magnetic–nonmagnetic, or conductive–insulating regions. Asymmetric core-shell systems, on the other hand, may exhibit a gradient in composition or an off-centered core, leading to anisotropic electronic, optical, or catalytic properties. These unique structural features allow precise control over interactions with external fields, fluids, or biological environments. The structural uniqueness of Janus and asymmetric core-shell systems arises from their controlled anisotropy. The spatial separation of different chemical functionalities allows for directional interactions, enabling selective self-assembly and orientation-dependent behavior. Some Janus particles possess a solid core with two distinct surface modifications, while others may have asymmetric shells with variations in thickness or composition. Several synthesis strategies have been developed to fabricate these asymmetric nanomaterials, including **phase separation techniques**, such as emulsion-based methods, where two immiscible materials segregate during particle formation [28]; **seeded growth methods**, where controlled deposition of different materials onto selective regions of the core leads to asymmetric structures [29]; **template-assisted fabrication**, which utilizes pre-designed templates to create spatially heterogeneous coatings [30]; **microfluidic synthesis**, enabling precise control over the shape, size, and composition of Janus nanoparticles for tailored applications [31].

Janus and asymmetric nanoparticles exhibit several distinctive properties compared to their symmetric counterparts. The dual nature of Janus particles allows them to align at interfaces selectively, making them ideal for stabilizing emulsions and directing self-assembly. Asymmetric core-shell catalysts can exhibit site-specific reactivity, improving efficiency in heterogeneous catalysis [32]. Variations in shell composition can lead to directional light absorption, tunable plasmonic responses, and anisotropic charge transport in electronic devices. Janus particles with catalytic coatings, such as platinum on one hemisphere, can undergo self-propulsion in fuel-containing environments, enabling applications in nanorobotics and microfluidics [33]. This property is crucial for electrophoretic displays, where asymmetric nanoparticles align under an electric field to modulate display brightness and contrast. The multifunctional nature of Janus and asymmetric nanostructures has led to their widespread adoption across various scientific and technological domains. Functionalized Janus nanoparticles are widely used in targeted drug delivery and imaging. The asymmetric surface chemistry allows selective binding to biological targets, improving therapeutic efficiency. Additionally, Janus nanoparticles with magnetic and fluorescent components enable multimodal imaging. Asymmetric core-shell catalysts exhibit site-selective activity, enhancing reaction kinetics in oxidation, reduction, and photocatalytic processes. For example, Janus Pt-TiO_2 catalysts enhance hydrogen evolution reactions (HER) due to their dual functional interfaces. Janus particles functionalized with catalytic coatings, such as Pt-Au systems, propel themselves by decomposing chemical fuels, making them promising for environmental remediation and microscale transport applications. Janus nanoparticles stabilize complex emulsions by selectively interacting with different solvent phases, making them valuable in cosmetics, food processing, and drug formulations. Asymmetric core-shell particles with tunable dielectric properties are utilized in waveguides, optical sensors, and high-performance display technologies.

The development of Janus and asymmetric core-shell systems continues to evolve with advances in synthesis techniques and theoretical modeling. Future research aims to design smart and responsive Janus materials capable of dynamically changing properties in response to external stimuli, such as temperature, pH, or light. Challenges remain in scaling up production while maintaining precise control over asymmetry, as well as in achieving long-term stability for practical applications. Nevertheless, the ability of Janus and asymmetric nanostructures to bridge the gap between multifunctionality and precision makes them an exciting frontier in nanotechnology, with significant implications for medicine, energy, and environmental science.

2.1.5 Core-shell Hybrid and Composite Nanostructures

Core-shell hybrid and composite nanostructures represent a class of advanced materials designed to integrate multiple functional components within a single nanoscale architecture. Unlike conventional core-shell systems, where the shell primarily serves as a protective or functionalizing layer, hybrid and composite core-shell structures are engineered to exploit the combined advantages of different materials. By incorporating diverse material classes such as metals, oxides, polymers, carbon nanostructures, or biomolecules, these nanostructures exhibit enhanced mechanical, optical, electronic, magnetic, and catalytic properties. Their multifunctional nature makes them highly valuable for applications in energy conversion, environmental remediation, biomedicine, and smart materials development.

Hybrid and composite core-shell materials can be classified based on their composition and functional interactions. Metal-metal core-shell structures, such as Au@Pt and Ag@Pd, are widely used in catalysis due to their enhanced electronic interactions and surface activity. Metal-oxide core-shell systems, including Fe_3O_4@TiO_2 and Co@SiO_2, combine magnetic and photocatalytic properties for environmental and biomedical applications. Polymer-inorganic hybrids, such as SiO_2@polyaniline, offer mechanical flexibility, stimuli-responsiveness, and biocompatibility, making them useful in drug delivery and wearable electronics. Carbon-based hybrid nanostructures, including graphene or carbon nanotubes (CNTs) combined with metals or oxides, provide exceptional electrical conductivity and chemical stability for energy storage and sensing applications [34]. Additionally, bioinspired core-shell materials mimic natural architectures, such as shell-protected enzymes or virus-like nanoparticles, for applications in drug delivery, biosensing, and tissue engineering. The synthesis of hybrid core-shell nanostructures requires precise control over material interfaces to maintain the distinct functionalities of each component while enabling synergistic interactions [7]. Sol–gel processing is commonly used to coat metallic or polymeric cores with oxide shells, providing thermal and chemical stability. Electrostatic self-assembly takes advantage of charge interactions to construct polymer-inorganic hybrids, while hydrothermal and solvothermal methods allow controlled crystallization of metal oxides onto various cores [35]. Layer-by-layer deposition provides precise control over shell thickness and composition, making it useful for smart coatings and biomedical applications. Advanced techniques such as chemical vapor deposition (CVD) and ALD enable conformal coating of semiconductor-based hybrid nanostructures, ensuring high-quality shell formation.

The unique properties of core-shell hybrid and composite nanostructures arise from the synergistic combination of different materials. The integration of metal-semiconductor hybrids, such as Au@TiO_2, enhances charge transport and photocatalytic efficiency by promoting electron-hole separation. Polymer-coated metal nanostructures improve mechanical strength and flexibility while preventing oxidation. Plasmonic metal cores combined with optically active shells enable tunable light absorption and scattering for photonic and imaging applications [36]. In catalysis, structures such as Pt@CeO_2 or Fe_3O_4@SiO_2 provide multiple active sites for oxidation and reduction reactions, enhancing overall efficiency. Additionally, stimuli-responsive hybrid materials exhibit changes in properties based on environmental conditions, making them useful for smart coatings and controlled drug release. Given their multifunctionality, core-shell hybrid and composite nanostructures are widely utilized in various scientific and industrial applications. In energy storage and conversion, carbon-based composites with metal oxides are used in supercapacitors, lithium-ion batteries, and fuel cells, enhancing charge storage and cycling stability. Environmental applications include Fe_3O_4@TiO_2 nanocomposites for wastewater treatment, which combine photocatalysis with magnetic recoverability. In biomedicine, hybrid nanoparticles with biocompatible coatings are employed for targeted drug delivery, bioimaging, and photothermal therapy [37]. Wearable and flexible electronics benefit from conductive polymer-based nanocomposites, enabling stretchable sensors and energy-harvesting devices. Catalysis applications take advantage of hybrid nanostructures to improve reaction selectivity and catalyst longevity in industrial chemical synthesis [38]. The continued advancement of core-shell hybrid and composite nanostructures depends on optimizing their structural integrity, interfacial interactions, and large-scale fabrication. Emerging research is focused on leveraging AI and computational modeling to design highly efficient hybrid materials tailored for specific applications. However, challenges such as scalability, reproducibility, and long-term stability must be addressed before these materials achieve widespread industrial adoption. Despite these challenges, the ability of hybrid core-shell nanostructures to integrate multiple functionalities within a single nanoscale entity makes them one of the most promising materials for next-generation nanotechnology.

2.2 Theoretical Basis of Core-shell Interactions

The functionality and performance of core-shell nanomaterials are fundamentally dictated by the interactions between the core and shell components. These interactions are governed by a combination of interfacial phenomena, charge transfer mechanisms, strain effects, and chemical compatibility between the materials.

Theoretical models describing these interactions provide insights into how the core and shell influence each other's properties, leading to tunable electronic, optical, mechanical, and catalytic behaviors. Understanding these interactions is crucial for the rational design of core-shell structures with optimized functionalities for applications in energy, electronics, catalysis, and biomedicine. The interfacial properties of core-shell materials determine their overall stability, electronic coupling, and charge transport efficiency. For example, in metal-semiconductor systems, the core-shell interface influences band alignment and charge carrier separation, which is essential for photocatalysis and optoelectronics [39]. Similarly, in catalytic applications, strain effects at the interface modify the electronic structure of active sites, thereby enhancing reaction kinetics. Additionally, mechanical and thermal stability considerations play a vital role in ensuring the longevity of core-shell materials, particularly in high-temperature or high-stress environments. Recent advances in computational modeling and simulation techniques have further enabled the prediction and optimization of core-shell interactions, paving the way for the design of next-generation multifunctional nanomaterials.

2.2.1 Interfacial Phenomena in Core-shell Materials

The interface between the core and shell in core-shell nanomaterials plays a decisive role in determining their structural integrity, physicochemical stability, and functional performance. Interfacial phenomena govern the extent of charge transfer, mechanical stress distribution, and chemical reactivity between the two components, ultimately influencing the material's electronic, optical, catalytic, and mechanical properties [40]. The nature of these interactions depends on factors such as the chemical compatibility between the core and shell, lattice mismatch, interfacial bonding strength, and defect formation at the interface. Interfacial interactions in core-shell materials can be broadly classified into covalent bonding, van der Waals forces, electrostatic interactions, and epitaxial strain effects. Covalent bonding occurs in systems where strong chemical bonds form between the core and shell, ensuring excellent structural stability [41]. In contrast, van der Waals forces dominate in weakly interacting systems such as polymeric or carbon-based core-shell structures, where the shell is physically adsorbed onto the core. Electrostatic interactions play a crucial role in the self-assembly of charged nanoparticles, especially in polyelectrolyte-based and biomolecular core-shell architectures. In crystalline core-shell structures, epitaxial strain arises due to lattice mismatch between the core and shell materials, leading to strain-induced modifications in electronic and optical properties.

The electronic structure of core-shell nanomaterials is strongly influenced by charge transfer processes occurring at the interface. In metal-semiconductor core-shell systems, for example, the alignment of Fermi levels dictates the movement of charge carriers, affecting conductivity and catalytic performance [42]. A well-matched interface can facilitate efficient electron transport, enhancing the material's performance in photovoltaics and photocatalysis. Conversely, an energetically unfavorable interface can lead to charge carrier recombination, reducing efficiency. Similarly, in metal-metal core-shell systems, such as Au@Pd or Pt@Ni, d-band interactions modify catalytic activity by shifting the adsorption energies of reaction intermediates. When the core and shell have different lattice parameters, lattice mismatch leads to strain formation at the interface, influencing the mechanical and electronic properties of the material. A small lattice mismatch can induce beneficial strain effects, tuning the band structure and catalytic activity, as observed in Pt@Ru and Au@Pt catalysts. However, excessive strain can lead to defect formation, shell delamination, or even structural collapse. Advanced deposition techniques, such as ALD, are often used to precisely engineer strain effects by controlling shell thickness and composition. Interfacial diffusion is another critical phenomenon that affects the stability and functionality of core-shell structures. Over time, thermal or chemical conditions may induce diffusion between the core and shell, leading to interfacial alloying or phase segregation. In metal-semiconductor systems, interfacial diffusion can alter band structures, affecting charge transport and optical behavior. Strategies such as the introduction of diffusion barrier layers (e.g., SiO_2 or Al_2O_3) and controlled synthesis conditions help in maintaining well-defined core-shell interfaces [43].

Defects such as vacancies, dislocations, and grain boundaries at the core-shell interface significantly impact material properties. For example, oxygen vacancies in oxide-core/oxide-shell structures can modify electronic conductivity and catalytic performance. Interface engineering techniques, including doping, surface passivation, and controlled oxidation, are employed to minimize defect-related performance degradation. Computational modeling approaches further aid in predicting defect formation and optimizing synthesis routes. Interfacial phenomena in core-shell materials are critical for diverse applications. In catalysis, optimized interfaces in metal-core/oxide-shell structures enhance reaction kinetics and selectivity. In optoelectronics, engineered core-shell interfaces improve charge carrier dynamics, benefiting LED and solar cell efficiency. Biomedical applications leverage biocompatible interfaces for targeted drug delivery and imaging. The continuous development of interface-modulated core-shell materials is expected to drive innovations in nanotechnology, enabling superior performance across multiple disciplines.

2.2.2 Electronic and Optical Coupling Between Core and Shell

The electronic and optical properties of core-shell nanomaterials are significantly influenced by the interactions between the core and shell components. These interactions determine charge carrier dynamics, exciton behavior, band structure modifications, and light-matter interactions, making core-shell architectures highly tunable for applications in optoelectronics, photovoltaics, plasmonics, and photocatalysis. The degree of electronic and optical coupling depends on factors such as material composition, band alignment, core-shell thickness ratio, and interface quality. Electronic coupling in core-shell nanomaterials is governed by the energy level alignment between the core and shell materials, as shown in Figure 2.2. When the core and shell have matching or complementary electronic structures, charge transfer can occur efficiently across the interface, influencing electrical conductivity, catalytic activity, and quantum efficiency [44].

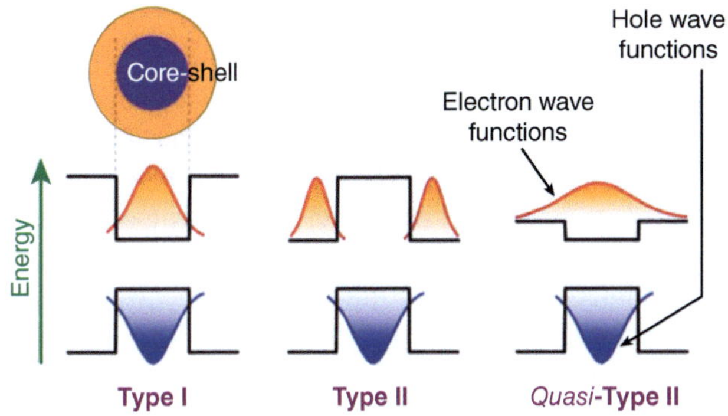

Figure 2.2 Energy band diagrams and carrier localization in different types of core-shell structures. Reproduced from [47]/Royal Society of Chemistry/CC BY 3.0.

a. **Type I Band Alignment:** In this configuration, both the conduction band minimum (CBM) and valence band maximum (VBM) of the shell are higher or lower in energy than those of the core, leading to charge carrier confinement within the core [45]. This effect is advantageous for applications requiring high exciton recombination efficiency, such as quantum dot–based LEDs and lasers.
b. **Type II Band Alignment:** Here, the CBM of one material is lower, and the VBM is higher than those of the other material, enabling spatial separation of electrons and holes [46]. This configuration is highly beneficial for photovoltaic and photocatalytic applications, where efficient charge separation is required for enhanced performance.
c. **Quasi Type II Band Alignment:** The hole is strongly confined in the core, while the electron is delocalized across core and shell. It's partially like Type I (hole strongly localized) and partially like Type II (electron spread out).

In metallic core-semiconductor shell systems (e.g., Au@TiO_2), electronic coupling at the interface leads to localized surface plasmon resonance (LSPR) effects, which enhance light absorption and photocatalytic efficiency. Similarly, in semiconductor core-shell quantum dots (e.g., CdSe@ZnS), the shell provides passivation, reducing surface defects and non-radiative recombination, thereby improving photoluminescence quantum yield. In nanoscale core-shell materials, exciton dynamics (electron–hole pair formation and recombination) are strongly influenced by quantum confinement effects. The core size and shell thickness determine the degree of confinement, affecting bandgap energy and optical absorption characteristics.

a. **Quantum Dot Core-shell Structures:** In CdSe@ZnS quantum dots, the ZnS shell passivates surface defects, enhancing fluorescence efficiency while also providing a protective barrier against oxidation and degradation [48].
b. **Core-shell Nanorods and Nanowires:** These structures offer anisotropic charge transport properties, useful in field-effect transistors and photodetectors. The core-shell configuration enables controlled exciton diffusion and recombination rates.

In metal-core/dielectric-shell systems (e.g., Au@SiO_2, Ag@TiO_2), plasmonic effects arise due to collective oscillations of free electrons in the metal core when excited by light. These plasmonic resonances enhance local electromagnetic fields, leading to increased light absorption, scattering, and energy transfer. Plasmonic core-shell nanostructures are widely used in SERS, biosensing, and photothermal therapy [49]. In Au@SiO_2 and Ag@TiO_2 nanoparticles, LSPR enhances electric field strength at the interface, improving photocatalytic activity by facilitating charge separation and electron injection into the semiconductor shell. In complex core-shell architectures, interactions between plasmonic and excitonic states lead to Fano resonances, which provide opportunities for tunable optical properties in photonic devices and biosensors.

By controlling core-shell composition, thickness, and interfacial interactions, the absorption and emission properties of nanomaterials can be precisely engineered. In core-shell semiconductor nanocrystals, increasing shell thickness can lead to either a redshift (lower energy absorption) or a blueshift (higher energy absorption) depending on the band alignment and strain effects. In fluorescent core-shell systems (e.g., CdTe@ZnS), the shell acts as a passivation layer, reducing surface trap states and enhancing photoluminescence. Conversely, in metal-core/fluorescent-shell structures, fluorescence quenching can occur due to Förster resonance energy transfer (FRET) or non-radiative energy dissipation [50]. The ability to engineer electronic and optical coupling in core-shell nanomaterials has led to significant advancements in various technological applications. Core-shell quantum dots and nanowires with tailored band alignment enhance charge separation and carrier transport in solar cells. Semiconductor core-shell structures improve color purity, brightness, and efficiency in LED applications. Plasmonic and fluorescent core-shell nanoparticles enable highly sensitive biosensing and targeted bioimaging. Plasmon-enhanced core-shell catalysts improve solar-to-fuel conversion efficiency in photocatalytic water splitting and CO_2 reduction.

2.2.3 Mechanical and Thermal Stability Considerations

Mechanical and thermal stability are critical factors in determining the performance and longevity of core-shell nanomaterials. The interplay between the core and shell materials significantly influences their resistance to mechanical deformation, thermal degradation, and environmental stress. These properties are particularly important for applications in structural nanocomposites, high-temperature catalysis, thermal insulation, and flexible electronics. Understanding the mechanical and thermal behavior of core-shell structures requires analyzing factors such as interfacial adhesion, mismatch in thermal expansion coefficients (TECs), and phase transitions at elevated temperatures [51]. The mechanical stability of core-shell structures is largely determined by the intrinsic strength of the core and shell materials, their interfacial bonding characteristics, and how stress is distributed across the interface. In many core-shell systems, the shell acts as a protective layer that redistributes mechanical stress, thereby enhancing structural integrity. For example, in metallic core-ceramic shell structures such as Au@TiO_2, the rigid ceramic shell provides mechanical reinforcement, preventing deformation of the softer metallic core under external stress. The adhesion between the core and shell plays a crucial role in mechanical stability, as strong interfacial bonding—whether covalent, ionic, or through van der Waals forces—prevents delamination or shell detachment. In contrast, weak interfacial interactions can lead to mechanical failure, reducing the durability of the material. Another aspect influencing mechanical behavior is the ability of core-shell materials to resist fracture and enhance toughness. Multi-shell or layered core-shell structures, such as SiO_2@TiO_2@ZrO_2, can efficiently distribute mechanical stress across multiple layers, preventing crack propagation and improving resistance to failure. This characteristic is particularly useful in protective coatings and impact-resistant materials. Additionally, in polymeric core-shell systems, the elasticity of the shell material determines the flexibility of the overall structure. Soft polymeric shells, such as polystyrene or polyvinyl alcohol, contribute to the mechanical robustness of nanostructures used in wearable electronics, stretchable sensors, and biomedical implants [52].

The thermal stability of core-shell materials is equally important, especially in applications involving high temperatures, such as catalysis, energy storage, and aerospace coatings. One key factor affecting thermal stability is the mismatch in TECs between the core and shell materials. When the core and shell expand at different rates upon heating, interfacial stress can develop, leading to structural failure such as cracking or delamination [53]. For example, metal-oxide core-shell structures like Al@Al_2O_3 are designed to mitigate such expansion mismatch, allowing the oxide shell to compensate for thermal stress and maintain structural integrity at high temperatures. Phase stability is another important consideration in high-temperature applications. Some core-shell materials undergo phase transitions at elevated temperatures, which can alter their structural and functional properties [20]. In Fe@Fe_3O_4 nanoparticles, for instance, the shell undergoes oxidation at high temperatures, which impacts its magnetic properties. To counteract such changes, researchers have focused on designing core-shell materials with thermally stable shells or incorporating barrier layers that prevent unwanted phase transitions. The thermal conductivity of core-shell structures can also be engineered for specific applications. In carbon-based core-shell structures such as CNT@SiO_2, the high thermal conductivity of CNTs enables efficient heat dissipation, while the insulating silica shell controls thermal transport. This tunability makes such materials suitable for applications requiring precise thermal management, including electronic devices and thermal interface materials. Additionally, ceramic-based core-shell structures such as ZrO_2@Al_2O_3 exhibit excellent resistance to thermal shock, allowing them to withstand rapid temperature fluctuations without undergoing structural degradation. These properties make them highly useful in aerospace thermal barrier coatings and high-temperature catalytic systems.

The enhanced mechanical and thermal stability of core-shell nanostructures has led to their adoption in various high-performance applications. In structural nanocomposites, reinforced core-shell nanoparticles improve toughness and mechanical resilience, making them valuable in aerospace and automotive industries. In high-temperature catalysis, core-shell catalysts remain stable under harsh reaction conditions, ensuring sustained catalytic activity. Polymeric core-shell structures play an essential role in flexible and wearable electronics, where mechanical durability is a key requirement. Furthermore, ceramic-based core-shell coatings provide excellent thermal protection, safeguarding aerospace components from extreme heat and oxidative degradation.

2.2.4 Core-shell Effects on Catalysis and Reactivity

Core-shell nanomaterials have garnered significant attention in catalysis due to their unique structural and interfacial properties that enable precise control over catalytic activity, selectivity, and stability. The interplay between the core and shell introduces synergistic effects that enhance reactivity, optimize electronic states, and improve resistance to catalyst deactivation. These advantages make core-shell nanostructures highly suitable for applications in heterogeneous catalysis, electrocatalysis, photocatalysis, and biocatalysis.

One of the most significant advantages of core-shell catalysts is their ability to modulate electronic and geometric properties through core-shell interactions. The electronic effects arise due to charge transfer between the core and shell, which influences the binding energy of reactants and intermediates on the catalyst surface [54]. For example, in Pt@Pd core-shell nanoparticles, the electron donation from the platinum core to the palladium shell alters the d-band structure of Pd, enhancing its catalytic efficiency in hydrogenation and oxygen reduction reactions (ORRs) [55]. Similarly, Au@Ag nanostructures exhibit improved catalytic activity in oxidation reactions due to the electronic coupling between gold and silver, which modifies the adsorption–desorption kinetics of reactants [56]. Geometrically, core-shell structures provide a platform for tuning active site exposure and coordination environment. By varying the shell thickness and composition, catalytic selectivity can be optimized. For instance, in Co@Pt core-shell nanoparticles, a thin Pt shell with a well-defined atomic arrangement ensures high catalytic performance while minimizing the use of expensive platinum. This design enhances the ORR efficiency in fuel cells while maintaining cost-effectiveness [57]. Additionally, bimetallic core-shell catalysts such as Ru@Pt and Pd@Au exhibit superior catalytic activity compared to their monometallic counterparts due to strain effects and ligand interactions at the core-shell interface.

The protective function of the shell is another crucial factor in catalytic applications. Many catalysts suffer from deactivation due to sintering, poisoning, or leaching under harsh reaction conditions. Core-shell architectures mitigate these issues by shielding the core from direct exposure to the reaction environment while preserving catalytic functionality. In Ni@SiO_2 catalysts, the silica shell prevents the aggregation of nickel nanoparticles at high temperatures, ensuring sustained catalytic performance in methane reforming reactions [58]. Similarly, Au@CeO_2 core-shell catalysts exhibit enhanced resistance to sulfur poisoning, making them highly effective in oxidation reactions for environmental remediation. In photocatalysis, core-shell nanostructures offer significant advantages by facilitating charge separation and enhancing light absorption. The band alignment between the core and shell materials determines the efficiency of electron-hole pair separation, which is crucial for photocatalytic reactions such as water splitting and CO_2 reduction. For example, TiO_2@CdS core-shell structures exhibit improved photocatalytic activity due to efficient charge transfer between the wide-bandgap TiO_2 core and the narrow-bandgap CdS shell, enabling effective utilization of visible light. Similarly, $ZnO@ZrO_2$ core-shell photocatalysts demonstrate enhanced photocatalytic stability, as the ZrO_2 shell prevents ZnO photocorrosion while maintaining high charge carrier mobility. In electrocatalysis, core-shell structures are widely used in fuel cells, metal-air batteries, and CO_2 electroreduction. The presence of a well-defined shell layer optimizes the catalytic interface and promotes faster charge transfer. For instance, Pt@Ru core-shell electrocatalysts exhibit superior performance in methanol oxidation reactions due to the synergistic interaction between the platinum shell and ruthenium core, which facilitates the removal of CO intermediates, preventing catalyst poisoning [59]. Similarly, Ni@Mo core-shell catalysts demonstrate high activity in HER due to their optimized electronic structure and improved conductivity.

Core-shell architectures also play a pivotal role in enzyme-mimicking catalysis (nanozymes) and biocatalysis. Enzyme-inspired core-shell nanostructures, such as $Fe_3O_4@C$ and $CeO_2@MnO_2$, mimic natural enzyme active sites, offering catalytic activity comparable to biological systems [21]. The controlled surface chemistry and functionalization of the shell allow for substrate-specific interactions, making core-shell nanomaterials valuable in biosensing, medical diagnostics, and targeted drug activation. Looking ahead, the future of core-shell catalysts lies in the rational design of multi-shell architectures, atomically precise interfaces, and defect-engineered surfaces to maximize catalytic efficiency. Advanced synthesis methods, such as ALD and molecular self-assembly, will enable precise tuning of core-shell compositions to achieve unprecedented catalytic performance. Additionally, machine learning (ML) and computational modeling will aid in the discovery of new core-shell catalysts with tailored electronic and surface properties for sustainable energy conversion, environmental remediation, and industrial catalysis.

2.2.5 Computational and Modeling Approaches in Core-shell Systems

Computational modeling and simulation play a crucial role in understanding, predicting, and optimizing the properties of core-shell nanomaterials. Given the complexity of core-shell architectures—where interactions at the interface dictate their electronic, mechanical, optical, and catalytic behavior—advanced theoretical approaches provide valuable insights into their fundamental mechanisms. Modeling techniques allow researchers to rationally design core-shell materials with tailored properties, minimizing costly trial-and-error experimental synthesis. These computational approaches range from quantum mechanical calculations to molecular dynamics (MD) simulations and ML-based predictions.

Density Functional Theory (DFT) is one of the most widely used computational methods for investigating the electronic structure of core-shell materials. DFT provides insight into charge transfer mechanisms, band structure modifications, and interfacial effects between the core and shell [60]. By calculating the density of states (DOS) and the Fermi level shift, researchers can predict how core-shell interactions influence catalytic activity, optical transitions, and electronic coupling. For example, DFT simulations have been used to study Pt@Au and Pd@Pt core-shell nanostructures, revealing how electronic charge redistribution affects their catalytic efficiency in hydrogen evolution and ORRs [61]. DFT also helps in understanding the stability of core-shell materials by evaluating formation energies, adsorption energies, and surface reactivity. In catalysis, these calculations aid in determining the optimal shell thickness and composition to achieve the best catalytic performance while maintaining stability under reaction conditions. For instance, in Pt@Ni and Pd@Cu core-shell electrocatalysts, DFT studies reveal that the shell's d-band center position relative to the Fermi level directly correlates with catalytic activity. MD simulations provide atomistic insights into the thermal, mechanical, and dynamic behavior of core-shell nanomaterials. MD simulations track the motion of atoms over time, enabling the study of diffusion, structural stability, and phase transitions at the core-shell interface [62]. These simulations are particularly useful in assessing the mechanical properties of core-shell nanoparticles, such as elastic modulus, fracture resistance, and TECs. For example, MD simulations have been used to study the sintering resistance of core-shell catalysts, such as Pt@SiO_2 and Au@TiO_2, under high-temperature conditions. These models predict how shell thickness and composition influence the suppression of nanoparticle coalescence, which is critical for improving the long-term stability of catalysts. Similarly, in polymeric core-shell materials, MD studies help in understanding solvent diffusion, polymer flexibility, and interfacial adhesion, which are essential for applications in drug delivery and smart coatings. Finite Element Modeling (FEM) is another powerful tool for studying the mechanical and thermal stability of core-shell systems, particularly in large-scale applications. FEM helps in predicting stress distribution, fracture mechanics, and heat dissipation in core-shell structures under different loading conditions [63]. This approach is particularly relevant for core-shell materials used in structural composites, battery electrodes, and thermal barrier coatings. For instance, FEM studies on Si@C core-shell anodes in lithium-ion batteries have revealed how stress accumulation at the silicon–carbon interface influences mechanical degradation during charge–discharge cycles. By optimizing shell thickness and material composition, computational models suggest ways to mitigate capacity fading and enhance battery lifespan. With the advent of AI and ML, computational approaches to core-shell nanomaterials have advanced significantly. ML algorithms analyze large datasets from experimental and theoretical studies to identify trends, predict material performance, and optimize synthesis parameters. In Figure 2.3, the schematic representation highlights the role of AI and ML in advancing core-shell nanostructures. These computational approaches enable precise material design, predictive modeling of properties, and optimization of structural modifications for enhanced performance. By using neural networks and decision-tree algorithms, researchers can screen thousands of potential core-shell combinations to identify those with the best electronic, catalytic, or optical properties [64]. For example, ML-driven models have been used to design core-shell photocatalysts for water splitting applications by predicting band alignments and charge separation efficiency. Similarly, in electrocatalysis, ML-based approaches have successfully identified optimal core-shell compositions for fuel cells and

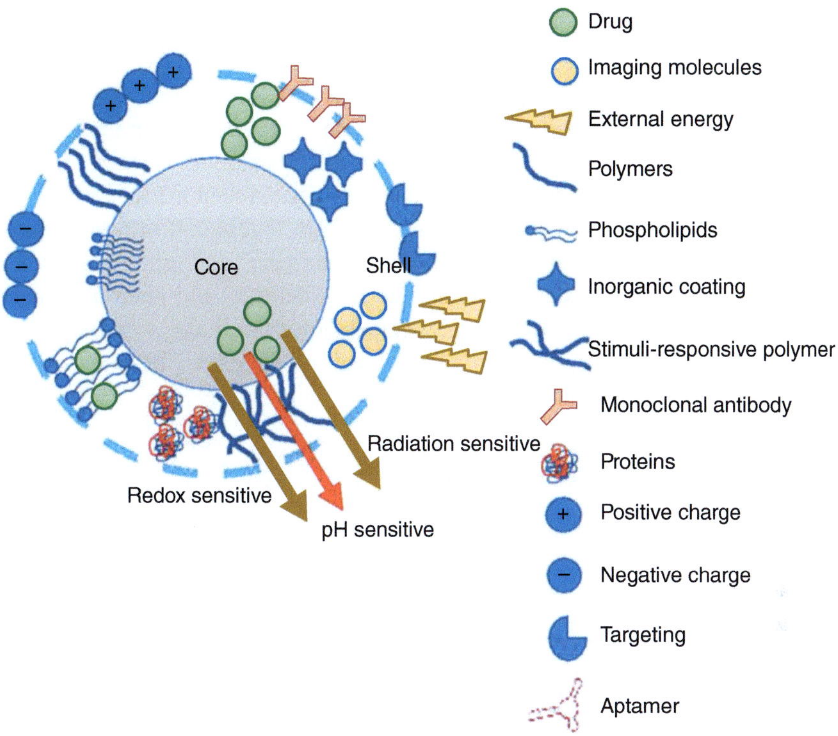

Figure 2.3 Schematic representation of core-shell nanostructures with AI and machine learning-driven modifications for future advancements. Reproduced from [65]/with permission of Elsevier.

CO_2 reduction reactions, accelerating the discovery of next-generation catalysts. Monte Carlo simulations provide statistical modeling of reaction kinetics, diffusion mechanisms, and adsorption–desorption processes in core-shell catalysts. These simulations are particularly useful in understanding how reactants and intermediates interact with the shell surface, which is critical for optimizing catalytic efficiency. Monte Carlo methods have been employed to model the diffusion of lithium ions in Si@C core-shell anodes and the adsorption dynamics of gases on Pd@Pt catalysts for hydrogen storage applications. As computational power increases and ML algorithms become more sophisticated, the future of core-shell nanomaterials research will increasingly rely on predictive modeling and high-throughput simulations. The integration of DFT, MD, FEM, and ML techniques will allow for multi-scale modeling, bridging the gap between atomic-level interactions and macroscopic material performance.

Additionally, real-time computational-experimental feedback loops will enable rapid prototyping of new core-shell architectures, reducing development time for energy storage, catalysis, and biomedical applications. In combination with in situ characterization techniques such as X-ray absorption spectroscopy (XAS) and transmission electron microscopy (TEM), computational models will provide deeper insights into the dynamic behavior of core-shell materials under operational conditions.

2.3 Core-shell Material Compositions

Core-shell nanomaterials exhibit a wide range of compositions, each tailored for specific properties and applications. The choice of core and shell materials dictates the functionality, stability, and reactivity of these nanostructures. These compositions can be broadly categorized into metallic, polymeric, metal oxide-based, and hybrid core-shell systems. Each category (as shown in Figure 2.4) brings unique advantages and plays a critical role in various technological and industrial fields, ranging from catalysis and energy storage to drug delivery and environmental remediation. The ability to fine-tune the composition, morphology, and interfacial characteristics of these materials provides a versatile platform for engineering next-generation functional materials.

2.3.1 Metallic Core-shell Nanostructures

Metallic core-shell nanostructures comprise a metallic core enveloped by a metallic or non-metallic shell. These structures are widely studied due to their tunable optical, electrical, and catalytic properties. Noble metal-based core-shell structures, such as gold and silver, are of particular interest due to their surface plasmon resonance (SPR) effects, which enable enhanced electromagnetic interactions. This property is crucial for applications in sensing, imaging, and photothermal therapy [66]. Additionally, transition metal core-shell nanomaterials, such as platinum or palladium shells over a nickel or cobalt core, are extensively employed in catalysis due to their high surface area and enhanced electron transfer capabilities. These materials are also instrumental in fuel cells, electrochemical energy storage, and environmental remediation. Apart from their electronic and optical applications, metallic core-shell structures also exhibit high stability and corrosion resistance [67]. By choosing an appropriate shell material, the core can be protected from oxidation, thereby extending the lifespan of the nanomaterial. For example, iron core nanoparticles encapsulated in gold shells prevent oxidation, ensuring prolonged magnetic functionality for biomedical applications such as targeted drug delivery and hyperthermia treatment. Another example is silver-core gold-shell nanostructures, which combine the high conductivity of silver with the chemical stability of gold, making them ideal for electronic and catalytic applications. Furthermore, bimetallic core-shell systems, such as Pt–Au or Pd–Ag, have been explored for advanced electrocatalysis, demonstrating improved reaction kinetics and resistance to poisoning effects.

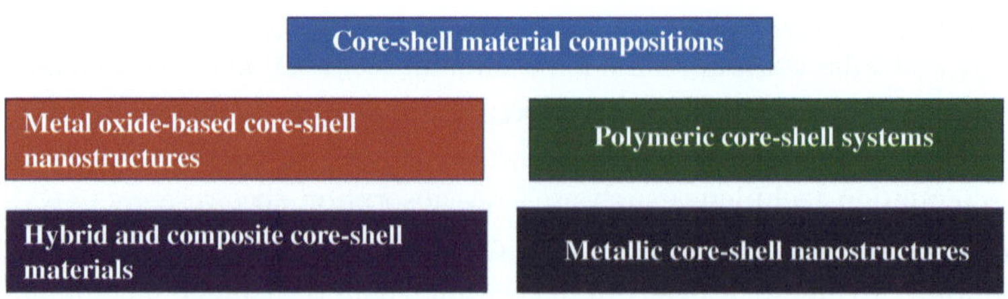

Figure 2.4 Different categories of core-shell material compositions.

2.3.2 Polymeric Core-shell Systems

Polymeric core-shell systems offer a unique class of nanomaterials designed for controlled release, biocompatibility, and environmental responsiveness. These structures typically consist of a polymeric core or shell, which can be engineered to exhibit specific chemical and physical behaviors. One of the most prominent advantages of polymer-based core-shell systems is their ability to form smart and stimuli-responsive materials. By incorporating polymers with temperature, pH, or light-sensitive properties, these nanostructures can dynamically alter their characteristics in response to environmental changes. Organic–inorganic hybrid polymeric core-shell architectures have found significant applications in drug delivery and tissue engineering [41]. For instance, biodegradable polymeric shells surrounding an inorganic core enable sustained drug release and enhance biocompatibility. Hydrogels and nanogels are common examples of polymeric core-shell systems utilized in pharmaceutical applications due to their ability to encapsulate drugs and release them in a controlled manner [53]. Additionally, polymeric coatings on metallic nanoparticles are often used to enhance dispersibility and stability in aqueous environments, making them suitable for biosensing and imaging applications. A prominent example includes polystyrene-core silica-shell nanoparticles, which are extensively used in bioimaging and biosensors due to their excellent stability and surface modification capabilities. Another widely studied system is polyethylene glycol (PEG)-coated nanoparticles, which offer stealth properties, reducing immune clearance and prolonging circulation time in drug delivery applications.

2.3.3 Metal Oxide-based Core-shell Nanostructures

Metal oxide-based core-shell nanostructures are crucial in numerous industrial and environmental applications. These materials, consisting of a metal oxide core or shell, exhibit diverse functionalities such as photocatalysis, magnetism, and dielectric properties. Semiconductor oxides like titanium dioxide (TiO_2) and zinc oxide (ZnO) are commonly employed as shell materials due to their excellent photocatalytic properties, which are highly beneficial in environmental remediation and solar energy conversion. Magnetic oxide-based core-shell systems, such as iron oxide (Fe_3O_4) cores with silica or polymeric shells, are widely used in biomedical applications, including magnetic resonance imaging (MRI) contrast agents and targeted drug delivery. These materials combine magnetic properties with enhanced biocompatibility and functionalization capabilities. Moreover, dielectric metal oxides such as aluminum oxide and silica are often used as protective shells to prevent degradation and agglomeration of core materials, thereby improving performance in catalysis and electronic applications [9]. An example of a widely used system is cobalt ferrite ($CoFe_2O_4$) core with silica shells, which exhibits enhanced magnetic properties useful for data storage and medical diagnostics. Another key feature of metal oxide-based core-shell structures is their ability to enhance charge separation in photovoltaic and photocatalytic applications. By carefully selecting the core and shell materials, researchers can design nanostructures that facilitate efficient electron–hole pair separation, thereby improving the efficiency of solar cells and water-splitting reactions. For instance, TiO_2-coated cadmium sulfide (CdS) core-shell structures have demonstrated improved photocatalytic activity for hydrogen generation from water splitting. Additionally, ZnO-based core-shell structures are being investigated for next-generation light-emitting diodes (LEDs) and transparent conductive films due to their excellent optical and electrical properties.

2.3.4 Hybrid and Composite Core-shell Materials

Hybrid and composite core-shell materials encompass a diverse range of structures that integrate different material classes, such as metals, oxides, polymers, and carbon-based nanostructures. The synergy between different material types enables the development of multifunctional nanostructures with enhanced performance and durability. Carbon-based core-shell systems, such as graphene and CNT-coated nanoparticles, offer remarkable electrical conductivity and mechanical strength. These materials are widely used in energy storage, supercapacitors, and flexible electronics. For example, CNT-coated silicon nanoparticles have been explored as anode materials in lithium-ion batteries, where the CNT shell enhances conductivity and prevents silicon degradation [68]. Bioinspired and biomimetic core-shell architectures represent another fascinating category of hybrid materials. Inspired by natural systems, such as biomineralization processes found in seashells and bones, these nanostructures are engineered to exhibit exceptional mechanical and functional properties. For example, calcium phosphate-based core-shell structures have been extensively investigated for bone tissue engineering and regenerative medicine applications. Additionally, chitosan-coated metal nanoparticles have been developed for antimicrobial and drug delivery applications due to their biocompatibility and controlled release properties. Composite core-shell systems also play a pivotal role in advanced catalysis. By combining different catalytic materials within a core-shell configuration, researchers can achieve enhanced activity, selectivity, and stability [69]. For instance, core-shell structures with a catalytic core encapsulated by a porous shell enable controlled diffusion of reactants, minimizing undesired side reactions and improving reaction efficiency in heterogeneous catalysis. A notable example is platinum-core ceria-shell nanostructures, which exhibit superior catalytic activity for fuel cell applications by leveraging the oxygen storage capacity of ceria and the catalytic efficiency of platinum. Another example includes metal-organic framework (MOF)-based core-shell materials, which offer high porosity and tunable chemistry for gas separation and storage [70]. Overall, core-shell material compositions offer an extensive platform for developing advanced nanomaterials with tailored properties. The ability to manipulate the composition, structure, and interfacial interactions of these materials has led to significant advancements in energy storage, biomedicine, environmental science, and electronics. As research continues to push the boundaries of core-shell nanotechnology, new compositions and functionalizations will emerge, further expanding their potential applications across multiple scientific and industrial domains.

2.4 Structure–Property Relationships in Core-shell Materials

The structure–property relationship in core-shell materials plays a fundamental role in determining their functionality and performance across various applications. By carefully designing the composition, thickness, morphology, and interfacial interactions of the core and shell, researchers can achieve highly tunable physical, chemical, and mechanical properties. The ability to manipulate these factors makes core-shell architectures a versatile platform for advancements in nanotechnology, catalysis, energy storage, optoelectronics, and biomedical applications. One of the most crucial aspects influencing the properties of core-shell structures is the geometric configuration, particularly the relative thickness of the core and shell. The dimensions of these components govern key attributes such as electronic band structure, optical response, magnetic behavior, and mechanical resilience. The selection of core and shell materials further dictates interfacial effects, which can enhance or suppress specific properties through charge transfer, strain modulation, or interdiffusion phenomena. Additionally, shell modifications enable surface functionalization, rendering the nanostructures responsive to external stimuli such as pH, temperature, or electromagnetic fields [71]. Beyond geometric and compositional factors, the core-shell interface plays a decisive role in dictating stability, charge transport, and catalytic activity. Strong interfacial bonding can enhance mechanical strength and thermal stability, whereas weak or mismatched interfaces may lead to structural degradation or phase separation under operational conditions. Understanding the interplay between interfacial energy, crystallographic orientation, and lattice mismatches is crucial for designing robust and high-performance core-shell materials. Moreover, core-shell structures offer a platform to tailor fundamental properties, such as bandgap engineering in semiconductors, plasmonic tuning in metallic nanostructures, and anisotropic behavior in Janus particles [72]. These modifications enable the development of materials with precisely controlled functionalities suited for targeted applications. The ability to integrate multiple material classes within a core-shell framework further broadens the scope for hybrid and multifunctional materials, paving the way for next-generation technologies in energy, environment, healthcare, and information processing.

2.4.1 Influence of Core and Shell Thickness on Properties

The thickness of both the core and shell in core-shell nanomaterials is a critical factor in determining their physical, chemical, and functional properties. By precisely controlling these dimensions, researchers can modulate key characteristics such as optical absorption, electronic conductivity, mechanical strength, and catalytic efficiency. The interplay between the core and shell thickness allows for the fine-tuning of nanomaterials to optimize their performance in specific applications, making them highly adaptable for use in fields such as optoelectronics, energy storage, medicine, and environmental remediation. One of the most significant impacts of core and shell thickness variation is observed in optical properties, particularly in plasmonic and semiconductor nanostructures. In metal-based core-shell systems, such as gold-core silver-shell nanoparticles, the LSPR shifts depending on the shell thickness [73]. A thicker shell generally results in redshifting of the plasmonic peak due to increased electron oscillation interactions within the structure. Similarly, in semiconductor core-shell nanostructures, shell thickness dictates quantum confinement effects, affecting bandgap energies and, consequently, optical absorption and emission characteristics. This property is particularly important for applications in photodetectors, LEDs, and photovoltaic devices, where precise control of light interaction is necessary.

In electronic and catalytic applications, core-shell thickness influences charge transfer dynamics and surface reactivity. A thin shell can facilitate efficient electron transport between the core and the surrounding medium, enhancing conductivity and catalytic performance. For example, in platinum-shell gold-core nanoparticles (Figure 2.5), an ultrathin platinum shell improves catalytic efficiency in fuel cells by exposing a high density of active sites while utilizing the gold core for structural stability [74]. Conversely, a thicker shell may act as an insulating layer, impeding charge transport but providing enhanced protection against oxidation or degradation. In battery electrodes and supercapacitors, core-shell nanomaterials with optimized shell thickness improve ionic diffusion and electrochemical stability, leading to superior energy storage capabilities. Mechanical and thermal stability are also highly dependent on the relative thickness of the core and shell. In high-temperature environments, a thicker shell can serve as a protective barrier, preventing thermal degradation of the core material [75]. This principle is widely utilized in ceramic core-shell systems, where an oxide shell protects a metallic core from oxidation and phase transformations. Additionally, the shell thickness plays

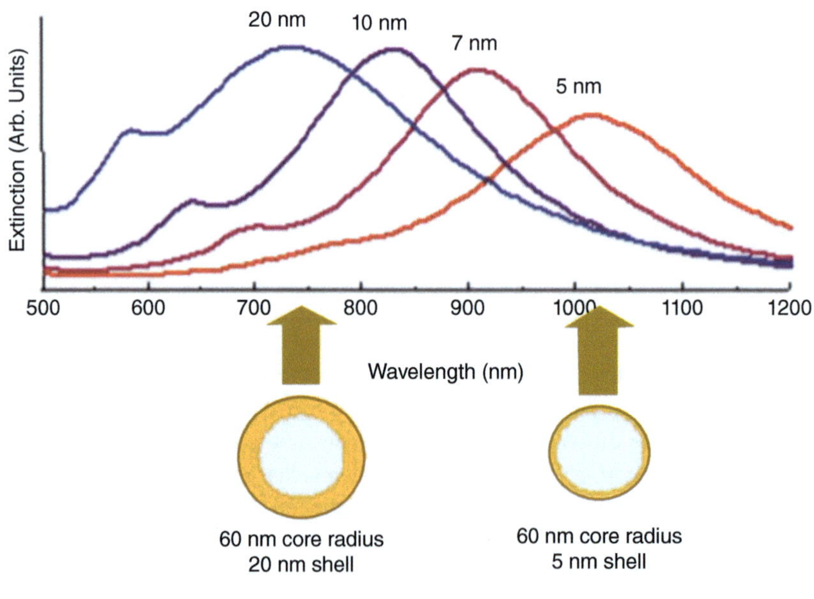

Figure 2.5 Optical resonance behavior of gold-shell/silica-core nanoshells as a function of their core-to-shell ratio. Reproduced from [76]/with permission of SAGE Publications.

a crucial role in stress distribution, affecting the mechanical strength of the overall nanostructure. For instance, core-shell polymer nanofibers with thick shells exhibit enhanced tensile strength and flexibility, making them ideal for wearable electronics and biomedical scaffolds. In the biomedical field, core-shell thickness impacts drug delivery and imaging performance. A thin polymeric shell over a magnetic or metallic core allows for rapid drug release, whereas a thicker shell provides sustained release profiles. This property is particularly useful for targeted cancer therapy, where controlled drug diffusion is essential for maximizing therapeutic efficacy while minimizing side effects. Similarly, in bioimaging applications, the shell thickness in quantum dot–based probes determines fluorescence lifetime and stability, affecting the resolution and duration of imaging capabilities.

Ultimately, optimizing the core and shell thickness requires a careful balance between competing effects. While a thicker shell may enhance stability and protection, it may also hinder functionality by reducing the active surface area or inhibiting charge transport. Conversely, a thinner shell may improve interaction dynamics but may compromise durability. Advances in synthesis techniques, such as ALD and controlled colloidal growth, now allow for precise manipulation of core and shell dimensions, enabling tailored nanomaterials for specific applications. By understanding the intricate effects of core and shell thickness, researchers can design next-generation core-shell materials with enhanced tunability and performance across diverse technological and industrial domains.

2.4.2 Effect of Core-shell Interfaces on Mechanical and Thermal Stability

The interface between the core and shell in core-shell nanomaterials plays a crucial role in determining their mechanical and thermal stability. The interactions at this interface significantly impact stress distribution, thermal expansion behavior, and overall structural integrity, making it an essential consideration in designing materials for high-performance applications. Properly engineered core-shell interfaces enhance mechanical strength, prevent structural degradation, and improve resistance to extreme temperatures, which is critical for applications in aerospace, energy storage, catalysis, and biomedical engineering [77]. The mechanical properties of core-shell materials depend largely on how the core and shell interact at the interface. Differences in material properties, such as elasticity, toughness, and adhesion strength, influence how stress is transferred between the two components. A well-integrated interface allows for even stress distribution, preventing fractures and structural failure under mechanical loading.

In composite materials, such as metal-ceramic or polymer-ceramic core-shell structures, the core-shell interface acts as a reinforcement site, enhancing mechanical resilience. For example, in silica-coated metal nanoparticles, the rigid silica shell protects the softer metal core from deformation, significantly enhancing structural durability [78]. Similarly, in polymer-core ceramic-shell systems, the outer shell imparts rigidity while the core maintains flexibility, providing a combination of toughness and elasticity useful in coatings, flexible electronics, and impact-resistant materials. However, poor adhesion or mismatched mechanical properties between the core and shell can lead to internal stress accumulation, resulting in delamination, cracks, or even catastrophic failure under repeated mechanical stress. To mitigate these issues, interfacial engineering techniques such as surface functionalization, gradient interfaces, and chemical bonding strategies are employed. For instance, functionalizing metal surfaces with organic linkers before shell deposition improves adhesion and ensures a mechanically robust core-shell interface.

Core-shell interfaces also play a fundamental role in defining the thermal stability of nanomaterials, particularly in high-temperature applications. When exposed to heat, different materials expand at different rates due to variations in their coefficients of thermal expansion (CTE) [79]. If the core and shell materials have significantly mismatched CTEs, internal stress develops at the interface, which can lead to structural distortion, cracking, or shell detachment. In high-temperature environments, such as catalytic reactors and thermal insulation coatings, core-shell nanomaterials must withstand extreme heat while maintaining their functional properties. For example, in ceramic-metal core-shell structures, an oxide shell can provide a protective barrier against oxidation and thermal degradation. Titanium carbide (TiC) or zirconia (ZrO_2) shells are commonly used to encapsulate metal cores, shielding them from high-temperature oxidation and extending their operational lifespan in harsh environments [80]. One strategy to counteract thermal expansion mismatch is using buffer layers or graded interfaces, where an intermediate layer gradually transitions between the core and shell materials. This design reduces abrupt thermal stress at the interface, enhancing durability. A notable example is alumina-coated nickel nanoparticles, where an intermediate oxide layer prevents stress accumulation and ensures long-term thermal stability in catalytic and electronic applications. Additionally, thermal conductivity is influenced by the core-shell interface, impacting heat dissipation efficiency. Core-shell nanostructures designed for thermal management applications, such as in heat sinks or thermal interface materials, require interfaces that promote efficient phonon transport. In cases where thermal resistance is a concern, optimizing the interfacial bonding through chemical modifications or nanostructuring can significantly improve heat dissipation and thermal performance.

The ability to tailor the mechanical and thermal properties of core-shell materials through interfacial engineering has enabled their widespread adoption in various technological fields. In lithium-ion batteries, core-shell structured cathode materials with thermally stable coatings enhance cycling stability and prevent capacity degradation. In structural materials, nanocomposites with robust core-shell interfaces provide superior mechanical strength, wear resistance, and impact absorption [81]. Future research in this domain focuses on improving interfacial adhesion and stability through advanced fabrication techniques, such as ALD, molecular self-assembly, and plasma-assisted coatings. By fine-tuning interfacial chemistry and structure, scientists aim to develop next-generation core-shell materials with enhanced durability, reliability, and performance in extreme environments. Understanding the complex interplay between core and shell interfaces is essential for designing mechanically and thermally stable materials that can withstand real-world operational challenges, paving the way for more robust and efficient nanotechnology solutions.

2.4.3 Optical and Electronic Properties of Core-shell Architectures

Core-shell nanomaterials exhibit unique optical and electronic properties that arise from the interaction between their core and shell components. These properties are highly tunable and can be engineered for a wide range of applications, including photovoltaics, optoelectronics, sensing, and plasmonic devices. The combination of different materials in a core-shell configuration allows for enhanced light absorption, energy transfer, charge separation, and emission properties, making them valuable for next-generation nanophotonic and electronic systems. The optical behavior of core-shell structures is primarily governed by their size, composition, refractive index contrast, and interfacial interactions. One of the most widely studied optical phenomena in metallic core-shell nanostructures is SPR, where conduction electrons in the metallic core oscillate in response to incident light [82]. The resonance frequency can be tuned by adjusting the core and shell dimensions, allowing for precise control over the material's optical response. For example, **gold-core silica-shell nanoparticles** exhibit tunable SPR peaks by varying the silica shell thickness, enabling their use in biosensing, imaging, and photothermal therapy [83]. Similarly, **silver-core gold-shell nanostructures** offer enhanced stability and broadened plasmonic responses, making them highly effective in SERS applications [84].

In semiconductor-based core-shell materials, such as **CdSe/ZnS quantum dots**, the shell serves as a passivation layer, reducing surface defects and enhancing photoluminescence quantum yield [85]. These structures are widely used in bioimaging and display technologies due to their narrow emission spectra and high color purity. Additionally, core-shell architectures in upconversion nanoparticles (UCNPs), such as **$NaYF_4$:Yb/Er@$NaYF_4$**, enhance energy transfer efficiency, enabling high-performance anti-Stokes luminescence for bioimaging and infrared-based security applications. Another key optical property is **fluorescence resonance energy transfer (FRET)**, which is significantly influenced by the core-shell interface. In certain dye-functionalized core-shell structures, such as **silica-coated fluorescent dyes**, FRET-based interactions can be tuned to improve sensing capabilities for biological and chemical detection [86]. The electronic behavior of core-shell materials depends on factors such as band alignment, charge transfer efficiency, and interfacial electronic coupling. Core-shell architectures offer significant advantages in controlling charge separation and recombination processes, making them highly useful in optoelectronic and energy-related applications.

In **semiconductor core-shell nanostructures**, bandgap engineering is a crucial aspect that determines electronic performance. For example, in **TiO_2-coated CdSe quantum dots**, the wide-bandgap TiO_2 shell facilitates electron transport while preventing charge recombination, enhancing the efficiency of photovoltaic and photocatalytic processes. Similarly, **GaAs/AlGaAs core-shell nanowires** are used in high-efficiency solar cells and LEDs due to their optimized charge carrier dynamics [87]. In core-shell **dielectric-metal structures**, the shell can act as an insulating barrier, preventing leakage currents and improving charge retention in memory storage devices. An example of this is **SiO_2-coated gold nanoparticles**, which show enhanced electron storage capabilities in nanostructured capacitors [88].

Figure 2.6 Illustration of the tunability of metal nanoshells. Reproduced with permission from [76]/SAGE Publications.

Another important aspect of electronic behavior in core-shell materials is the **quantum confinement effect**, where reducing the core size leads to discrete electronic states and tunable bandgaps. **CdTe/CdS core-shell nanostructures** exploit this effect to improve the performance of quantum dot–based solar cells and photodetectors. In Figure 2.6, core-shell nanomaterials exhibit tunable optical properties, as evident from the color variations in colloidal solutions. These variations arise due to changes in shell composition, thickness, and core/shell ratio, which influence plasmonic resonance and light absorption. Such tunability makes core-shell structures highly desirable for applications in sensing, imaging, and photonic devices.

The ability to manipulate optical and electronic properties in core-shell materials has enabled breakthroughs in diverse fields. Core-shell quantum dots and nanowires improve charge separation and light absorption, enhancing solar cell efficiencies. Tunable bandgaps in core-shell semiconductors make them ideal for LEDs, lasers, and photodetectors. Core-shell metallic nanostructures are widely used in biosensors, SERS, and plasmon-enhanced spectroscopy. Fluorescent and plasmonic core-shell nanoparticles provide enhanced imaging contrast and targeted therapeutic capabilities. Future research will focus on integrating advanced computational models, ML, and AI to design next-generation core-shell materials with even greater control over optical and electronic behavior. As material synthesis techniques advance, new hybrid core-shell architectures will emerge, pushing the boundaries of optoelectronic performance and leading to revolutionary applications in nanotechnology and energy conversion.

2.4.4 Magnetic and Catalytic Performance Optimization

Core-shell nanomaterials offer significant advantages in both magnetic and catalytic applications due to their tailored interfaces, controlled structural properties, and synergistic effects between the core and shell materials. By engineering the composition, thickness, and morphology of core-shell structures, researchers can optimize their performance in various technological applications, including magnetic data storage, biomedical imaging, targeted drug delivery, fuel cells, and heterogeneous catalysis. The magnetic properties of core-shell materials are influenced by several key factors, including the nature of the core and shell materials, their dimensions, interfacial interactions, and crystallographic structure [89]. These materials are particularly valuable in applications such as spintronics, magnetoresistive devices, hyperthermia therapy, and magnetic separation. One of the primary advantages of core-shell magnetic nanostructures is their ability to exhibit **enhanced thermal and chemical stability** compared to their bare magnetic counterparts. For instance, **Fe_3O_4@SiO_2 core-shell nanoparticles** (Fe_3O_4 core with a silica shell) demonstrate improved dispersibility in biological environments and reduced oxidation, making them highly effective in biomedical imaging and targeted drug delivery. The silica shell prevents aggregation and degradation while maintaining the magnetic response necessary for controlled manipulation using external magnetic fields. Additionally, **exchange coupling effects** between the core and shell materials can significantly modify the magnetic behavior of core-shell nanostructures [90]. In **hard–soft magnetic core-shell systems**, such as **FePt@Fe_3O_4**, the soft magnetic shell enhances the magnetic moment while the hard magnetic core maintains high coercivity. This property is critical in high-density data storage and permanent magnet applications.

Another crucial aspect of magnetic performance optimization is the **control of superparamagnetic behavior** in core-shell nanoparticles. Superparamagnetic materials exhibit strong magnetic properties in the presence of an external field but show no remanence once the field is removed, preventing unwanted particle aggregation. This is particularly useful in **biomedical applications** such as MRI contrast agents and targeted drug delivery. Superparamagnetic core-shell nanoparticles, such as **Fe_3O_4@Au**, provide a balance between magnetic response and biocompatibility, allowing for effective diagnostic and therapeutic functionalities [91]. In addition to biomedical applications, core-shell magnetic nanostructures have been extensively studied for **water purification and environmental remediation**. **Magnetic core-shell catalysts**, such as **Fe_3O_4@TiO_2**, enable easy recovery after catalytic reactions, providing sustainable solutions for pollutant degradation, wastewater treatment, and oil spill cleanup. Catalysis is another domain where core-shell nanomaterials demonstrate superior performance due to their well-controlled interface chemistry, high surface area, and tailored electronic interactions. Core-shell structures allow for precise modulation of catalytic activity, selectivity, and stability, making them ideal candidates for applications in fuel cells, hydrogen production, CO_2 reduction, and chemical synthesis.

One of the key advantages of core-shell catalysts is the ability to **reduce the amount of expensive noble metals while maintaining high catalytic efficiency**. For example, **Pt@Pd core-shell nanoparticles** utilize a platinum core with a palladium shell to maximize catalytic activity in fuel cells while minimizing the overall use of costly platinum. Similarly, **Au@Pt core-shell nanostructures** enhance catalytic efficiency in HER by leveraging the electronic synergy between gold and platinum. Core-shell nanocatalysts also exhibit **enhanced stability and resistance to poisoning**, a common challenge in catalytic processes [4]. In fuel cell applications, **Pt@Ru core-shell catalysts** are known to improve the tolerance to carbon monoxide (CO) poisoning, ensuring prolonged operational efficiency. The Ru shell provides active sites that mitigate CO adsorption, thereby extending the catalyst's durability. Another widely studied catalytic application of core-shell nanostructures is **photocatalysis**, where the combination of semiconductors in a core-shell arrangement improves charge separation and light absorption. For instance, **TiO_2@CdS core-shell nanostructures** enhance photocatalytic hydrogen production by effectively separating photogenerated electron-hole pairs, thereby increasing reaction efficiency. Similarly, **ZnO@MoS_2 nanostructures** have shown remarkable photocatalytic degradation capabilities for organic pollutants in wastewater treatment.

The design of **hollow core-shell nanocatalysts** further enhances catalytic activity by exposing a higher surface area and providing active sites for chemical reactions. **Hollow Pd@Pt core-shell nanoparticles**, for example, demonstrate exceptional activity in electrocatalytic oxidation reactions due to their increased active surface and improved mass transport properties. Additionally, **hierarchical multi-shell catalysts**, such as **Pt@CeO_2@SiO_2**, offer unique advantages in catalytic reactions by incorporating multiple functional layers that promote efficient reaction kinetics and stability. The CeO_2 shell acts as an oxygen storage material, improving redox performance, while the SiO_2 layer enhances structural integrity. Core-shell nanomaterials optimized for magnetic and catalytic applications have paved the way for numerous real-world advancements.

a. **Magnetic Hyperthermia Therapy:** Superparamagnetic core-shell nanoparticles for targeted cancer treatment.
b. **Energy Storage and Conversion:** Core-shell catalysts for improving the efficiency of fuel cells and batteries.
c. **Environmental Remediation:** Magnetic core-shell nanoparticles for removing pollutants and heavy metals from water.
d. **Chemical Synthesis:** Core-shell catalysts for selective hydrogenation, oxidation, and CO_2 reduction reactions.

By integrating advanced synthesis techniques with computational modeling, researchers aim to create next-generation core-shell systems with even greater precision, efficiency, and sustainability.

2.4.5 Tunability of Properties Through Shell Modification

Core-shell nanomaterials offer unparalleled opportunities for tailoring material properties by modifying the composition, thickness, and structure of the shell. The shell plays a crucial role in dictating the nanomaterial's physicochemical behavior, influencing properties such as optical response, electrical conductivity, magnetic behavior, thermal stability, and catalytic efficiency. By carefully selecting and engineering the shell material, researchers can achieve enhanced performance and functionality for applications in energy storage, biomedicine, environmental remediation, and nanotechnology. The choice of shell material significantly impacts the overall performance of core-shell nanostructures. By varying the shell composition, it is possible to introduce novel functionalities and improve stability, dispersibility, and reactivity.

a. **Metallic Shells:** Encapsulating a core material with a metallic shell can enhance electrical and thermal conductivity while providing protective and catalytic benefits [92]. For instance, **gold (Au) shells** are commonly used to enhance chemical stability and biocompatibility in biomedical applications. **Fe_3O_4@Au nanoparticles**, for example, combine the magnetic properties of Fe_3O_4 with the chemical inertness and bio-functionalization potential of gold, making them suitable for targeted drug delivery and bioimaging.

b. Oxide Shells: Metal oxide shells such as **SiO_2, TiO_2, and Al_2O_3** are widely used to improve chemical stability, prevent oxidation, and modify surface interactions. **SiO_2 shells** are particularly effective in protecting magnetic and catalytic cores from aggregation and degradation, extending their operational lifespan in harsh environments [93]. A common example is **Fe_3O_4@SiO_2**, where the silica shell enhances colloidal stability and biocompatibility while preserving the magnetic functionality of the Fe_3O_4 core.

c. Polymeric Shells: Polymers provide versatility in functionalization, allowing for stimuli-responsive behavior, biocompatibility, and controlled release properties. Polymeric coatings such as **PEG, polystyrene, and poly(N-isopropylacrylamide) (PNIPAM)** can render nanoparticles hydrophilic, improving their dispersibility in aqueous environments [94]. For instance, **PNIPAM-coated nanoparticles** exhibit temperature-sensitive behavior, making them valuable for drug delivery systems that release payloads at specific temperatures.

d. Carbon-based Shells: Graphene and CNT coatings improve electrical conductivity, mechanical strength, and stability in electrochemical applications. **Si@C core-shell nanostructures** are widely employed as anode materials in lithium-ion batteries, where the carbon shell enhances electrical conductivity while buffering volume expansion during charge–discharge cycles.

The thickness of the shell is another crucial factor that determines the functional performance of core-shell nanostructures. Thin shells allow for efficient electron and ion transfer, whereas thicker shells provide enhanced protection and surface passivation. In catalytic and electronic applications, thin shells ensure minimal resistance to charge transport. **Pt@Pd core-shell catalysts** with an ultrathin palladium shell exhibit superior catalytic efficiency in fuel cells due to improved electron transfer and active surface exposure. A thicker shell enhances the material's durability, chemical resistance, and mechanical robustness. **ZnO@SiO_2 nanoparticles**, for example, benefit from a thicker silica shell, which protects the ZnO core from photocorrosion, extending their usability in UV protection coatings and photocatalysis. In some applications, a gradient shell thickness provides a balance between reactivity and protection. **Fe_3O_4@TiO_2 nanoparticles** with controlled shell thickness offer tunable photocatalytic activity, allowing optimization for specific environmental and energy applications. Beyond altering the shell composition and thickness, surface functionalization provides an additional level of tunability, enabling precise control over interactions with the surrounding environment. Modifying the shell with functional ligands, such as amines, carboxyl groups, or thiols, enhances biocompatibility, solubility, and selective binding. **SiO_2-coated nanoparticles functionalized with amino (—NH_2) or carboxyl (—COOH) groups** improve targeting in biomedical imaging and drug delivery. In advanced applications, multi-shell architectures introduce layered functionalities. **Pt@CeO_2@SiO_2 core-shell structures** integrate multiple functionalities, where the CeO_2 shell provides oxygen storage capability, and the SiO_2 shell ensures structural stability, making them highly efficient in fuel cell catalysis. Smart core-shell systems with **pH-, temperature-, or light-responsive shells** exhibit tunable behavior, allowing applications in controlled drug release and environmental sensing. **Thermo-responsive PNIPAM-coated nanoparticles** have been developed to release therapeutic agents at specific temperatures, optimizing drug delivery efficiency.

The ability to tailor properties through shell modification has led to transformative advancements across multiple disciplines. Core-shell materials with conductive shells, such as **Si@C anodes**, enhance lithium-ion battery performance. Functionalized polymer and gold shells improve targeted therapy and bioimaging capabilities. Tunable shells enable optimized photocatalytic degradation of pollutants and enhanced catalytic efficiency in green chemistry.

2.5 Conclusion

Core-shell nanomaterials represent a groundbreaking advancement in nanotechnology, offering unprecedented control over material properties through structural and compositional engineering. Their unique architectures, consisting of a distinct core encapsulated by a carefully chosen shell, allow for tailored enhancements in optical, electronic, mechanical, magnetic, and catalytic performance. By manipulating the core and shell compositions, thickness, and interfacial interactions, researchers have developed highly functional materials with applications spanning energy storage, catalysis, medicine, and environmental remediation. A key advantage of core-shell nanomaterials is their ability to combine complementary properties from different materials, leading to multifunctional capabilities. For instance, metallic core-shell structures leverage plasmonic and catalytic properties, polymeric core-shell systems enhance drug delivery and biocompatibility, and metal oxide-based core-shell architectures optimize photocatalysis and electronic applications. Furthermore, hybrid and composite core-shell materials integrate multiple material classes, providing synergistic benefits such as improved conductivity, structural stability, and chemical resistance. The relationship between structure and properties in core-shell nanomaterials underscores the importance of precise control at the nanoscale. The thickness of the shell significantly influences the material's stability, reactivity, and performance, while the nature of the core-shell interface dictates mechanical robustness and thermal resilience. Moreover, optical, electronic, magnetic, and catalytic behaviors can be finely tuned through shell modifications, enabling tailored functionality for diverse applications. Advanced synthetic strategies and computational modeling have further accelerated the rational design of core-shell materials, allowing for the development of highly efficient and application-specific nanostructures. Looking ahead, continued research in core-shell nanotechnology is expected to drive innovation in energy, healthcare, and environmental sciences. The integration of ML and AI in material design, along with advances in in situ characterization techniques, will further refine our ability to engineer core-shell structures with precise functionalities. Additionally, sustainable and scalable fabrication methods will be crucial for transitioning these materials from laboratory research to industrial applications.

References

1 Wang J, Wan J, Wang D. Hollow multishelled structures for promising applications: Understanding the structure-performance correlation. Acc Chem Res [Internet]. 2019 Aug 20 [cited 2025 Mar 7];52(8):2169–78. Available from: https://pubs.acs.org/doi/full/10.1021/acs.accounts.9b00112

2 Walther A, Müller AHE. Janus particles: Synthesis, self-assembly, physical properties, and applications. Chem Rev [Internet]. 2013 Jul 10 [cited 2025 Mar 7]; 113(7):5194–261. Available from: https://pubs.acs.org/doi/full/10.1021/cr300089t

3 Mohamed El-Toni A, Habila MA, Labis JP, Alothman ZA, Alhoshan M, Elzatahry AA, et al. Design, synthesis and applications of core–shell, hollow core, and nanorattle multifunctional nanostructures. Nanoscale. 2016;8:2510–31. https://doi.org/10.1039/C5NR07004J

4 Gawande MB, Goswami A, Asefa T, Guo H, Biradar AV, Peng DL, et al. Core–shell nanoparticles: Synthesis and applications in catalysis and electrocatalysis. Chem Soc Rev [Internet]. 2015 Oct 19 [cited 2025 Mar 7];44(21):7540–90. Available from: https://pubs.rsc.org/en/content/articlehtml/2015/cs/c5cs00343a

5 Starr JD. Janus-type ceramic nanomaterials: Anisotropic building blocks for the formation of new composites [Dissertation]. University of Florida. 2014.

6 Hayes R, Ahmed A, Edge T, Zhang H. Core–shell particles: Preparation, fundamentals and applications in high performance liquid chromatography. J Chromatogr A. 2014 Aug 29;1357:36–52.

7 El-Toni AM, Habila MA, Labis JP, Alothman ZA, Alhoshan M, Elzatahry AA, et al. Design, synthesis and applications of core–shell, hollow core, and nanorattle multifunctional nanostructures. Nanoscale [Internet]. 2016 Jan 29 [cited 2025 Mar 7];8(5):2510–31. Available from: https://pubs.rsc.org/en/content/articlehtml/2016/nr/c5nr07004j

8 Rajabifar N, Rostami A, Afshar S, Mosallanezhad P, Zarrintaj P, Shahrousvand M, et al. Wound dressing with electrospun core-shell nanofibers: From material selection to synthesis. Polymers [Internet]. 2024 Sep 5 [cited 2025 Mar 7];16(17):2526. Available from: https://www.mdpi.com/2073-4360/16/17/2526/htm

9 Zou Y, Sun Z, Wang Q, Ju Y, Sun N, Yue Q, et al. Core-shell magnetic particles: Tailored synthesis and applications. Chem Rev [Internet]. 2025 Jan 22 [cited 2025 Mar 7];125(2):972–1048. Available from: https://pubs.acs.org/doi/full/10.1021/acs.chemrev.4c00710

10 Wei S, Wang Q, Zhu J, Sun L, Lin H, Guo Z. Multifunctional composite core–shell nanoparticles. Nanoscale [Internet]. 2011 Nov 2 [cited 2025 Mar 7];3(11):4474–502. Available from: https://pubs.rsc.org/en/content/articlehtml/2011/nr/c1nr11000d

11 Liu D, Xue C, Liu D, Xue C. Plasmonic coupling architectures for enhanced photocatalysis. Adv Mater [Internet]. 2021 Nov 1 [cited 2025 Mar 7];

33(46):2005738. Available from: https://onlinelibrary.wiley.com/doi/full/10.1002/adma.202005738

12 Yao Q, Yu Z, Li L, Huang X. Strain and surface engineering of multicomponent metallic nanomaterials with unconventional phases. Chem Rev [Internet]. 2023 Aug 9 [cited 2025 Mar 7];123(15):9676–717. Available from: https://pubs.acs.org/doi/full/10.1021/acs.chemrev.3c00252

13 Serda M, Becker FG, Cleary M, Team RM, Holtermann H, The D, et al. Fabrication of functionalized SiO_2/TiO_2 hollow nanoparticles for drug delivery and neuronal development applications. In: Balint G, Antala B, Carty C, Mabieme JMA, Amar IB, Kaplanova A, editors. Uniwersytet śląski [Internet]. 2017 [cited 2025 Mar 7];7(1):343–54. Available from: https://s-space.snu.ac.kr/handle/10371/119827

14 Awiaz G, Lin J, Wu A. Recent advances of Au@Ag core–shell SERS-based biosensors. Exploration [Internet]. 2023 Feb 1 [cited 2025 Mar 7];3(1):20220072. Available from: https://onlinelibrary.wiley.com/doi/full/10.1002/EXP.20220072

15 Zhang J, Li Y, Cao K, Chen R. Advances in atomic layer deposition. Nanomanuf Metrol [Internet]. 2022 Apr 8 [cited 2025 Mar 7];5(3):191–208. Available from: https://link.springer.com/article/10.1007/s41871-022-00136-8

16 Prieto G, Tüysüz H, Duyckaerts N, Knossalla J, Wang GH, Schüth F. Hollow nano- and microstructures as catalysts. Chem Rev [Internet]. 2016 Nov 23 [cited 2025 Mar 7];116(22):14056–119. Available from: https://pubs.acs.org/doi/full/10.1021/acs.chemrev.6b00374

17 Wang X, Feng J, Bai Y, Zhang Q, Yin Y. Synthesis, properties, and applications of hollow micro-/nanostructures. Chem Rev [Internet]. 2016 Sep 28 [cited 2025 Mar 7];116(18):10983–1060. Available from: https://pubs.acs.org/doi/full/10.1021/acs.chemrev.5b00731

18 He Y, Matthews B, Wang J, Song L, Wang X, Wu G. Innovation and challenges in materials design for flexible rechargeable batteries: From 1D to 3D. J Mater Chem A Mater [Internet]. 2018 Jan 16 [cited 2025 Mar 7];6(3):735–53. Available from: https://pubs.rsc.org/en/content/articlehtml/2018/ta/c7ta09301b

19 Saha R, Mondal B, Mukherjee PS. Molecular cavity for catalysis and formation of metal nanoparticles for use in catalysis. Chem Rev [Internet]. 2022 Jul 27 [cited 2025 Mar 7];122(14):12244–307. Available from: https://pubs.acs.org/doi/full/10.1021/acs.chemrev.1c00811

20 Ma X, Li Y, Hussain I, Shen R, Yang G, Zhang K, et al. Core–shell structured nanoenergetic materials: Preparation and fundamental properties. Adv Mater [Internet]. 2020 Jul 1 [cited 2025 Mar 7];32(30):2001291. Available from: https://onlinelibrary.wiley.com/doi/full/10.1002/adma.202001291

21 Kesici-Meco E, Unur-Yilmaz E. Enzymatic electrochemical biosensor for glucose detection based on biomass-derived Fe_3O_4/C and PEDOT:PSS modification. J Mater Sci: Mater Electron [Internet]. 2024 Nov 1 [cited 2025 Mar 7];35(33):1–18. Available from: https://link.springer.com/article/10.1007/s10854-024-13885-7

22 Cialla D, März A, Böhme R, Theil F, Weber K, Schmitt M, et al. Surface-enhanced Raman spectroscopy (SERS): Progress and trends. Anal Bioanal Chem [Internet]. 2012 Apr 29 [cited 2025 Mar 7];403(1):27–54. Available from: https://link.springer.com/article/10.1007/s00216-011-5631-x

23 Wang J, Wan J, Yang N, Li Q, Wang D. Hollow multishell structures exercise temporal–spatial ordering and dynamic smart behaviour. Nat Rev Chem [Internet]. 2020 Feb 11 [cited 2025 Mar 7];4(3):159–68. Available from: https://www.nature.com/articles/s41570-020-0161-8

24 Kumar R, Mondal K, Panda PK, Kaushik A, Abolhassani R, Ahuja R, et al. Core–shell nanostructures: Perspectives towards drug delivery applications. J Mater. Chem. B [Internet]. [cited 2025 Mar 7]. Available from: https://pubs.rsc.org/en/content/articlehtml/2020/tb/d0tb01559h

25 Zhuo M, Wang X, Horizons LLM. Construction and optoelectronic applications of organic core/shell micro/nanostructures. Materials Horizons. 2020;7:3161–75. https://doi.org/10.1039/D0MH00967A.

26 Du J, Reilly RKO. Anisotropic particles with patchy, multicompartment and Janus architectures: Preparation and application. Chem Soc Rev [Internet]. 2011 Apr 18 [cited 2025 Mar 7];40(5):2402–16. Available from: https://pubs.rsc.org/en/content/articlehtml/2011/cs/c0cs00216j

27 Le TC, Zhai J, Chiu WH, Tran PA, Tran N. Janus particles: Recent advances in the biomedical applications. Int J Nanomedicine. 2019;14:6749–77.

28 Swisher JH, Jibril L, Petrosko SH, Mirkin CA. Nanoreactors for particle synthesis. Nat Rev Mater [Internet]. 2022 Jan 13 [cited 2025 Mar 7];7(6):428–48. Available from: https://www.nature.com/articles/s41578-021-00402-z

29 Fan N, Yang Y, Wang W, Zhang L, Chen W, Zou C, et al. Selective etching induces selective growth and controlled formation of various platinum nanostructures by modifying seed surface free energy. ACS Nano [Internet]. 2012 May 22 [cited 2025 Mar 7];6(5):4072–82. Available from: https://pubs.acs.org/doi/full/10.1021/nn3004668

30 Chakarvarti SK. Track-etch membranes enabled nano-/microtechnology: A review. Radiat Meas. 2009 Oct 1;44(9–10):1085–92.

31 Yang S, Guo F, Kiraly B, Mao X, Lu M, Leong KW, et al. Microfluidic synthesis of multifunctional Janus particles for biomedical applications. Lab Chip [Internet]. 2012 May 22 [cited 2025 Mar 7];12(12):2097–102. Available from: https://pubs.rsc.org/en/content/articlehtml/2012/lc/c2lc90046g

32 Yan M, Xie L, Tang J, Liang K, Mei Y, Kong B. Recent advances in heterosilica-based micro/nanomotors: Designs, biomedical applications, and future perspectives. Chem Mater [Internet]. 2021 May 11 [cited 2025 Mar 7];33(9):3022–46. Available from: https://pubs.acs.org/doi/full/10.1021/acs.chemmater.1c00192

33 Guo J, Yang J, Xiang Z, Zhai H, Chen W. An asymmetrically coordinated Zn–Co diatomic site catalyst for efficient hydrogen evolution reactions. Chem Commun [Internet]. 2024 Jun 13 [cited 2025 Mar 7];60(49):6320–23. Available from: https://pubs.rsc.org/en/content/articlehtml/2024/cc/d4cc01832j

34 Etesami M, Nguyen MT, Yonezawa T, Tuantranont A, Somwangthanaroj A, Kheawhom S. 3D carbon nanotubes-graphene hybrids for energy conversion and storage applications. Chem Eng J. 2022 Oct 15;446:137190.

35 Sui R, Charpentier PA, Marriott RA. Metal oxide-related dendritic structures: Self-assembly and applications for sensor, catalysis, energy conversion and beyond. Nanomaterials [Internet]. 2021 Jun 27 [cited 2025 Mar 7];11(7):1686. Available from: https://www.mdpi.com/2079-4991/11/7/1686/htm

36 Abass NAHA, Jawad MF, Haider AJ, Taha BA. Exploring random laser characteristics in core@ shell nano-scatter centers: Trends and opportunities. Opt Quantum Electron [Internet]. 2024 Nov 24 [cited 2025 Mar 7];56(12):1–30. Available from: https://link.springer.com/article/10.1007/s11082-024-06881-y

37 Malehmir S, Esmaili MA, Khaksary Mahabady M, Sobhani-Nasab A, Atapour A, Ganjali MR, et al. A review: Hemocompatibility of magnetic nanoparticles and their regenerative medicine, cancer therapy, drug delivery, and bioimaging applications. Front Chem. 2023 Aug 29;11:1249134.

38 Ye R, Zhao J, Wickemeyer BB, Toste FD, Somorjai GA. Foundations and strategies of the construction of hybrid catalysts for optimized performances. Nat Catal [Internet]. 2018 Apr 9 [cited 2025 Mar 7];1(5):318–25. Available from: https://www.nature.com/articles/s41929-018-0052-2

39 Mondal K, Sharma A. Recent advances in the synthesis and application of photocatalytic metal–metal oxide core–shell nanoparticles for environmental remediation and their recycling process. RSC Adv [Internet]. 2016 Sep 2 [cited 2025 Mar 7];6(87):83589–612. Available from: https://pubs.rsc.org/en/content/articlehtml/2016/ra/c6ra18102c

40 Al-Amiery AA, Fayad MA, Abdul Wahhab HA, Al-Azzawi WK, Mohammed JK, Majdi HS. Interfacial engineering for advanced functional materials: Surfaces, interfaces, and applications. Results Eng. 2024 Jun 1;22:102125.

41 Distefano JG, Murthy AA, Hao S, Dos Reis R, Wolverton C, Dravid VP. Topology of transition metal dichalcogenides: The case of the core–shell architecture. Nanoscale [Internet]. 2020 Dec 18 [cited 2025 Mar 7];12(47):23897–919. Available from: https://pubs.rsc.org/en/content/articlehtml/2020/nr/d0nr06660e

42 Yang T, Lu B, Zuo Y, Huang J. Configuration engineering of plasmonic-metal/semiconductor nanohybrids for solar fuel production†. Chem Mater [Internet]. 2025 [cited 2025 Mar 7]. Available from: https://pubs.acs.org/doi/full/10.1021/acs.chemmater.4c03170

43 Serda M, Becker FG, Cleary M, Team RM, Holtermann H, The D, et al. A study on the synthesis and electrical properties of one-dimensional Zinc-based semiconductor structures. In: Balint G, Antala B, Carty C, Mabieme JMA, Amar IB, Kaplanova A, editors. Uniwersytet śląski [Internet]. 2017 [cited 2025 Mar 7]; 7(1):343–54. Available from: https://s-space.snu.ac.kr/handle/10371/136761

44 Li W, Elzatahry A, Aldhayan D, Zhao D. Core–shell structured titanium dioxide nanomaterials for solar energy utilization. Chem Soc Rev [Internet]. 2018 Nov 12 [cited 2025 Mar 7];47(22):8203–37. Available from: https://pubs.rsc.org/en/content/articlehtml/2018/cs/c8cs00443a

45 Yang X, Shu H, Chen X. Band structures and spatial carrier confinement in GaAs/GaP core-shell nanowires: Core/shell composition and size effects. J Alloys Compd. 2016 Oct 15;682:571–8.

46 Zhang QK, Zhao WH, Zhou ZP, Cao LM, Yin WJ, Wei XL, et al. 2D Janus MoSSe/$MoGeSiN_4$ vdW heterostructures for photovoltaic and photocatalysis applications. J Alloys Compd. 2023 Mar 25;938:168708.

47 Mulla R, Dunnill CW. Core–shell nanostructures for better thermoelectrics. Mater Adv [Internet]. 2022 Jan 4 [cited 2025 Mar 7];3(1):125–41. Available from: https://pubs.rsc.org/en/content/articlehtml/2022/ma/d1ma00955a

48 Yeh CW, Chen GH, Ho SJ, Chen HS. Inhibiting the surface oxidation of low-cadmim-content ZnS:(Cd,Se) quantum dots for enhancing application reliability. ACS Appl Nano Mater [Internet]. 2019 Aug 23 [cited 2025 Mar 7];2(8):5290–301. Available from: https://pubs.acs.org/doi/full/10.1021/acsanm.9b01213

49 Li JF, Zhang YJ, Ding SY, Panneerselvam R, Tian ZQ. Core-shell nanoparticle-enhanced Raman spectroscopy. Chem Rev [Internet]. 2017 Apr 12 [cited 2025 Mar 7];117(7):5002–69. Available from: https://pubs.acs.org/doi/full/10.1021/acs.chemrev.6b00596

50 Ricardo J, Martins G. Study of the Förster Resonance Energy Transfer in Ensembles of Colloidal PbS Quantum Dots,emitting in the Near-Infrared Spectral Range. Universidade do Minho Escola de Ciências. 2020.

51 Ghosh Chaudhuri R, Paria S. Core/shell nanoparticles: Classes, properties, synthesis mechanisms, characterization, and applications. Chem Rev [Internet]. 2012 Apr 11 [cited 2025 Mar 7];112(4):2373–433. Available from: https://pubs.acs.org/doi/full/10.1021/cr100449n

52 Llerena Zambrano B, Renz AF, Ruff T, Lienemann S, Tybrandt K, Vörös J, et al. Soft electronics based on stretchable and conductive nanocomposites for biomedical applications. Adv Healthc Mater [Internet]. 2021 Feb 1 [cited 2025 Mar 7];10(3):2001397. Available from: https://onlinelibrary.wiley.com/doi/full/10.1002/adhm.202001397

53 Liu J, Xiang L, Kan T. The effect of temperature on the bending properties and failure mechanism of composite truss core sandwich structures. Compos Part A Appl Sci Manuf. 2015 Dec 1;79:146–54.

54 Das S, Pérez-Ramírez J, Gong J, Dewangan N, Hidajat K, Gates BC, et al. Core–shell structured catalysts for thermocatalytic, photocatalytic, and electrocatalytic conversion of CO_2. Chem Soc Rev [Internet]. 2020 May 26 [cited 2025 Mar 7];49(10):2937–3004. Available from: https://pubs.rsc.org/en/content/articlehtml/2020/cs/c9cs00713j

55 Tao L, Huang B, Jin F, Yang Y, Luo M, Sun M, et al. Atomic PdAu interlayer sandwiched into Pd/Pt core/shell nanowires achieves superstable oxygen reduction catalysis. ACS Nano [Internet]. 2020 Sep 22 [cited 2025 Mar 7];14(9):11570–8. Available from: https://pubs.acs.org/doi/full/10.1021/acsnano.0c04061

56 Xu J, Qu Z, Ke G, Wang Y, Huang B. Catalytic activity of gold-silver nanoalloys for HCHO oxidation: Effect of hydroxyl and particle size. Appl Surf Sci. 2020 May 30;513:145910.

57 Chen H, Liu J, Wu X, Ye C, Zhang J, Luo JL, et al. Pt–Co electrocatalysts: Syntheses, morphologies, and applications. Small [Internet]. 2022 Oct 1 [cited 2025 Mar 7];18(40):2204100. Available from: https://onlinelibrary.wiley.com/doi/full/10.1002/smll.202204100

58 Medina OE, Amell AA, López D, Santamaría A. Comprehensive review of nickel-based catalysts advancements for CO_2 methanation. Renew Sustain Energy Rev. 2025 Jan 1;207:114926.

59 Wang H, Gao C, Liu Z, Li B, Dok Kim Y, Feng J, et al. Pt-Ru atomic alloys confined in mesoporous carbon hollow spheres for accelerating methanol oxidation. J Colloid Interface Sci. 2025 Jan 15;678:1004–11.

60 Besley NA. Density functional theory based methods for the calculation of X-ray spectroscopy. Acc Chem Res [Internet]. 2020 Jul 21 [cited 2025 Mar 7];53(7): 1306–15. Available from: https://pubs.acs.org/doi/full/10.1021/acs.accounts.0c00171

61 Laghrissi A, Es-Souni M. Au-nanorods supporting Pd and Pt nanocatalysts for the hydrogen evolution reaction: Pd is revealed to be a better catalyst than Pt. Nanomaterials [Internet]. 2023 Jul 1 [cited 2025 Mar 7];13(13):2007. Available from: https://www.mdpi.com/2079-4991/13/13/2007/htm

62 Zhang H, Su YC, Han Y, Jiang S. Molecular dynamics study of melting behavior of planar stacked Ti–Al core–shell nanoparticles. J Compos Sci [Internet]. 2022 Apr 26 [cited 2025 Mar 7];6(5):126. Available from: https://www.mdpi.com/2504-477X/6/5/126/htm

63 Luo L, Zhang F, Wang L, Liu Y, Leng J. Recent advances in shape memory polymers: Multifunctional materials, multiscale structures, and applications. Adv Funct Mater [Internet]. 2024 Apr 1 [cited 2025 Mar 7];34(14):2312036. Available from: https://onlinelibrary.wiley.com/doi/full/10.1002/adfm.202312036

64 Yang Z, Gao W. Applications of machine learning in alloy catalysts: Rational selection and future development of descriptors. Adv Sci [Internet]. 2022 Apr 1 [cited 2025 Mar 7];9(12):2106043. Available from: https://onlinelibrary.wiley.com/doi/full/10.1002/advs.202106043

65 Mendes M, Sousa J, Vitorino C, Pais A. Clinical applications of nanostructured drug delivery systems: From basic research to translational medicine. In: Core-Shell Nanostructures for Drug Delivery and Theranostics: Challenges, Strategies and Prospects for Novel Carrier Systems. 2018 Jan 1;43–116.

66 Hang Y, Wang A, Wu N. Plasmonic silver and gold nanoparticles: Shape- and structure-modulated plasmonic functionality for point-of-caring sensing, bio-imaging and medical therapy. Chem Soc Rev [Internet]. 2024 Mar 18 [cited 2025 Mar 7];53(6):2932–71. Available from: https://pubs.rsc.org/en/content/articlehtml/2024/cs/d3cs00793f

67 Yang C, Lin ZI, Chen JA, Xu Z, Gu J, Law WC, et al. Organic/inorganic self-assembled hybrid nano-architectures for cancer therapy applications. Macromol Biosci [Internet]. 2022 Feb 1 [cited 2025 Mar 7];22(2):2100349. Available from: https://onlinelibrary.wiley.com/doi/full/10.1002/mabi.202100349

68 Terranova ML, Orlanducci S, Tamburri E, Guglielmotti V, Rossi M. Si/C hybrid nanostructures for Li-ion anodes: An overview. J Power Sources. 2014 Jan 15;246:167–77.

69 Zhang Q, Lee I, Joo JB, Zaera F, Yin Y. Core-shell nanostructured catalysts. Acc Chem Res [Internet]. 2013 Aug 20 [cited 2025 Mar 7];46(8):1816–24. Available from: https://pubs.acs.org/doi/full/10.1021/ar300230s

70 Wang D, Li T. Toward MOF@Polymer core-shell particles: Design principles and potential applications. Acc Chem Res [Internet]. 2023 Feb 21 [cited 2025 Mar 7];56(4):462–74. Available from: https://pubs.acs.org/doi/full/10.1021/acs.accounts.2c00695

71 Comanescu C. Recent advances in surface functionalization of magnetic nanoparticles. Coatings [Internet]. 2023 Oct 15 [cited 2025 Mar 7];13(10):1772. Available from: https://www.mdpi.com/2079-6412/13/10/1772/htm

72 Wan X, Liu J, Zhang J. Construction of plasmonic Metal@Semiconductor core–shell photocatalysts: From epitaxial to nonepitaxial strategies. Small redStruct [Internet]. 2022 Sep 1 [cited 2025 Mar 7];3(9):2200045. Available from: https://onlinelibrary.wiley.com/doi/full/10.1002/sstr.202200045

73 Yilmaz A. Comparison of core-shell alloy plasmonic nanoparticles with 3 different models [MS thesis]. The Graduate School of Natural and Applied Sciences of Middle East Technical University.

74 Banerjee I, Kumaran V, Santhanam V. Synthesis and characterization of Au@Pt nanoparticles with ultrathin platinum overlayers. J Phy Chem C [Internet]. 2015 Mar 19 [cited 2025 Mar 7];119(11):5982–87. Available from: https://pubs.acs.org/doi/full/10.1021/jp5113284

75 Li Z, Li M, Bian Z, Kathiraser Y, Kawi S. Design of highly stable and selective core/yolk–shell nanocatalysts—A review. Appl Catal B. 2016 Jul 5;188:324–41.

76 Loo C, Lin A, Hirsch L, Lee MH, Barton J, Halas N, et al. Nanoshell-enabled photonics-based imaging and therapy of cancer. Technol Cancer Res Treat [Internet]. 2004 Feb 1 [cited 2025 Mar 8];3(1):33–40. Available from: https://journals.sagepub.com/doi/abs/10.1177/153303460400300104

77 Khan NR, Sharmin T, Bin Rashid A. Exploring the versatility of aerogels: Broad applications in biomedical engineering, astronautics, energy storage, biosensing, and current progress. Heliyon [Internet]. 2024 Jan 15 [cited 2025 Mar 7];10(1):e23102. Available from: https://www.cell.com/action/showFullText?pii=S2405844023103100

78 Hanske C, Sanz-Ortiz MN, Liz-Marzán LM. Silica-coated plasmonic metal nanoparticles in action*. In: Colloidal Synthesis of Plasmonic Nanometals [Internet]. 2020 Apr 23 [cited 2025 Mar 7];755–820. Available from: https://www.taylorfrancis.com/chapters/edit/10.1201/9780429295188-25/silica-coated-plasmonic-metal-nanoparticles-action-christoph-hanske-marta-sanz-ortiz-luis-liz-marz%C3%A1n

79 Haynes JA, Pint BA, Porter WD, Wright IG. Comparison of thermal expansion and oxidation behavior of various high-temperature coating materials and superalloys. Mater High Temp [Internet]. 2004 [cited 2025 Mar 7];21(2):87–94. Available from: https://www.tandfonline.com/doi/abs/10.1179/mht.2004.012

80 Ajayan PM, Schadler LS, Braun PV. Nanocomposite Science and Technology. Nanocomposite Science and Technology [Internet]. 2003 Aug 26 [cited 2025 Mar 7]. Available from: https://onlinelibrary.wiley.com/doi/book/10.1002/3527602127

81 Chen H, Zhang L, Li M, Xie G. Synthesis of core–shell micro/nanoparticles and their tribological application: A review. Materials [Internet]. 2020 Oct 15 [cited 2025 Mar 7];13(20):4590. Available from: https://www.mdpi.com/1996-1944/13/20/4590/htm

82 Zhang C, Chen BQ, Li ZY, Xia Y, Chen YG. Surface plasmon resonance in bimetallic core-shell nanoparticles. J Phy Chem C [Internet]. 2015 Jul 23 [cited 2025 Mar 7];119(29):16836–45. Available from: https://pubs.acs.org/doi/full/10.1021/acs.jpcc.5b04232

83 Moreira AF, Rodrigues CF, Reis CA, Costa EC, Correia IJ. Gold-core silica shell nanoparticles application in imaging and therapy: A review. Micropor Mesopor Mater. 2018 Nov 1;270:168–79.

84 Yilmaz A, Yilmaz M. Bimetallic core–shell nanoparticles of gold and silver via bioinspired polydopamine layer as surface-enhanced Raman spectroscopy (SERS) platform. Nanomaterials [Internet]. 2020 Apr 5 [cited 2025 Mar 7];10(4):688. Available from: https://www.mdpi.com/2079-4991/10/4/688/htm

85 Huang B, Yang H, Zhang L, Yuan Y, Cui Y, Zhang J. Effect of surface/interfacial defects on photo-stability of thick-shell CdZnSeS/ZnS quantum dots. Nanoscale [Internet]. 2018 Oct 4 [cited 2025 Mar 7];10(38):18331–40. Available from: https://pubs.rsc.org/en/content/articlehtml/2018/nr/c8nr04224a

86 Ali Z. Surface Modification and Functionalization of Colloidal Nanoparticles [Doctoral thesis]. Philipps-Universität Marburg. 2011 [cited 2025 Mar 7]. Available from: https://archiv.ub.uni-marburg.de/diss/z2011/0083

87 Vettori M, Piazza V, Cattoni A, Scaccabarozzi A, Patriarche G, Regreny P, et al. Growth optimization and characterization of regular arrays of GaAs/AlGaAs core/shell nanowires for tandem solar cells on silicon. Nanotechnology [Internet]. 2018 Dec 31 [cited 2025 Mar 7];30(8):084005. Available from: https://iopscience.iop.org/article/10.1088/1361-6528/aaf3fe

88 Interfacing Bioelectronics and Biomedical Sensing - Google Books [Internet]. [cited 2025 Mar 7]. Available from: https://books.google.co.in/books?hl=en&lr=&id=LgDQDwAAQBAJ&oi=fnd&pg=PA29&dq=SiO%E2%82%82-coated+gold+nanoparticles,+which+show+enhanced+electron+storage+capabilities+in+nanostructured+capacitors.+&ots=qSbDxgzsNK&sig=d-TjBqumo4Ad3Pdhx5nXoCDz7b4&redir_esc=y#v=onepage&q&f=false

89 Mélinon P, Begin-Colin S, Duvail JL, Gauffre F, Boime NH, Ledoux G, et al. Engineered inorganic core/shell nanoparticles. Phys Rep. 2014 Oct 20;543(3):163–97.

90 Lottini E, López-Ortega A, Bertoni G, Turner S, Meledina M, Tendeloo G Van, et al. Strongly exchange coupled core|shell nanoparticles with high magnetic anisotropy: A strategy toward rare-earth-free permanent magnets. Chem Mater [Internet]. 2016 Jun 28 [cited 2025 Mar 7];28(12):4214–22. Available from: https://pubs.acs.org/doi/full/10.1021/acs.chemmater.6b00623

91 Shah A, Sun H, Qiao Z, Umair Khan M, Ali N, Gao C, et al. Synthesis of iron oxide nanoparticles and its antimicrobial, anticancer, anti-inflammatory, wound healing, and immunomodulatory activities –A review. Phytopharmacol Res J [Internet]. 2024 [cited 2025 Mar 7];3(3):1–28. Available from: https://ojs.prjn.org/index.php/prjn/article/view/75

92 Deng J, Deng D, Bao X, Deng J, Deng D, Bao X. Robust catalysis on 2D materials encapsulating metals: Concept, application, and perspective. Adv Mater [Internet]. 2017 Nov 1 [cited 2025 Mar 7];29(43):1606967. Available from: https://onlinelibrary.wiley.com/doi/full/10.1002/adma.201606967

93 Guo Y, Tan K, Guo X, Li H, Jian X. Multi-core/shell $SiO_2@Al_2O_3$ nanostructures deposited on $Ti_3AlC_2$to enhance high-temperature stability and microwave absorption properties. Nanotechnol Rev [Internet]. 2023 Jan 1 [cited 2025 Mar 7];12(1):20220545. Available from: https://www.degruyter.com/document/doi/10.1515/ntrev-2022-0545/html

94 Schubert J, Chanana M. Coating matters: Review on colloidal stability of nanoparticles with biocompatible coatings in biological media, living cells and organisms. Curr Med Chem. 2018 Jun 4;25(35):4553–86.

3

Synthesis of Core-shell Nanomaterials

The synthesis of core-shell nanomaterials represents a sophisticated and strategically guided process that enables the precise assembly of materials with hierarchical structures and tailored functionalities. At the heart of this endeavor lies the challenge of achieving controlled nucleation, shell growth, and interface formation between the core and shell constituents without compromising their individual or collective properties. As these materials exhibit highly structure-dependent behavior, the choice of synthesis method plays a pivotal role in determining their performance across applications ranging from catalysis to biomedicine and electronics. The synthetic strategies for core-shell nanomaterials are as diverse as the materials themselves, broadly classified into physical, chemical, and green or sustainable approaches. Each method offers distinct advantages and challenges in terms of precision, scalability, cost, environmental impact, and compatibility with different material systems. Physical methods typically provide excellent control over thickness and morphology but may involve high-energy or vacuum environments, making them less accessible for large-scale production. In contrast, chemical approaches offer versatility in compositional tuning and are suitable for synthesizing complex architectures under relatively mild conditions. Emerging green synthesis methods, on the other hand, are gaining prominence as researchers aim to minimize environmental impact and move toward safer, more sustainable material fabrication.

The choice of synthetic route also directly influences interfacial integrity, crystallinity, defect density, and surface functionality—the factors that critically affect the overall stability and application potential of core-shell nanostructures. Achieving uniform shell coverage, avoiding unwanted secondary nucleation, and controlling the interface between dissimilar materials are among the key challenges encountered during synthesis. Furthermore, as the field progresses from laboratory-scale proof-of-concept studies to commercial and industrial-scale applications, emphasis is shifting toward synthesis techniques that are not only reproducible and cost-effective but also amenable to automation and high-throughput production. The chapter provides a detailed exploration of the major synthesis techniques employed for fabricating core-shell nanomaterials. It begins with physical synthesis methods, progresses through chemical routes, and discusses green and sustainable

Core-Shell Nanomaterials: From Fundamentals to Applications, First Edition. Shreya Sharma et al.

approaches. Finally, the chapter addresses the important aspects of scalability, industrial feasibility, and emerging trends such as AI-guided synthesis, aiming to provide a comprehensive understanding of how these materials are brought to life from the nanoscale up.

3.1 Physical Methods for Core-shell Synthesis

The physical synthesis of core-shell nanomaterials encompasses a diverse range of top-down and vapor-phase approaches that capitalize on physical phenomena such as evaporation, ion bombardment, mechanical force, or electric fields to form well-defined nanostructured shells around preexisting cores. Unlike chemical methods, which depend on solution-based reactions or molecular precursors, physical methods often operate in vacuum, inert, or controlled environments and are characterized by their high precision, cleanliness, and suitability for materials with high melting points or sensitivity to chemical degradation. These techniques are particularly valuable in applications where shell composition and structural perfection are critical, such as in nanoelectronics, spintronics, photonics, and catalysis, due to their ability to yield ultra-thin, uniform coatings with low defect densities. Many physical methods offer exquisite control over shell thickness, morphology, and crystallinity, enabling the design of materials with fine-tuned interfacial properties and enhanced functional performance [1]. Additionally, certain physical approaches such as mechanical milling and electrospinning lend themselves to large-scale production with minimal chemical waste, aligning with sustainability goals in manufacturing. The key physical synthesis techniques include physical vapor deposition (PVD), thermal evaporation, sputtering, pulsed laser deposition (PLD), mechanical milling, electrospinning, and template-assisted methods.

3.1.1 PVD and Thermal Evaporation

PVD is a class of vacuum-based coating techniques widely used for the fabrication of core-shell nanomaterials, particularly when uniform, high-purity, and defect-free shell layers are required. PVD processes rely on the physical transfer of atoms or molecules from a condensed phase (solid or liquid source) to the vapor phase, followed by condensation onto a target substrate or core particle to form a thin shell [2]. The absence of chemical reactions during deposition is a distinguishing feature of PVD methods, offering greater control over purity, composition, and crystallinity of the resulting nanostructures. This is especially advantageous in applications involving catalysis, optics, energy storage, and microelectronics, where contamination or uncontrolled stoichiometry can significantly impact performance.

Among the various PVD techniques shown in Figure 3.1, thermal evaporation is one of the most straightforward and extensively utilized methods. In this process, the source material, typically a metal, semiconductor, or oxide, is heated in a vacuum chamber until it reaches its vaporization point. The vapor then travels through the vacuum and condenses onto a cooler surface, which, in the context of core-shell synthesis, is often a preexisting nanoparticle or a batch of dispersed core materials

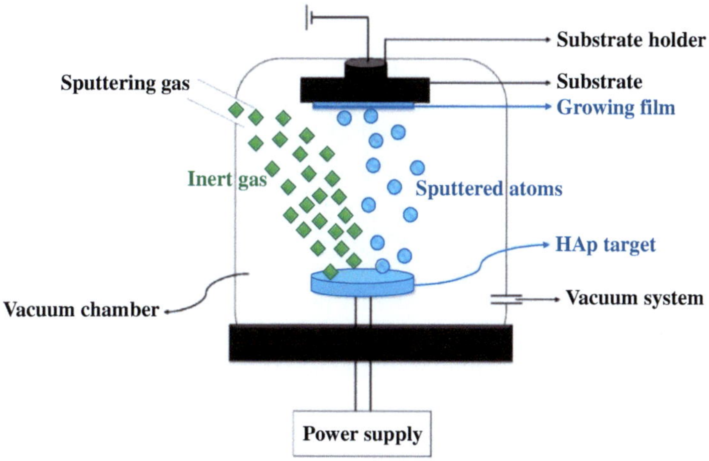

Figure 3.1 Schematic diagram of the physical vapor deposition (PVD) process for thin film growth. [5]/MDPI/CC BY 4.0.

supported on a substrate. The vacuum environment minimizes contamination and facilitates the deposition of high-purity, adherent films. By adjusting parameters such as source temperature, deposition rate, vacuum pressure, and substrate positioning, researchers can tailor the thickness and uniformity of the shell layer with great precision [3]. Thermal evaporation is particularly effective for producing metal or oxide shells on dielectric or polymeric cores. For example, silica (SiO_2) or polystyrene nanospheres can be coated with thin metallic films such as gold (Au), silver (Ag), or aluminum (Al), creating nanoshells with tunable optical properties, especially surface plasmon resonance effects that are highly desirable in photonic and sensing applications. Similarly, zinc oxide (ZnO) or titanium dioxide (TiO_2) shells can be deposited for photocatalytic or electronic uses. One notable advantage of thermal evaporation is its compatibility with a wide variety of core materials, as long as the substrate can withstand moderate heat without degradation or deformation.

Another important aspect of PVD, especially thermal evaporation, is the directional nature of vapor flux, often referred to as the "line-of-sight" limitation [4]. This can pose challenges when attempting to uniformly coat irregular or non-spherical core geometries, as shadowing effects may result in partial or non-uniform shell formation. To mitigate this, techniques such as substrate rotation, planetary motion stages, and use of rotating crucibles are employed to improve coating uniformity. Advanced modifications, including electron-beam evaporation and laser-assisted evaporation, offer additional control over energy input and deposition rate, expanding the applicability of thermal evaporation to more complex or sensitive material systems. Despite these innovations, there are still constraints to thermal evaporation when scaling up the process or dealing with materials with extremely high melting points. In such cases, other PVD variants like sputtering or PLD may be more suitable. Nevertheless, thermal evaporation remains a cornerstone technique in nanomaterials research due to its simplicity, versatility, and effectiveness in producing conformal coatings with desirable functional properties.

3.1.2 Sputtering and PLD

Sputtering and PLD are advanced PVD techniques that offer superior control over film thickness, composition, and adhesion when synthesizing core-shell nanostructures. Both techniques are vacuum-based and allow the deposition of a wide range of materials, including metals, oxides, nitrides, and complex compounds. They are particularly effective in fabricating conformal and high-performance shell layers over various nanoscale cores, even those with intricate geometries. In sputtering, energetic ions, typically from an inert gas such as argon, are generated in a plasma and accelerated toward a target (source) material [6]. The bombardment causes atoms to be ejected (sputtered) from the target's surface. These atoms then travel through the chamber and condense onto the core material, forming the shell layer. Depending on the target and chamber environment, different sputtering methods can be used: DC sputtering for conductive materials, RF sputtering for insulating targets, and magnetron sputtering for improved deposition rates and film uniformity [7]. In reactive sputtering, reactive gases such as oxygen or nitrogen are introduced to form compound films like metal oxides or nitrides on the core surface. Sputtering is widely used for coating nanoparticles with metals such as Pt, Au, and Ag, or metal oxides like TiO_2 and ZnO. It allows for fine-tuning of deposition parameters such as ion energy, pressure, substrate bias, and temperature—critical for achieving desired shell characteristics, including thickness, crystallinity, and adhesion strength. Importantly, it offers excellent coverage over both planar and moderately complex surfaces, and the use of rotating substrates and magnetic field assistance further enhances uniformity. Its compatibility with industrial-scale processes also makes sputtering a preferred choice for fabricating core-shell materials in electronics, magnetic storage, and sensor technologies.

PLD is another versatile PVD method that utilizes short. In Figure 3.2, a high-energy laser pulses, often from an excimer or Nd:YAG laser to ablate material

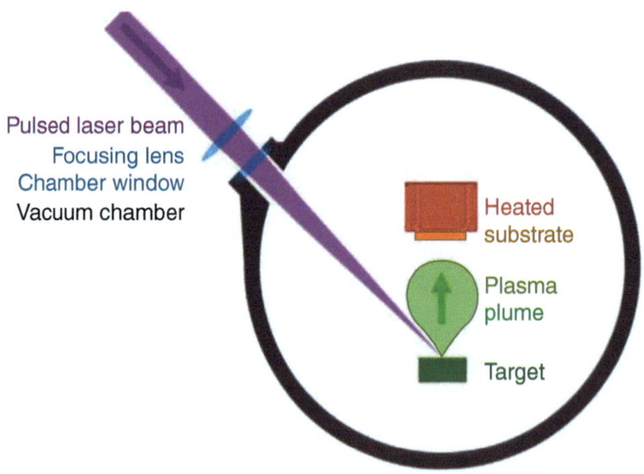

Figure 3.2 Schematic diagram of the pulsed laser deposition (PLD) process illustrating laser ablation of the target material, plume formation, and thin film deposition onto the substrate. [10]/Frontiers Media S.A/CC BY 4.0.

from a solid target. The intense laser energy generates a transient plasma plume containing ions, atoms, and clusters, which rapidly deposit onto the substrate or nanocore material [8]. The dynamic nature of the laser-material interaction enables the deposition of complex and stoichiometrically accurate films, even for multicomponent systems such as perovskites, doped oxides, or mixed metal chalcogenides. One of PLD's key advantages is its exceptional ability to transfer the exact composition of the target to the deposited film, making it ideal for core-shell structures requiring precise functional properties, such as high-temperature superconductors, multiferroic materials, or catalytic layers. The plume dynamics can be manipulated by varying laser fluence, repetition rate, background gas pressure, and target-substrate distance. Moreover, PLD allows for sequential or multi-target deposition, enabling the creation of layered shells or compositionally graded coatings within a single synthesis cycle [9]. PLD has been successfully used to deposit uniform shell layers of oxides like ZnO, TiO_2, and $YBa_2Cu_3O_7$ over cores such as carbon nanotubes, polymer beads, and metal nanoparticles. Its applications span thin-film electronics, energy harvesting, catalysis, and optical coatings. However, the technique is more suited for laboratory-scale production due to limitations in uniform large-area coverage and slower deposition rates compared to sputtering. Additionally, the high cost of lasers and vacuum equipment presents challenges for commercial-scale adoption.

Both sputtering and PLD represent powerful tools for engineering core-shell nanomaterials with high precision and material versatility. While sputtering is more adaptable to industrial-scale coating processes, PLD excels in laboratory-scale fabrication of complex and high-performance shell layers. Together, they serve as indispensable techniques in advancing the frontier of nanoscale material design and functionalization.

3.1.3 Mechanical Milling and Ball Milling Approaches

Mechanical milling, including high-energy ball milling, is a top-down, solid-state synthesis technique that has gained significant traction in fabricating core-shell nanomaterials due to its simplicity, cost-effectiveness, and scalability [11]. Unlike vapor-phase deposition techniques, mechanical milling relies on the application of mechanical force to induce repeated fracture, cold welding, and re-welding of powder particles in a milling medium. This dynamic process allows for intimate contact between core and shell precursor materials, leading to the formation of core-shell structures through physical encapsulation, diffusion, or chemical interaction during subsequent processing steps.

In high-energy ball milling (Figure 3.3), the milling media (typically steel or ceramic balls) are set in motion within a rotating or vibrating chamber, generating substantial kinetic energy that is transferred to the powder mixture. When core material and shell precursors are milled together, the continuous impact causes the shell material to deform plastically and coat the core particles. For instance, aluminum particles have been successfully coated with an alumina (Al_2O_3) shell via ball milling in an oxygen-rich atmosphere. This oxide shell enhances the particles' resistance to oxidation, thermal degradation, and premature reaction—properties that are critical in energetic materials and composite applications [12]. One of the key advantages of this

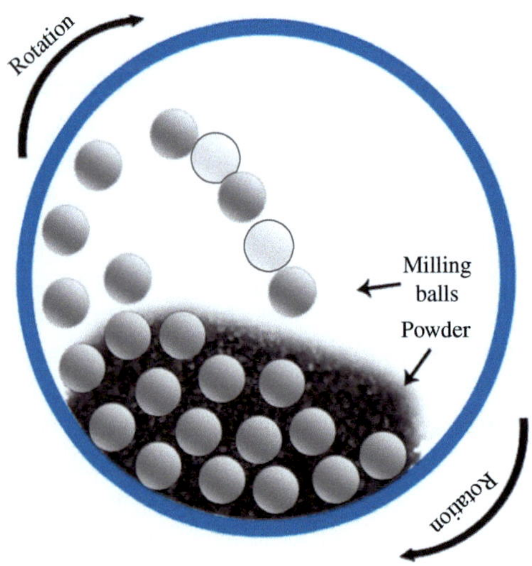

Figure 3.3 Ball milling approaches: overview of different techniques for mechanical grinding and particle size reduction. [14]/MDPI/CC BY 4.0.

approach is its ability to produce core-shell structures in bulk, which is essential for industrial applications such as structural materials, thermites, and magnetic composites. The method can also accommodate a wide variety of material systems, including metal-metal, metal-ceramic, and metal-polymer combinations. For example, iron (Fe) particles coated with silica (SiO_2), carbon, or polymeric shells have been synthesized using ball milling followed by controlled annealing. These core-shell nanostructures exhibit enhanced corrosion resistance, magnetic properties, and thermal stability, making them suitable for data storage, catalysis, and biomedical applications [13]. However, the mechanical nature of milling can introduce challenges. The high-energy impacts may cause crystallographic defects, grain refinement, or even partial amorphization of the materials, which could either enhance or degrade functional performance depending on the application. Furthermore, controlling shell thickness and uniformity is non-trivial, as it depends on several interdependent parameters, including milling time, ball-to-powder weight ratio, rotation speed, milling atmosphere, and process temperature. Shorter milling times may lead to incomplete coating, while extended durations can cause agglomeration or damage to the core particles.

To overcome some of these issues, advanced milling strategies have been explored. These include cryomilling (where milling is conducted at cryogenic temperatures to prevent thermal degradation), mechanochemical synthesis (which leverages mechanical energy to drive chemical reactions between core and shell precursors), and post-milling annealing (which helps in diffusing the shell material uniformly and restoring crystallinity). Moreover, combining mechanical milling with other techniques such as sol–gel coating or spray drying has been reported to enhance shell conformity and functional integration [15]. Despite its limitations, mechanical milling remains a practical and versatile method for synthesizing core-shell nanomaterials, especially where large-scale production and material compatibility are key considerations. Its continued development is expected to contribute significantly to the fabrication of advanced materials for energy storage, catalysis, structural reinforcement, and environmental remediation.

3.1.4 Electrospinning and Physical Encapsulation Techniques

Electrospinning is a versatile and powerful physical technique for fabricating one-dimensional nanostructures, particularly core-shell fibers. As shown in Figure 3.4, it utilizes a high-voltage electric field to draw a charged polymer solution or melt from a capillary spinneret, forming ultrafine fibers that solidify and deposit on a grounded collector. In the context of core-shell nanomaterials, coaxial electrospinning plays a crucial role. This advanced form of electrospinning involves a concentric dual-capillary setup that enables the simultaneous ejection of two different fluids, typically a core solution and a shell solution, resulting in nanofibers with distinct core-shell morphology. The core-shell arrangement within electrospun fibers offers synergistic properties, combining the mechanical strength, biocompatibility, or conductivity of one phase with the functional or responsive behavior of the other. Electrospun core-shell fibers have been extensively used in biomedical applications such as controlled drug delivery, where the drug is encapsulated in the core and protected by a biodegradable polymer shell that regulates release kinetics [16]. In tissue engineering, core-shell scaffolds can mimic the extracellular matrix, providing both structural support and localized delivery of growth factors. In energy and environmental fields, electrospun core-shell nanofibers have been employed in battery separators, supercapacitor electrodes, and air filtration membranes, thanks to their high surface area, tunable porosity, and ability to host active materials within the core or shell.

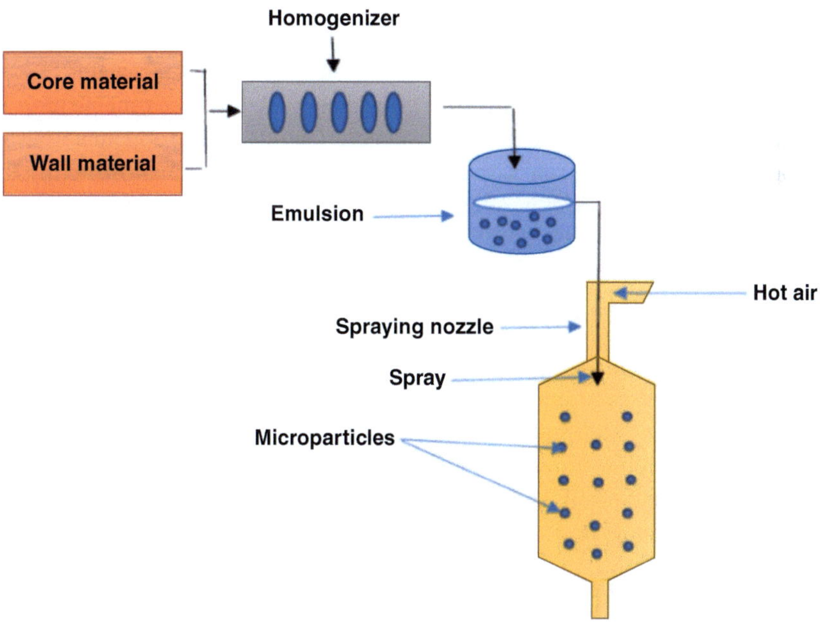

Figure 3.4 Electrospinning combined with spray nozzle technique for producing ultrafine nanofibers alongside microparticles, enabling controlled fabrication of composite micro- and nano-scale structures. [19]/MDPI/CC BY 4.0.

The process offers considerable control over the diameter, alignment, and morphology of the fibers, but achieving uniform core-shell structures demands precise tuning of parameters such as viscosity, conductivity, flow rate, and applied voltage. One of the main advantages of electrospinning is its scalability and compatibility with a wide range of polymers, composites, and hybrid systems [17]. Moreover, it enables the inclusion of functional nanoparticles, catalysts, or therapeutic agents in either the core or shell, expanding its utility in multifunctional applications. Despite these advantages, the technique faces challenges related to solvent compatibility, phase separation, and long-term stability of the resulting fibers. Beyond electrospinning, other physical encapsulation techniques offer additional routes to core-shell structures, especially for particulate systems. Spray drying, for instance, involves atomizing a core-shell precursor solution into a hot gas stream, causing rapid solvent evaporation and formation of microcapsules with shell coatings [18]. This method is widely used in food processing, pharmaceuticals, and agriculture for encapsulating flavors, drugs, and pesticides to enhance their stability and control release. Freeze drying or lyophilization, on the other hand, is used to physically encapsulate sensitive biomolecules or phase change materials (PCMs) within polymeric or inorganic shells, providing structural integrity and protection under storage or operational conditions.

Phase separation and emulsion-based encapsulation are also key techniques under the umbrella of physical encapsulation. These involve forming emulsions or dispersions where the shell-forming material precipitates or solidifies around the dispersed core, producing stable core-shell micro- or nanostructures. For example, paraffin wax as a core material can be encapsulated within a polymeric shell to yield microcapsules for thermal energy storage, capable of repeated melting and solidification cycles without leakage or degradation [20]. Collectively, electrospinning and other physical encapsulation methods offer flexible, scalable, and relatively low-cost strategies for fabricating core-shell materials with tunable properties. These techniques are particularly valuable for integrating soft materials, biological agents, and phase-sensitive compounds into functional nanostructures, thereby broadening the spectrum of applications in healthcare, energy, and environmental technologies.

3.1.5 Template-assisted Physical Synthesis

Template-assisted synthesis is a strategically guided approach for fabricating core-shell nanomaterials, where pre-formed templates—ranging from hard spheres to porous membranes—are employed to dictate the shape, size, and structural integrity of the resulting architectures. This method relies on the deposition or growth of shell materials onto or within the template structure, followed by selective removal or retention of the template to yield the desired core-shell configuration. Unlike random or isotropic synthesis techniques, template-assisted approaches enable superior morphological control and reproducibility, which are crucial for achieving uniformity and function-specific design in nanomaterials.

Hard templates, such as silica spheres, polystyrene beads, and anodic aluminum oxide (AAO) membranes, are commonly used to create hollow or yolk-shell nanostructures. In one typical route, a solid core material (e.g., SiO_2 or PS spheres) is coated with the desired shell material through techniques such as PVD, atomic layer deposition (ALD), or sol–gel coating [21]. After the shell has been adequately formed, the core template is removed using chemical etching agents (e.g., hydrofluoric acid for silica or solvents for polystyrene), resulting in a hollow shell. These hollow core-shell nanostructures have immense utility in catalysis, drug delivery, energy storage, and photonic applications due to their low density, high surface area, and void space for hosting active agents. In another variant, AAO templates—characterized by their ordered porous channels—are used for synthesizing nanowire and nanotube arrays with core-shell geometries [22]. By infiltrating the pores with core material via electrodeposition or vapor-phase techniques, followed by coating with a shell material, researchers can fabricate vertically aligned, uniform nanostructures. After removing the AAO matrix, free-standing nanowire arrays with tailored core-shell features are obtained. These materials find significant application in biosensors, battery electrodes, field emitters, and photovoltaic devices, owing to their anisotropic properties and directional charge transport capabilities.

Soft templates, such as micelles, vesicles, and emulsions, are also employed in template-assisted synthesis. These templates, while more transient and dynamic, offer the advantage of self-assembly and ease of removal. In such systems, the amphiphilic molecules form structured domains in solution that act as scaffolds for shell material deposition. For instance, surfactant-based micelles can encapsulate hydrophobic cores while the shell is formed via polymerization or sol–gel processes around the micelle structure [23]. After solvent evaporation or thermal decomposition, the soft template is removed, yielding hollow or composite nanostructures.

A key advantage of template-assisted methods is the precision and uniformity in shell thickness, porosity, and surface functionality that can be achieved. Moreover, this technique allows the incorporation of multiple layers or gradient shells, enabling multifunctionality in a single nanostructure. However, challenges persist in terms of template removal, potential contamination, and scalability. The etching or calcination steps involved in removing templates must be carefully controlled to avoid damaging the delicate shell structure [24]. Additionally, the cost and environmental impact of template materials and solvents must be considered when transitioning from laboratory to industrial-scale production. Despite these limitations, template-assisted physical synthesis remains a cornerstone technique for engineering nanostructures with hierarchical and highly ordered features. Its ability to fine-tune geometry and composition at the nanoscale renders it especially attractive for high-performance applications in targeted drug delivery, advanced catalysis, optical metamaterials, and electrochemical systems.

3.2 Chemical Methods for Core-shell Synthesis

Chemical synthesis methods form a broad and versatile category for fabricating core-shell nanomaterials, offering molecular-level control over composition, structure, and surface functionality. Unlike physical methods that often require high-energy inputs or vacuum environments, chemical approaches typically operate under milder conditions, making them more suitable for incorporating sensitive or functional species into the core-shell structure. These methods rely on wet-chemical reactions, precipitation processes, or chemical vapor transformations to form shell layers around preexisting cores or during simultaneous core-shell formation. A significant advantage of chemical methods is their adaptability to various material systems, including metals, metal oxides, semiconductors, polymers, and hybrid organic-inorganic frameworks. Moreover, the scalability of many solution-based processes makes them attractive for large-scale manufacturing [25]. Chemical techniques are particularly advantageous when precise control over shell thickness, crystallinity, porosity, and composition is needed, as in the design of functional nanostructures for catalysis, drug delivery, sensing, or energy storage.

Another important feature of chemical synthesis is the possibility of tailoring interfacial interactions between the core and shell. Through judicious selection of precursors, surfactants, solvents, and reaction parameters, researchers can engineer strong or weak adhesion at the interface, modulate charge transfer properties, and manipulate surface energy. This tunability is essential for optimizing core-shell performance in targeted applications. Furthermore, many chemical methods enable the encapsulation of active or labile materials within protective or functional shells, extending their stability, biocompatibility, or reactivity. As a result, chemical synthesis routes have become integral to the fabrication of advanced nanomaterials with hierarchical architectures and multifunctional behavior. In the following subsections, various chemical methods such as sol–gel processing, hydrothermal and solvothermal techniques, coprecipitation, chemical vapor deposition (CVD), and colloidal synthesis will be discussed in detail, each offering unique advantages and considerations in the context of core-shell nanomaterial fabrication.

3.2.1 Sol–Gel Method and Controlled Precipitation

The sol–gel method is a widely adopted chemical approach for synthesizing core-shell nanomaterials, particularly when oxide-based shells are desired. It involves the hydrolysis and condensation of metal alkoxides or inorganic salts to form a colloidal suspension or "sol," which progressively evolves into a gel-like network. This transition from sol to gel provides a unique advantage in encapsulating various core materials, including nanoparticles, polymers, or biomolecules, within a uniform and highly adherent shell matrix. The mild reaction conditions, typically carried out at or near room temperature, make the sol–gel process compatible with a broad range of core materials, including temperature-sensitive or biologically active components. In a typical sol–gel-based core-shell synthesis shown in Figure 3.5, the preformed core material is dispersed in a sol containing the desired shell precursor. Controlled hydrolysis and condensation reactions then promote the nucleation and growth of a shell layer directly on the surface of the cores [26]. The resulting core-shell structures can subsequently be dried and heat-treated to enhance shell crystallinity or remove residual organics. The thickness and porosity of the shell can be finely tuned by varying precursor concentration, pH, reaction time, and the use of surfactants or stabilizers.

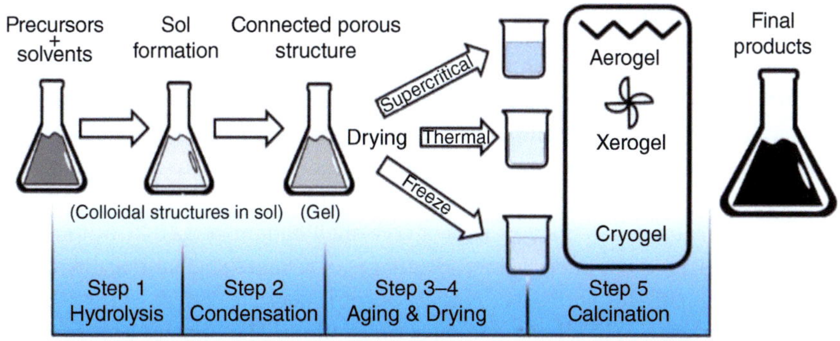

Figure 3.5 Illustration of the sol–gel method, a versatile chemical process used for synthesizing nanomaterials and thin films through the transition of a solution (sol) into a solid gel phase. [28]/MDPI/CC BY 4.0.

For example, silica shells can be grown on gold or magnetic nanoparticles using tetraethyl orthosilicate (TEOS) in an ethanol–water–ammonia mixture, a process known for producing highly uniform and monodisperse core-shell particles. Such silica-coated nanoparticles find extensive applications in bioimaging, diagnostics, and catalysis, owing to their chemical inertness, optical transparency, and ease of surface functionalization.

Controlled precipitation, on the other hand, offers an alternative route to chemical encapsulation where the shell material is deposited from a supersaturated solution onto the core particles. Unlike the sol–gel method that relies on hydrolysis-condensation, precipitation-based approaches involve the gradual addition of reagents to induce the controlled nucleation of an insoluble compound around the core surface. This technique is particularly effective in synthesizing core-shell nanoparticles composed of metals, metal oxides, sulfides, or carbonates. It is especially useful for constructing complex heterostructures, such as Fe_3O_4@ZnO or TiO_2@Ag, where the core and shell materials exhibit distinct functionalities. In controlled precipitation, achieving homogeneous shell coverage without agglomeration remains a critical challenge. To address this, stabilizing agents such as surfactants, polymers, or ligands are employed to control the growth rate and prevent uncontrolled nucleation in the bulk solution [27]. Process parameters such as temperature, stirring rate, pH, and ion concentration must be precisely managed to favor heterogeneous nucleation on the core surface over homogeneous nucleation in the solution. One prominent example is the synthesis of calcium carbonate-coated core particles for drug delivery applications, where the mild pH and biocompatibility of the shell are crucial for protecting sensitive therapeutic agents.

Overall, both sol–gel and controlled precipitation methods offer flexible, scalable, and cost-effective routes to tailor the composition, structure, and functionality of core-shell nanomaterials. Their compatibility with a wide range of cores and ability to create complex architectures make them indispensable tools in the chemical synthesis toolbox for nanotechnology applications.

3.2.2 Hydrothermal and Solvothermal Approaches

Hydrothermal and solvothermal methods are versatile, solution-based synthesis techniques that operate under high pressure and elevated temperature in sealed vessels, typically autoclaves. These approaches enable the controlled crystallization of inorganic materials from aqueous (hydrothermal) or non-aqueous (solvothermal) solutions, often resulting in highly crystalline and morphologically well-defined nanostructures. When applied to the synthesis of core-shell nanomaterials, hydrothermal and solvothermal methods offer distinct advantages in terms of shell uniformity, crystallinity, and compatibility with a wide variety of core materials [29]. In hydrothermal synthesis, water acts as the solvent and reaction medium. The elevated temperature and pressure enhance the solubility of precursors and promote slow, controlled nucleation and growth of the shell layer onto the core surface. This facilitates the formation of well-adhered, dense shells with minimal defects. For example, magnetic nanoparticles such as Fe_3O_4 can be encapsulated with crystalline TiO_2 or ZnO shells via hydrothermal treatment, yielding hybrid nanostructures with combined magnetic and photocatalytic properties. These core-shell structures are particularly valued in applications like pollutant degradation, drug delivery, and magnetic separation.

Solvothermal synthesis follows similar principles but employs organic solvents such as ethanol, ethylene glycol, or oleylamine, allowing for higher reaction temperatures and expanded chemical compatibility. This approach is ideal for forming shells composed of materials that are unstable or insoluble in water, such as metal-organic frameworks (MOFs), chalcogenides, or certain metal oxides. For instance, solvothermal methods can be used to synthesize carbon-coated metal nanoparticles or encapsulate luminescent cores within organometallic or polymeric shells, enhancing their chemical stability and functionality for use in sensors, photonics, and biomedical applications [30]. The key to successful core-shell synthesis using these methods lies in the careful tuning of reaction parameters, including temperature, precursor concentration, solvent type, reaction duration, and pH. The use of surfactants, ligands, or capping agents can further aid in directing shell growth exclusively on the core surface and preventing the formation of free shell particles. Additionally, one-pot synthesis strategies, where both the core and shell components are formed sequentially or simultaneously in a single reaction, are increasingly explored to simplify processing and improve scalability.

Hydrothermal and solvothermal methods are particularly attractive for producing anisotropic or hierarchical core-shell nanostructures, such as rods, plates, or flower-like morphologies, which can exhibit enhanced optical, catalytic, or electronic properties. Their ability to yield high-quality, crystalline coatings under relatively benign conditions (compared to solid-state or high-temperature gas-phase methods) makes them an essential part of the synthetic toolkit for core-shell nanomaterials.

3.2.3 Coprecipitation and Layer-by-layer Assembly

Coprecipitation and layer-by-layer (LbL) assembly are two widely employed chemical techniques for the synthesis of core-shell nanomaterials. Both methods offer low-cost, scalable, and adaptable routes to engineer complex nanoscale architectures with tunable composition and functionality. While coprecipitation is particularly effective for simultaneously forming cores and shells in a single reaction medium, the LbL approach allows the stepwise construction of multilayer shells with exceptional precision in thickness and structure. Coprecipitation involves the simultaneous precipitation of multiple ionic species from a homogeneous solution, often triggered by changes in pH, temperature, or the addition of a precipitating agent such as ammonia or sodium hydroxide [31]. This method is particularly suited for synthesizing oxide-based core-shell nanostructures, especially magnetic or dielectric systems like $Fe_3O_4@SiO_2$, $CoFe_2O_4@MnFe_2O_4$, or $TiO_2@CeO_2$. In typical protocols, the core nanoparticles are either formed in situ or added to the reaction mixture before shell precursor ions are introduced. Upon precipitation, the shell material deposits directly onto the surface of the core particles due to favorable surface interactions and nucleation energetics.

The success of coprecipitation depends strongly on factors such as ionic concentration, mixing speed, reaction temperature, and pH stability. Surfactants or capping agents are often used to control particle agglomeration and direct the nucleation to the core surface, ensuring uniform shell growth. One of the major advantages of coprecipitation is its simplicity and suitability for large-scale synthesis. However, challenges remain in achieving precise control over shell thickness and avoiding the formation of undesired homogeneous nucleates, especially in multi-component systems.

LbL assembly, on the other hand, is a sequential deposition technique that relies on the alternate adsorption of oppositely charged species, typically polyelectrolytes, nanoparticles, or other functional molecules onto the surface of a core material [32]. This electrostatic interaction-driven process can be used to deposit multiple shell layers with nanometer-level precision. The method is particularly powerful when precise control over shell architecture, porosity, or functionalization is required, as in the case of drug delivery systems, sensors, and stimuli-responsive nanomaterials. The LbL technique begins with the surface functionalization of the core nanoparticle to create a charge or affinity for the first layer. Subsequent layers are deposited by alternately immersing or mixing the particles with positively and negatively charged precursors, followed by rinsing steps to remove loosely bound species. This process can be repeated for multiple cycles to build shells of desired thickness and functionality. A diverse range of materials, including polyanions, polycations, enzymes, DNA, carbon nanotubes, and metal oxides can be incorporated into LbL shells, enabling the fabrication of multifunctional and responsive nanostructures. LbL assembly is inherently modular and allows the incorporation of responsive or functional moieties at each layer, which can be triggered by external stimuli such as pH, temperature, or light. Despite its precision and versatility, the technique can be labor-intensive and time-consuming, especially for multilayer structures [33]. Advances in automation and microfluidic-assisted LbL processing are currently being explored to enhance throughput and reproducibility.

Together, coprecipitation and LbL assembly represent complementary strategies for fabricating core-shell nanomaterials. While coprecipitation is well-suited for rapid, large-scale production of relatively simple oxide or hydroxide shells, LbL assembly excels in fabricating multilayered, functional, and hierarchically structured shells with high customization potential.

3.2.4 Chemical Vapor Deposition

Chemical vapor deposition (CVD) is a widely utilized gas-phase technique for the controlled synthesis of thin films and coatings, and it has been extensively adapted for the fabrication of core-shell nanostructures. In CVD, volatile precursors are introduced into a reaction chamber, where they undergo thermal decomposition or chemical reactions on a heated substrate or core material, leading to the deposition of a solid shell layer. This technique offers excellent control over the thickness, composition, and crystallinity of the shell, making it particularly advantageous for high-performance applications in electronics, energy storage, catalysis, and biomedical devices. For core-shell synthesis demonstrated in Figure 3.6, CVD enables the uniform and conformal coating of core nanoparticles, nanowires, or nanotubes with materials such as carbon, silicon, silicon dioxide, boron nitride, or various metal oxides [34]. A classic example is the fabrication of carbon-coated $LiFePO_4$ particles for lithium-ion battery cathodes. In this case, a hydrocarbon gas like acetylene or methane is decomposed at elevated temperatures in the presence of $LiFePO_4$ particles, forming a uniform carbon shell that enhances electrical conductivity and protects the active material from degradation. Similarly, SiO_2 and TiO_2 shells can be deposited onto metallic or magnetic cores using suitable organometallic precursors such as TEOS or titanium isopropoxide (TTIP), respectively.

The major advantage of CVD lies in its ability to produce highly pure, dense, and adherent coatings with excellent control over growth parameters such as temperature, gas flow rate, and precursor concentration. These parameters can be optimized to tune the morphology, thickness, and microstructure of the shell, including the formation of amorphous, polycrystalline, or single-crystalline coatings. Moreover,

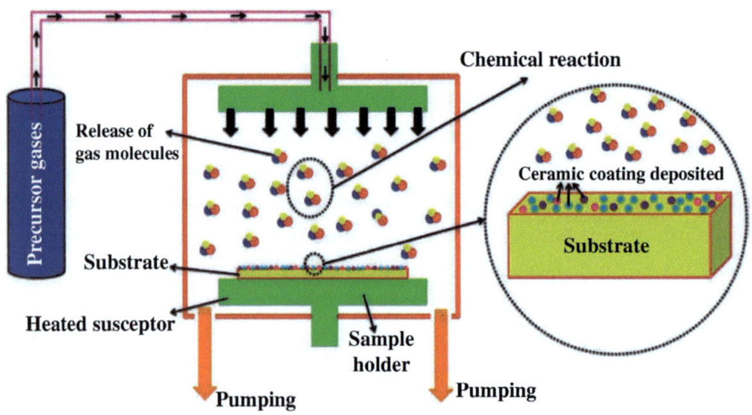

Figure 3.6 CVD process illustrating the controlled chemical reactions of gaseous precursors on a heated substrate, resulting in the formation of a high-quality thin film or coating. [35]/MDPI/CC BY 4.0.

advanced variants such as Plasma-Enhanced CVD (PECVD), Atomic Layer CVD (ALCVD), and Metal-Organic CVD (MOCVD) allow for deposition at lower temperatures and offer even greater control over nanoscale architectures.

ALD, a subclass of CVD, deserves special mention for its LbL growth mechanism based on self-limiting surface reactions. ALD is particularly useful for depositing ultra-thin and conformal shells on high-aspect-ratio nanostructures or porous materials [36]. This has led to its widespread adoption in fields requiring atomic-scale precision, such as microelectronics, catalysis, and protective coatings. Despite its many advantages, the limitations of CVD include high equipment costs, the need for elevated temperatures (which may not be compatible with thermally sensitive cores), and the potential for precursor toxicity or hazardous byproducts. Additionally, maintaining uniform coating in large-scale production or on irregularly shaped core particles remains a technical challenge. Nonetheless, ongoing advances in reactor design, precursor development, and hybrid synthesis methods are continuously expanding the capabilities of CVD in core-shell nanomaterials research. CVD offers a powerful, versatile, and highly tunable platform for fabricating core-shell nanostructures with superior uniformity, purity, and structural control. Its ability to deposit a broad spectrum of materials makes it indispensable in cutting-edge applications spanning semiconductors, batteries, sensors, and biomedical systems.

3.2.5 Colloidal Synthesis and Wet-chemical Techniques

Colloidal synthesis and wet-chemical approaches are among the most versatile and widely used methods for fabricating core-shell nanomaterials. These bottom-up techniques operate in liquid media under controlled conditions, enabling the production of a broad range of materials with tunable size, morphology, and composition. Their adaptability, cost-effectiveness, and compatibility with large-scale production make them especially attractive for both academic research and industrial applications. In colloidal synthesis, precursor materials are dissolved or dispersed in a suitable solvent, where nucleation and growth of nanoparticles occur under the influence of parameters such as temperature, pH, reaction time, and surfactant concentration [37]. For core-shell structures, the synthesis typically involves two sequential stages: the formation of the core particles followed by the deposition of the shell material. The shell is often grown through a controlled chemical reaction that ensures heterogeneous nucleation on the existing core surface rather than the formation of new particles. This can be achieved by carefully tuning the reactivity of precursors, applying selective surface functionalization, or introducing capping agents that promote shell growth. An exemplary case is the synthesis of gold-core/silver-shell (Au@Ag) nanoparticles, which are commonly fabricated by first reducing gold salts to form stable Au nanoparticles, followed by the slow reduction of silver ions in the presence of a mild reducing agent. This process ensures that silver is preferentially deposited onto the gold surface, forming a coherent shell. Similar strategies are employed in the fabrication of other metallic, semiconductor, or dielectric core-shell combinations, such as SiO_2@Au, Fe_3O_4@SiO_2, and CdSe@ZnS, each designed to exploit unique optical, magnetic, or electronic properties. The success of colloidal synthesis hinges on several critical factors. The choice of solvent affects solubility and reaction kinetics, while surfactants or ligands help stabilize the growing nanoparticles and control their surface chemistry. pH and ionic strength of the solution can also influence nucleation behavior and interfacial stability. Furthermore, temperature plays a crucial role in determining particle size and crystallinity; higher temperatures generally promote faster reactions and more crystalline products.

Wet-chemical techniques also include precipitation reactions, sol–gel transformations, microemulsion methods, and redox-assisted deposition. These methods allow for precise control over shell thickness, composition, and functionalization, often enabling the integration of dopants or bioactive molecules. For instance, silica shells can be grown on magnetic nanoparticles via a modified Stöber process, enhancing their dispersibility and biocompatibility for drug delivery or magnetic resonance imaging (MRI) contrast applications [38]. Similarly, polymer shells can be formed around inorganic cores to impart stimuli-responsiveness or improve processability in composite formulations. Despite their advantages, colloidal and wet-chemical methods can suffer from challenges such as batch-to-batch variability, difficulty in achieving perfect monodispersity, and the need for careful purification to remove unreacted precursors or surfactants. Additionally, surface defects and incomplete shell coverage may arise if reaction conditions are not meticulously optimized.

Nevertheless, the inherent flexibility of these approaches, combined with the possibility of post-synthetic modifications and surface functionalization, makes them indispensable in the development of next-generation core-shell nanomaterials. Whether for catalysis, imaging, sensing, or environmental remediation, colloidal synthesis continues to provide a robust and scalable route for tailoring core-shell architectures with high precision.

3.3 Green and Sustainable Synthesis Approaches

The advancement of core-shell nanomaterials has historically been driven by performance optimization, often at the cost of environmental sustainability. Figure 3.7 shows the advantages of using green synthesis as, with the growing emphasis on green chemistry and sustainable engineering, there is a concerted effort to develop synthesis methodologies that are not only efficient but also environmentally benign. Green synthesis approaches seek to minimize the ecological footprint of nanomaterials by reducing energy consumption, eliminating toxic reagents, employing renewable resources, and minimizing waste generation. This shift is critical in aligning materials science with global sustainability goals and regulatory frameworks. A green synthesis approach is generally characterized by the use of non-toxic, biodegradable, and renewable inputs, including plant extracts, biopolymers, or naturally occurring reducing agents, as well as water or other benign solvents as the reaction medium [39]. Additionally, sustainable methods aim to operate under mild conditions such as ambient temperature and pressure, thereby significantly reducing energy demands and associated carbon

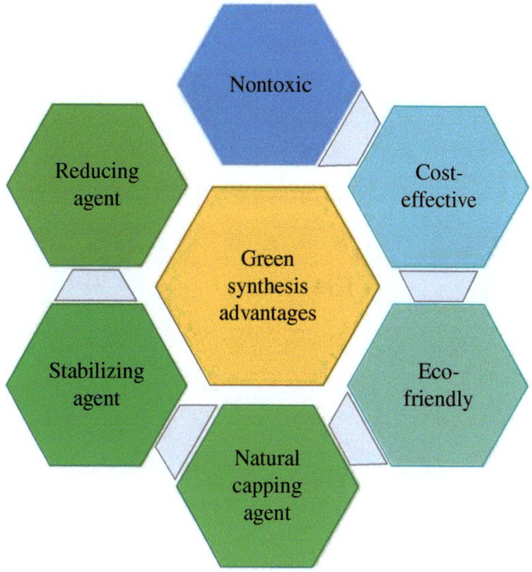

Figure 3.7 Advantages of green synthesis. Reproduced from [40]/ John Wiley & Sons/CC BY 4.0.

emissions. The principles of atom economy, resource efficiency, and life-cycle thinking are central to these methodologies, promoting the responsible design of both the process and the product.

Sustainability in core-shell synthesis also extends to the post-synthesis phase, where efficient purification, solvent recovery, and waste management practices are integral to the overall environmental performance of the process. The concept of circularity, including the reuse of reagents and recycling of byproducts or spent materials, is increasingly being explored to ensure a closed-loop approach to nanomaterials fabrication. Moreover, green synthesis routes often offer additional benefits such as improved biocompatibility, safer handling, and lower processing costs, making them attractive for applications in biomedicine, agriculture, and environmental remediation. As the field matures, the integration of green metrics—such as the Environmental Factor (E-factor), process mass intensity (PMI), and energy intensity—into the design and assessment of synthetic routes will be essential [41]. These metrics can provide quantitative insights into the sustainability of a given method, helping researchers make informed decisions about trade-offs between performance, cost, and environmental impact.

3.3.1 Biogenic and Plant-based Synthesis of Core-shell Nanomaterials

Biogenic and plant-based synthesis approaches represent an innovative and eco-conscious route for fabricating core-shell nanomaterials, leveraging the inherent chemical diversity and reducing potential of biological systems. This method employs natural resources such as plant extracts, microorganisms (bacteria, fungi, algae), and biomolecules (proteins, polysaccharides, enzymes) as reducing, stabilizing, and capping agents, as shown in Figure 3.8. These biological entities often contain a rich array of functional groups such as hydroxyl, carboxyl, amine, and phenolic moieties, which

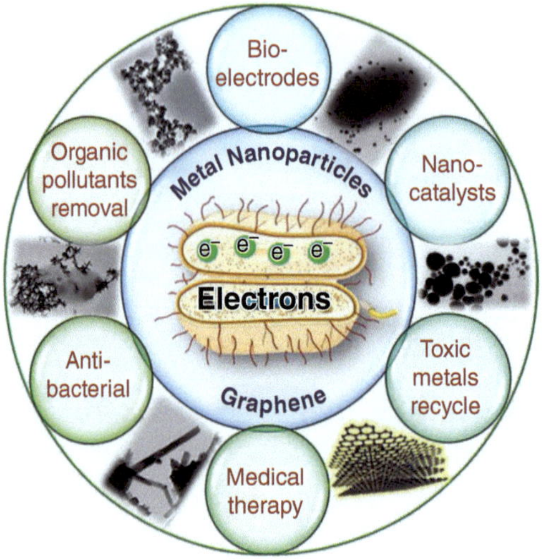

Figure 3.8 Green synthesis of core-shell nanomaterials using biogenic and plant-based sources: eco-friendly routes for core-shell synthesis. [43]/Springer Nature/ CC BY 4.0.

can mediate the nucleation and growth of nanoparticles and facilitate the formation of well-defined core-shell structures under mild conditions.

Plant-mediated synthesis is particularly attractive due to its simplicity, scalability, and minimal requirement for specialized equipment. Aqueous extracts from various parts of plants—leaves, roots, bark, seeds, or fruits—have been successfully used to synthesize metallic, oxide, and hybrid core-shell nanostructures [42]. For example, gold or silver cores encapsulated in silica, chitosan, or other biopolymer shells have been fabricated using green tea, neem, and aloe vera extracts. These phytochemicals act as both reducing and capping agents, enabling the formation of uniform and stable nanoparticles without the need for toxic chemicals or high-temperature processing.

Microbial synthesis offers another promising route, wherein live cells or cell-free extracts of bacteria, fungi, or algae are employed to mediate nanoparticle formation. The biological metabolism of these microorganisms can drive redox reactions that precipitate metal cores or shells, while their secreted proteins and metabolites can aid in templating and stabilization. For instance, fungi such as *Fusarium oxysporum* have been reported to synthesize core-shell structures like Fe_3O_4@Ag or Au@ZnO with high monodispersity and excellent biocompatibility. This technique is advantageous for biomedical applications, as the resulting nanomaterials often exhibit reduced toxicity and enhanced biological functionality.

The major advantages of biogenic and plant-based approaches lie in their sustainability, non-toxicity, and ability to produce biofunctionalized surfaces inherently suitable for medical and environmental applications. Moreover, these methods often result in nanomaterials with unique morphologies and surface chemistries that are difficult to achieve using conventional chemical techniques. However, one of the key challenges is the reproducibility and consistency of these biological systems, as variations in plant species, age, extraction method, and growth conditions can lead to variability in nanoparticle size, shape, and shell uniformity [44]. Continued research in this domain is focusing on optimizing process parameters, understanding reaction mechanisms, and integrating biogenic routes with other green technologies to enhance control and scalability. By harnessing the chemistry of nature, plant-based and microbial synthesis strategies offer a viable pathway toward the greener fabrication of core-shell nanomaterials with significant potential for sustainable innovation across diverse fields.

3.3.2 Use of Non-toxic and Eco-friendly Precursors

The use of non-toxic and eco-friendly precursors in the synthesis of core-shell nanomaterials is a fundamental principle of green chemistry that aims to minimize the environmental and health risks associated with nanomaterial production. Traditional synthesis methods often involve hazardous reagents such as heavy metal salts, volatile organic solvents, or strong reducing agents like hydrazine, sodium borohydride, or ammonia. These substances can pose significant risks in terms of toxicity, flammability, and environmental persistence. In contrast, green synthesis strategies seek to replace such harmful chemicals with safer, biodegradable, and renewable alternatives without compromising the structural integrity or functional performance of the resulting nanomaterials. One approach involves the use of benign metal precursors, such as metal acetates, citrates, and nitrates, which exhibit lower toxicity compared to their chloride or perchlorate counterparts [45]. These compounds are not only safer to handle but also offer better solubility in water, facilitating aqueous-phase synthesis routes that avoid the use of harmful organic solvents. Additionally, the use of plant-derived or biomolecule-based stabilizers and surfactants—such as starch, gelatin, cellulose, chitosan, and alginate—can provide effective control over nanoparticle morphology while being non-toxic and biodegradable.

Reducing agents are another critical component of nanomaterial synthesis. Conventional reductants can be replaced with naturally occurring compounds like ascorbic acid (vitamin C), glucose, tannic acid, and polyphenols, which are not only eco-friendly but can also contribute additional functionality to the synthesized nanostructures. For example, ascorbic acid has been widely used in the reduction of gold and silver salts to produce metallic cores that are subsequently coated with biocompatible shells, forming stable and non-toxic core-shell assemblies suitable for medical and sensing applications [46]. Shell materials can also be synthesized using green precursors. Silica shells, for instance, can be formed using non-toxic alkoxysilanes like TEOS in ethanol–water mixtures under mild pH conditions. Similarly, metal oxide shells such as ZnO and TiO_2 can be generated using metal alkoxides or nitrate precursors in combination with plant extracts or low-temperature hydrolysis methods. In cases where polymeric shells are desired, biopolymers such as polylactic acid (PLA), polyethylene glycol (PEG), and polydopamine (PDA) provide a sustainable alternative to petroleum-based polymers.

The shift toward non-toxic precursors also aligns with regulatory and safety standards for nanomaterials, particularly in biomedical, food, and environmental applications. Nanomaterials synthesized using eco-friendly routes exhibit enhanced biocompatibility and reduced risk of bioaccumulation or ecotoxicity. Moreover, these materials are increasingly favored in public and industrial perception, as they conform to the growing demand for green and sustainable technologies. Despite their numerous benefits, the transition to non-toxic and eco-friendly precursors is often challenged by issues of reactivity, availability, and cost. Additionally, maintaining precise control over core-shell structure, crystallinity, and functional performance remains an active area of research. Nevertheless, continuous advancements in green chemistry and material science are making it increasingly feasible to design and produce high-quality core-shell nanomaterials using safe and sustainable inputs, paving the way for environmentally responsible nanotechnology.

3.3.3 Energy-efficient and Low-temperature Processing

Energy consumption is a critical factor in the sustainable synthesis of core-shell nanomaterials, as traditional high-temperature and energy-intensive processes often have significant environmental footprints and operational costs. To address these concerns, energy-efficient and low-temperature processing techniques have gained substantial attention in recent years. These approaches aim to reduce the overall energy input during synthesis while maintaining control over nanoparticle size, shell thickness, crystallinity, and functional properties. Low-temperature synthesis strategies typically operate at or near ambient conditions, avoiding the need for prolonged heating, high pressures, or vacuum environments that are typical of conventional thermal or CVD processes. By doing so, these methods not only reduce energy consumption but also limit unwanted side reactions, phase transformations, and thermal degradation of sensitive core or shell materials.

One common energy-efficient approach is the use of microwave-assisted synthesis, where microwave radiation rapidly heats the reaction medium volumetrically and uniformly. This results in shorter reaction times and lower temperatures compared to conventional heating, enabling the formation of core-shell nanomaterials with well-defined morphologies and crystalline phases. Microwave methods also facilitate solvent-free or aqueous-based synthesis, further contributing to sustainability [47]. Another promising technique is sonochemical synthesis, which utilizes ultrasound waves to induce cavitation—rapid formation and collapse of microbubbles in liquids. This creates localized hot spots with extremely high temperatures and pressures, promoting rapid nucleation and growth of core-shell structures at low bulk temperatures. Sonochemistry has been applied to prepare metal–metal oxide core-shell particles, polymer-coated quantum dots, and catalyst-supported nanostructures with fine control over shell uniformity.

Solvothermal and hydrothermal methods conducted at moderate temperatures (typically below 200 °C) provide another energy-efficient route to high-quality core-shell materials. These methods leverage the unique solvent properties at elevated pressures to promote controlled nucleation and growth while preserving delicate core structures. Lower temperatures reduce energy input and minimize thermal stress, making these processes suitable for temperature-sensitive materials such as organic cores or biomolecules [48]. Additionally, room-temperature wet-chemical routes utilizing mild reducing agents and stabilizers have been developed to form core-shell nanomaterials without the need for external heating. Such ambient conditions not only save energy but also allow integration of biologically active or polymeric shells that would degrade at elevated temperatures. Despite the benefits, low-temperature and energy-efficient methods sometimes face challenges related to slower reaction kinetics, incomplete shell coverage, or lower crystallinity compared to high-temperature routes. These limitations can often be overcome through process optimization, such as adjusting precursor concentrations, pH, reaction time, or using hybrid methods that combine mild heating with microwave or sonochemical assistance.

3.3.4 Waste Utilization and Recycling in Core-shell Synthesis

The increasing emphasis on sustainable manufacturing has brought waste utilization and recycling to the forefront of nanomaterial synthesis, including the fabrication of core-shell nanostructures. Traditional synthesis methods often generate chemical waste, leftover reagents, and byproducts that pose environmental and economic challenges. Addressing these concerns, waste valorization and recycling strategies are being integrated into the core-shell synthesis processes to minimize resource consumption and environmental pollution.

Waste utilization in core-shell synthesis involves repurposing industrial, agricultural, or biological waste streams as raw materials or precursor sources. For example, metallic ions extracted from electronic waste (e-waste) or mining residues can serve as economical and eco-friendly core or shell precursors. Plant biomass ash, food-processing byproducts, and wastewater sludge have also been explored as sources of silica-, alumina-, or carbon-based materials that form functional shell layers. This approach not only reduces the reliance on virgin raw materials but also mitigates the disposal problems associated with waste accumulation. Recycling strategies focus on recovering and reusing valuable components from spent or end-of-life core-shell nanomaterials. Magnetic core-shell particles, for instance, can be magnetically separated and regenerated for multiple catalytic or environmental remediation cycles, significantly lowering production costs and material consumption [49]. Additionally, chemical methods have been developed to reclaim precious metals like gold and silver from core-shell composites, allowing reintegration into new synthesis batches. Closed-loop synthesis processes are gaining traction, where solvent recycling, recovery of unreacted precursors, and reuse of reaction byproducts are systematically implemented. Techniques such as membrane filtration, centrifugation, and solvent extraction enable purification and recycling of reaction media, reducing water and chemical use. This reduces the overall environmental footprint of the synthesis and improves the economic viability of large-scale production.

Moreover, innovative green solvents like supercritical fluids and ionic liquids facilitate waste minimization by enhancing precursor solubility and reaction efficiency, often enabling milder conditions and reducing side product formation. These solvents can often be recycled and reused multiple times without significant loss of activity. Despite these advances, challenges remain in fully integrating waste utilization and recycling into core-shell synthesis workflows. Issues such as variability in waste composition, contamination, and process scalability need further investigation. Nonetheless, the continued development of effective waste valorization and recycling approaches is vital for sustainable nanomaterial manufacturing, fostering circular economy principles and reducing environmental burdens. The strategic incorporation of waste utilization and recycling techniques within core-shell nanomaterial synthesis represents a promising direction toward greener, more sustainable nanotechnology. It bridges material innovation with environmental stewardship, enabling resource-efficient production aligned with global sustainability goals.

3.3.5 Water-based and Solvent-free Synthesis Strategies

Water-based and solvent-free synthesis approaches have emerged as pivotal strategies in the pursuit of greener and more sustainable core-shell nanomaterial fabrication. Traditional chemical syntheses often rely on organic solvents that can be toxic, flammable, and environmentally harmful, creating challenges for waste disposal and process safety. In contrast, water as a solvent offers a non-toxic, abundant, and eco-friendly medium, while solvent-free processes aim to eliminate solvents altogether, minimizing waste and environmental impact. Water-based synthesis techniques exploit the unique properties of water, such as high dielectric constant, excellent heat capacity, and hydrogen bonding capability, which can facilitate controlled nucleation and growth of core-shell nanostructures [50]. Aqueous media enable environmentally benign reactions at relatively mild conditions, often improving the biocompatibility of the resulting materials—a crucial factor for biomedical and environmental applications. For instance, water-based sol–gel, hydrothermal, and coprecipitation methods have been successfully used to synthesize metal oxide and polymer-coated core-shell nanoparticles with high uniformity and stability [51]. Beyond environmental benefits, water-based synthesis can simplify post-synthesis processing, as water is easy to remove by evaporation or centrifugation without leaving hazardous residues. Additionally, surfactants and stabilizing agents compatible with water are often employed to control particle size and prevent agglomeration during shell formation, enabling precise tuning of core-shell architectures.

Solvent-free synthesis techniques, on the other hand, often involve mechanochemical or solid-state processes such as ball milling, thermal treatment, or vapor-phase deposition without the use of liquid solvents. These approaches minimize chemical waste generation, reduce processing steps, and enhance atom economy. For example, mechanochemical methods can directly induce solid-state reactions between core and shell precursors under ambient or mild conditions, leading to intimate interfacial bonding and unique structural properties [52]. In some solvent-free approaches, vapor-phase reactions—such as ALD or CVD—enable thin and conformal shell formation on cores without the need for liquid media. These methods also contribute to high purity and crystallinity of the final nanostructures. Despite their advantages, water-based and solvent-free methods face challenges, including controlling reaction kinetics, preventing particle agglomeration, and ensuring uniform shell coverage. The limited solubility of some precursors in water or the difficulty of scaling solid-state processes to industrial levels are ongoing areas of research. However, advances in surfactant chemistry, green additives, and process engineering are progressively overcoming these limitations. The water-based and solvent-free synthesis strategies represent promising directions to produce core-shell nanomaterials with minimal environmental impact. Their development aligns with global sustainability goals by reducing hazardous waste, improving safety, and enabling cost-effective, scalable production of advanced functional nanostructures.

3.4 Scalability and Cost-efficient Manufacturing

The transition from laboratory-scale synthesis of core-shell nanomaterials to commercial-scale manufacturing presents a complex set of challenges and opportunities. While many synthesis methods are well-established in controlled research environments, scaling these processes without compromising material quality, performance, and reproducibility requires careful optimization. Achieving cost efficiency is equally critical to enable widespread adoption across industries such as energy, electronics, catalysis, and biomedicine. Scalability involves not only increasing the production volume but also ensuring that the physicochemical properties of the core-shell nanomaterials remain consistent and within specifications. This entails precise control over parameters such as particle size distribution, shell thickness, composition, and morphology during scale-up. The choice of synthesis technique plays a significant role in this regard, as some methods naturally lend themselves to large-scale production due to their simplicity and robustness, while others require specialized equipment and conditions that are difficult to replicate on an industrial scale. Cost considerations extend beyond raw materials to encompass energy consumption, labor, equipment investment, waste management, and process throughput. The economic viability of core-shell nanomaterial production is tightly linked to minimizing expensive reagents, reducing synthesis times, and maximizing yield. Additionally, environmental regulations and sustainability mandates increasingly influence manufacturing approaches, favoring green and energy-efficient processes that reduce emissions and hazardous waste. Ensuring product quality and uniformity in mass production introduces further complexities. High-volume synthesis can lead to batch-to-batch variations, defects, or contamination, which can adversely affect material performance. Therefore, robust quality control protocols, in-line monitoring, and process automation are essential to maintain consistency and traceability throughout production cycles. The advent of automated manufacturing and artificial intelligence (AI)-driven process optimization offers promising avenues to overcome many scale-up challenges. These technologies can enable real-time control, predictive maintenance, and adaptive parameter tuning, thereby enhancing efficiency and reliability in core-shell nanomaterial fabrication. Together, these aspects define the roadmap for scalable, cost-effective, and sustainable manufacturing

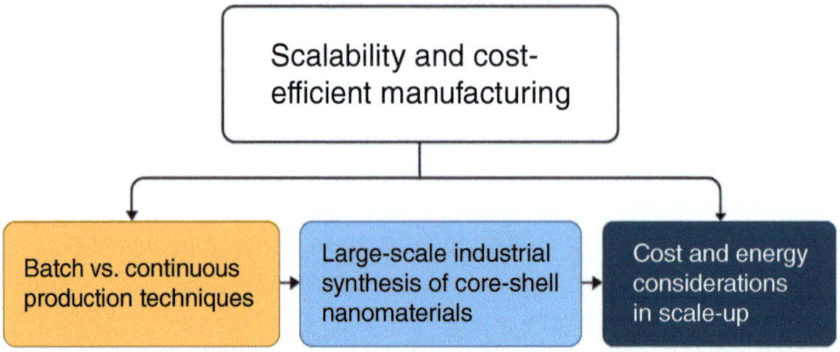

Figure 3.9 Flowchart illustrating key aspects of scalability and cost-efficient manufacturing of core-shell nanomaterials, covering batch vs. continuous production, large-scale industrial synthesis, cost and energy considerations during scale-up. Generated with AI using ChatGPT (version GPT-4).

of core-shell nanomaterials, which is crucial to meet the growing industrial demand and realize their full commercial potential. Figure 3.9 shows the flowchart illustrating the key aspects of scalability and cost-efficient manufacturing of core-shell nanomaterials.

3.4.1 Batch vs. Continuous Production Techniques

The choice between batch and continuous production techniques plays a pivotal role in the scalability and efficiency of core-shell nanomaterial synthesis. Each approach offers distinct advantages and challenges that impact product quality, throughput, process control, and economic feasibility. Batch production is traditionally the most common method used in laboratory and pilot-scale synthesis. In batch processes, reactants are combined in a fixed volume reactor, and the synthesis proceeds for a defined duration before the product is harvested. This approach provides high flexibility, allowing researchers to precisely control reaction parameters such as temperature, time, concentration, and mixing. Batch synthesis is well-suited for producing small quantities or multiple variants of core-shell nanomaterials with varying compositions or morphologies. It also facilitates easy modifications and rapid optimization of reaction conditions. However, batch methods often suffer from limitations when scaled up for industrial production [53]. Larger batch sizes can introduce challenges related to mixing efficiency, heat and mass transfer, and uniformity of shell coating, which may lead to inconsistencies in particle size distribution and shell thickness. Additionally, batch processes typically have lower throughput and involve downtime between cycles for cleaning, loading, and unloading, which can reduce overall productivity and increase labor costs.

Continuous production techniques, in contrast, are designed to operate non-stop, with reactants fed into the system and product continuously harvested. This approach is highly advantageous for large-scale manufacturing as it enables consistent process conditions, better heat and mass transfer, and improved reproducibility. Continuous flow reactors, microfluidic systems, and spray pyrolysis are examples of continuous synthesis methods that have been adapted for core-shell nanomaterial production. The steady-state operation in continuous synthesis allows for tighter control over particle nucleation and growth, resulting in more uniform core-shell structures. Moreover, continuous methods can significantly increase throughput while minimizing waste and energy consumption, enhancing overall cost efficiency [54]. The ability to integrate real-time monitoring and automation further improves process stability and quality assurance. Despite these benefits, continuous production requires significant upfront investment in equipment and process development. The design and optimization of continuous reactors can be complex, especially when dealing with multiphase reactions or materials sensitive to temperature and shear stress. Transitioning from batch to continuous synthesis also demands thorough understanding of reaction kinetics and scale-up dynamics to maintain product quality.

The selection between batch and continuous production hinges on factors such as production scale, material complexity, required flexibility, and cost targets. While batch methods remain indispensable for research and niche applications, continuous synthesis is increasingly preferred for industrial-scale manufacture of core-shell nanomaterials due to its scalability, efficiency, and consistency. Hybrid approaches that combine the strengths of both methods are also emerging as promising solutions to meet diverse production demands.

3.4.2 Large-scale Industrial Synthesis of Core-shell Nanomaterials

Core-shell nanomaterials are increasingly recognized for their potential in environmental applications, particularly in areas related to sustainability, pollution control, and resource management. Their unique properties enable them to address key environmental challenges, including water purification, air quality management, waste treatment, and energy efficiency. The multifunctional nature of core-shell structures allows for the development of materials that not only improve environmental outcomes but also enhance the overall sustainability of industrial processes and everyday technologies. One of the most significant applications of core-shell nanomaterials in environmental protection is in water treatment and purification. Water scarcity and contamination are critical global issues, and the demand for effective water purification technologies is growing [55]. Core-shell materials, especially those with photocatalytic or adsorptive properties, are highly effective in breaking down harmful contaminants such as heavy metals, organic pollutants, and microbial pathogens. For instance, core-shell nanoparticles made of metal oxide cores (e.g., titanium dioxide) with a shell of functionalized carbon or polymer coatings can serve as highly active photocatalysts under UV light. These nanomaterials can degrade organic pollutants and kill bacteria, providing a sustainable approach to water decontamination [56]. The core-shell structure allows for optimal exposure of the catalytic surface, enhancing the material's reactivity and stability and ensuring long-lasting performance in purification processes. In addition to water treatment, core-shell nanomaterials are gaining attention for their role in air pollution control. Nanocatalysts, composed of a reactive core material surrounded by a protective shell, are used in the catalytic conversion of harmful gases such as carbon monoxide (CO), nitrogen oxides (NOx), and volatile organic compounds (VOCs) [57]. The shell material, often a metal oxide or carbon, acts as a stabilizer, protecting the active core and facilitating the adsorption and transformation of pollutants into less harmful substances. This is particularly beneficial in automotive exhaust systems and industrial applications, where air quality regulations require efficient removal of toxic gases. The enhanced surface area and tunable chemical properties of core-shell nanomaterials make them highly effective in these applications, offering better catalytic efficiency and longer operational lifetimes compared to conventional catalysts.

Another promising application is in the field of environmental remediation and waste treatment. Core-shell nanomaterials can be engineered to capture and degrade hazardous substances, such as heavy metals and pesticides, from contaminated soils or industrial effluents. For example, core-shell nanoparticles with magnetic cores and functionalized shells can be used for the selective removal of toxic metals from wastewater [58]. The magnetic properties of the core allow for easy separation and recycling of the nanoparticles after the cleanup process, ensuring that the materials can be reused, which adds to the sustainability aspect of this technology. Additionally, core-shell materials can be employed to remove oils and hydrocarbons from water and soil, addressing oil spills and other environmental disasters with enhanced efficiency and lower environmental impact [59]. Core-shell nanomaterials are also being explored for energy-efficient environmental solutions, such as the development of energy-efficient coatings for buildings and other infrastructure. The use of core-shell composites in thermal insulating materials allows for better heat retention and temperature regulation, reducing the need for external energy inputs in heating and cooling systems [60]. These energy-saving applications contribute to the reduction of carbon footprints and energy consumption in both residential and industrial buildings, making them an important component of sustainable construction and urban planning.

In the field of renewable energy, core-shell nanomaterials contribute to environmental sustainability by improving the efficiency and performance of solar energy devices and other renewable technologies. For example, in solar cells, core-shell quantum dots and nanostructures can enhance light absorption and improve energy conversion efficiency, reducing the overall carbon footprint of energy production [61]. The integration of core-shell materials in solar technologies can also lower the cost of energy production, making renewable sources more competitive with traditional fossil fuels. Beyond direct applications in pollution control and resource management, core-shell nanomaterials contribute to the broader concept of sustainability through their potential in recycling and waste reduction. For instance, in e-waste recycling, core-shell nanoparticles can be used to selectively extract valuable metals such as gold, silver, and copper from discarded electronic devices [62]. The shell material can be designed to interact with specific metals, ensuring efficient extraction and minimizing the release of hazardous substances. This capability not only supports the recovery of valuable materials but also reduces the environmental burden associated with e-waste disposal. The ability to engineer core-shell nanomaterials with specific functional properties makes them highly suitable for a wide range of environmental applications [63]. By addressing key challenges such as water and air pollution, waste treatment, energy efficiency, and renewable energy development, these materials are paving the way for a more sustainable and environmentally friendly future. As research and development continue, core-shell nanomaterials are expected to play an even more significant role in advancing global sustainability efforts, providing solutions to some of the most pressing environmental issues of our time.

3.4.3 Cost and Energy Considerations in Scale-up

Core-shell nanomaterials are rapidly advancing biomedical applications, particularly in drug delivery, imaging, and therapeutic treatments. The unique structural and chemical properties of core-shell nanoparticles make them ideal candidates for targeted drug delivery, enhanced bioimaging, and the development of new therapeutic modalities [64–66]. These materials can be engineered to offer highly controllable release profiles, biocompatibility, and multifunctionality, positioning them at the forefront of modern medicine. The ability to design core-shell structures with tailored properties enables the development of nanomaterials that can address the complex demands of biomedical applications, improving treatment efficacy and reducing side effects. One of the most significant applications of core-shell nanomaterials in biomedicine is in targeted drug delivery [67]. The core-shell architecture allows for the encapsulation of therapeutic agents within the core, while the shell can be functionalized to provide specific targeting capabilities. For instance, the shell can be decorated with ligands, antibodies, or peptides that bind selectively to receptors on the surface of target cells or tissues, such as cancer cells or inflamed tissue [68]. This selective targeting reduces the systemic distribution of the drug, thereby minimizing off-target effects and enhancing the therapeutic efficacy of the treatment. The core material can be designed to encapsulate a wide range of therapeutic agents, including small molecules, proteins, RNA, and even gene-editing tools such as Clustered Regularly Interspaced Short Palindromic Repeats (CRISPR) [69]. The shell can provide stability and protection to these fragile agents, preventing their premature degradation and ensuring their efficient delivery to the target site.

The controlled release of drugs is another major advantage of core-shell nanomaterials in drug delivery. By altering the composition of the core and shell, the release rate of the encapsulated drug can be precisely regulated, enabling sustained or triggered release depending on the needs of the treatment. For example, the core can be made from a biodegradable polymer that breaks down over time, allowing for the gradual release of the drug. Alternatively, the shell can be responsive to external stimuli such as pH, temperature, or light, enabling the release of the drug in response to the specific conditions at the disease site [70]. This "smart" drug delivery approach allows for more efficient treatments by ensuring that drugs are delivered only when and where they are needed, improving patient outcomes and reducing the frequency of dosing. Core-shell nanomaterials are also playing a pivotal role in the development of nanocarriers for gene therapy [71]. The core-shell structure can encapsulate genetic material, such as DNA or RNA, and protect it from degradation in the bloodstream, while the shell can be modified to enhance cellular uptake and facilitate endosomal escape. This capability is especially important for the development of non-viral gene delivery systems, which offer a safer alternative to viral vectors that can carry the risk of immune reactions and insertional mutagenesis.

Core-shell nanoparticles can also improve the stability of genetic material, providing an effective vehicle for gene editing tools and therapies [72]. As gene therapies continue to advance, core-shell nanomaterials will play an increasingly vital role in their translation from research to clinical application.

Beyond drug delivery, core-shell nanomaterials are making significant contributions to biomedical imaging and diagnostic applications. The core-shell structure allows for the incorporation of various imaging agents, such as fluorescent dyes, magnetic nanoparticles, or radioisotopes, within the core, while the shell can be designed to enhance the stability and biocompatibility of the particles. These nanoparticles can be used in a variety of imaging modalities, including MRI, computed tomography (CT), positron emission tomography (PET), and fluorescence imaging. By integrating both therapeutic and diagnostic functions into a single nanoparticle (a strategy known as "theranostics"), core-shell nanomaterials enable more precise diagnosis, real-time monitoring of treatment efficacy, and the delivery of targeted therapies. Core-shell nanomaterials are also being explored for their potential in regenerative medicine [73]. The core can be made from biocompatible materials that promote tissue regeneration, while the shell can be functionalized with growth factors, peptides, or biomolecules that stimulate cell proliferation and tissue repair. For instance, core-shell nanoparticles have been used in bone tissue engineering to deliver osteogenic growth factors to sites of bone injury or degeneration. By combining the regenerative properties of the core with the bioactive molecules on the shell, core-shell nanomaterials can accelerate the healing process and improve the outcomes of tissue repair and regeneration [74]. The development of core-shell nanomaterials for biomedical applications requires careful consideration of factors such as biocompatibility, biodegradability, and toxicity. However, advances in material design and surface functionalization have led to the creation of nanoparticles that are not only effective but also safe for use in medical applications. The versatility and multifunctionality of core-shell nanomaterials provide significant opportunities for improving healthcare outcomes, from more effective drug delivery and gene therapy to enhanced diagnostic imaging and regenerative medicine.

3.4.4 Challenges in Mass Production and Quality Control

Core-shell nanomaterials are making significant strides in the fields of electronics, photonics, and wearable technologies due to their ability to provide enhanced properties such as conductivity, optical response, and mechanical flexibility. Their unique architecture, with a functional core and a shell that can be tailored for specific functions, enables the design of materials with a combination of characteristics that are difficult to achieve with conventional single-component nanomaterials. As a result, core-shell structures are increasingly being explored for applications in flexible electronics, optoelectronics, sensors, and next-generation wearable devices [75]. In electronics, core-shell nanomaterials offer exceptional advantages in the development of high-performance conductors and semiconductors. For example, the core can consist of a material with high electrical conductivity, such as gold, copper, or carbon nanotubes, while the shell can be made from a semiconducting or insulating material [76]. This combination allows for the design of nanoparticles that exhibit both high conductivity and tailored electronic properties, making them ideal candidates for use in transistors, integrated circuits, and other electronic components. Additionally, the core-shell structure can be used to enhance the stability of nanomaterials in electronic devices, protecting the conductive core from oxidation or degradation while maintaining efficient charge transport properties.

Core-shell nanomaterials also hold great potential in photonics, particularly in applications such as light-emitting diodes (LEDs), solar cells, and photodetectors [77]. The optical properties of core-shell nanoparticles can be finely tuned by adjusting the composition of the core and shell, allowing for the design of materials with specific absorption and emission characteristics. For example, core-shell quantum dots are widely used in photonics due to their ability to emit light at precisely controlled wavelengths, which is crucial for applications in displays, lighting, and solar energy conversion [78]. By altering the shell material, the surface states of the core can be controlled, leading to improvements in photoluminescence efficiency, stability, and optical properties. In the field of solar energy, core-shell nanomaterials are being explored for use in next-generation solar cells. The core can be made from a highly efficient light-absorbing material, such as cadmium telluride or silicon, while the shell can be engineered to improve charge separation, enhance stability, and reduce recombination losses [61]. The shell can also protect the core material from environmental degradation, such as moisture and oxidation, which is a significant challenge in the long-term performance of solar devices. Core-shell structures have the potential to increase the efficiency and durability of photovoltaic devices, making them a promising solution for renewable energy applications. Wearable technologies represent one of the most exciting areas where core-shell nanomaterials are making an impact. As demand grows for lightweight, flexible, and durable devices that can be worn on the body, core-shell nanomaterials offer a unique solution due to their ability to combine strength, flexibility, and conductivity. For example, core-shell nanoparticles can be incorporated into flexible substrates to create stretchable electronics, such as sensors, displays, and communication devices [79]. The shell material can be designed to provide mechanical flexibility, while the core material can impart conductivity and electronic functionality, allowing for the creation of devices that can withstand deformation without compromising performance. This combination of properties is critical for applications in health monitoring, smart clothing, and other wearable technologies.

Core-shell nanomaterials are also being used in the development of sensors for wearable applications. The core can be engineered to have specific interactions with target molecules or environmental conditions, such as temperature, humidity, or pH, while the shell can be functionalized with recognition elements or catalysts that enable sensitive and selective detection. For instance, core-shell nanomaterials have been used in wearable biosensors that monitor glucose levels, detect environmental pollutants, or track physiological parameters such as heart rate and sweat composition [80]. These sensors are essential for personalized healthcare, providing real-time monitoring and data collection that can inform treatment decisions and improve patient outcomes. In the rapidly growing field of optoelectronics, core-shell nanomaterials are being investigated for their potential in applications such as light-emitting devices, photodetectors, and lasers. By manipulating the core-shell architecture, researchers can control the band gap, optical absorption, and emission characteristics of the material, enabling the design of more efficient devices for use in communication systems, imaging, and sensing technologies. For example, core-shell quantum dots are being explored for use in LEDs, where they offer improved color purity and energy efficiency compared to traditional semiconductor materials [81]. Similarly, in photodetector applications, core-shell structures can be engineered to enhance light absorption and improve the device's sensitivity, paving the way for advancements in optical communication and sensing technologies. Core-shell nanomaterials are also integral to the development of flexible, stretchable, and highly efficient batteries and energy storage devices for wearable technologies [82]. The core-shell structure allows for improved charge storage capacity and cycling stability, which is essential for portable electronics that require long battery life. By designing core-shell electrodes with tailored ion diffusion paths and optimized structural stability, it is possible to increase the efficiency and longevity of wearable energy storage systems. Additionally, the mechanical flexibility of core-shell materials makes them ideal for integration into wearable devices without sacrificing performance, enabling the development of lightweight, long-lasting energy solutions for the growing demand in wearable electronics.

3.4.5 Automated and AI-driven Synthesis Methods

The integration of automation and AI into the synthesis of core-shell nanomaterials is revolutionizing the field by enabling enhanced precision, reproducibility, and scalability. Automated synthesis platforms reduce human error and variability by precisely controlling reaction parameters such as temperature, pressure, precursor dosing, and mixing rates. This not only improves consistency across batches but also accelerates the exploration of synthesis conditions by enabling rapid iteration and data collection. AI-driven approaches, including machine learning (ML) algorithms, are increasingly employed to analyze large datasets generated from synthesis experiments and characterization results. These tools can identify hidden patterns, optimize synthesis protocols, and predict the effects of varying parameters on material properties. For example, ML models can recommend optimal shell thickness or composition to achieve desired optical, magnetic, or catalytic performance, significantly shortening development cycles. Robotic platforms combined with AI facilitate autonomous experimentation where the system designs, conducts, and analyzes synthesis runs with minimal human intervention. Such "self-driving" laboratories enable high-throughput screening of precursor combinations and process conditions, which is particularly valuable for complex multi-component core-shell systems. Moreover, feedback loops can be established where real-time characterization data are used to dynamically adjust synthesis parameters to maintain quality and tailor material properties.

In addition to optimizing synthesis, AI methods support predictive maintenance and process control in industrial manufacturing. By continuously monitoring equipment performance and product quality, AI can anticipate faults, reduce downtime, and ensure efficient resource utilization, thereby lowering production costs. The challenges in implementing automated and AI-driven synthesis include the need for extensive high-quality data to train algorithms, integration of diverse instrumentation, and the development of user-friendly interfaces for chemists and engineers. Nonetheless, ongoing advances in sensor technology, data science, and robotics are rapidly overcoming these barriers. Overall, automated and AI-enhanced synthesis represents a transformative direction for core-shell nanomaterials manufacturing. By combining precision control with intelligent decision-making, these technologies pave the way for scalable, cost-effective, and highly tailored production processes that can meet the increasing demand for advanced nanomaterials in energy, environment, medicine, and beyond.

3.5 Emerging Trends in Synthesis Strategies

The synthesis of core-shell nanomaterials is witnessing rapid innovation, driven by the need for enhanced performance, multifunctionality, and environmental sustainability across various application domains. As traditional techniques mature, emerging synthesis strategies are increasingly focused on achieving atomic-level precision, reducing environmental impact, and enabling complex architectures that go beyond simple core-shell geometries. One of the prominent trends is the development of stimuli-responsive synthesis routes, where the formation of core-shell structures is triggered or controlled by external stimuli such as light, pH, magnetic fields, or electric fields. These methods offer dynamic control over shell formation and have shown promise in producing smart nanomaterials for targeted drug delivery, catalysis, and adaptive sensing applications. Another key direction is the integration of hybrid synthesis techniques, combining physical, chemical, and green methods to harness the advantages of each. For example, coupling ALD with sol–gel processes or using microwave-assisted precipitation in combination with biosynthetic agents has allowed researchers to fabricate highly uniform, defect-free shells while minimizing energy consumption and solvent use. Atomic-level engineering of core-shell structures is also gaining momentum, enabled by advancements in in situ characterization tools and computational modeling. Techniques such as atomic layer epitaxy and molecular beam epitaxy (MBE) allow researchers to precisely control the deposition of shell layers at the monolayer level, resulting in unprecedented control over interface quality, band alignment, and lattice strain. These developments are crucial for next-generation optoelectronic and quantum devices. In parallel, bottom-up self-assembly methods are being refined to construct hierarchical and multi-shell architectures with spatially controlled functionalities. Sophisticated templating strategies and biomimetic approaches are helping to realize intricate structures that mimic natural systems, such as virus-like nanoparticles or layered inorganic shells with gradient composition.

The adoption of digital and data-driven synthesis planning, leveraging ML, high-throughput experimentation, and automation, continues to transform how new materials are discovered and synthesized. Predictive models are being trained to recommend synthesis pathways and conditions for achieving specific core-shell configurations, drastically reducing trial-and-error experimentation. Lastly, the field is moving toward modular and reconfigurable synthesis platforms, where universal reactor systems can be programmed to handle multiple types of materials, reactants, and synthesis routes. This adaptability is particularly important for industrial translation and scale-up, ensuring flexibility in meeting diverse application demands. These emerging trends are not only expanding the capabilities of core-shell nanomaterials but also redefining the way researchers approach material synthesis. As interdisciplinary approaches become more integrated, the future of core-shell nanomaterial synthesis will likely be defined by intelligent, sustainable, and highly customizable methodologies tailored to address complex global challenges in energy, environment, and health.

3.6 Conclusion

The synthesis of core-shell nanomaterials represents a critical foundation for advancing functional nanostructures with tailored properties and enhanced performance across a wide spectrum of applications. The chapter has comprehensively reviewed the diverse array of synthesis strategies available, ranging from physical and chemical methods to green and scalable approaches, each offering distinct advantages and facing unique challenges. Physical methods such as PVD, sputtering, and mechanical milling provide excellent control over structural integrity and crystallinity but often require sophisticated infrastructure and may face scalability limitations. Chemical synthesis techniques, including sol–gel, CVD, and colloidal methods, offer versatility in material composition and shell architecture, making them suitable for fabricating complex nanostructures with tunable functionalities. Meanwhile, green and sustainable strategies are increasingly gaining momentum, addressing critical environmental and safety concerns by leveraging eco-friendly precursors, low-temperature conditions, and waste-minimizing processes. Equally important are the developments in scalable and cost-effective manufacturing techniques, which bridge the gap between laboratory-scale innovation and industrial-scale production. The integration of automation, continuous flow systems, and AI-driven optimization is expected to transform the synthesis landscape by improving consistency, reducing cost, and accelerating discovery cycles. Emerging trends in synthesis, including atomic-level control, hybrid methodologies, and data-assisted design, underscore the dynamic and interdisciplinary nature of this field. As researchers continue to innovate at the intersection of materials science, engineering, and sustainability, the synthesis of core-shell nanomaterials will remain a pivotal area enabling next-generation technologies in energy, environment, and biomedicine.

References

1 Harish V, Ansari MM, Tewari D, Yadav AB, Sharma N, Bawarig S, et al. Cutting-edge advances in tailoring size, shape, and functionality of nanoparticles and nanostructures: A review. J Taiwan Inst Chem Eng [Internet]. 2023 Aug 1 [cited 2025 Jun 10];149:105010. Available from: https://www.sciencedirect.com/science/article/pii/S1876107023003395?casa_token=qC4kwXj

mh4kAAAAA:q3qRBosfwxcgw2kfzn_drSQZtAZjuhU36L4WBBKAMe7Lvya D9IkfCEPHDqma6EYpw8oTO29H2qJT
2 Mattox DM. Handbook of Physical Vapor Deposition (PVD) processing - Google books [Internet]. Elsevier. 2014 [cited 2025 Jun 10].
3 Swady R. Development of vacuum technologies for the preparation of high-purity thin films in simple systems [Doctoral thesis]. Loughborough University of Technology. 1992 [cited 2025 Jun 10]. Available from: https://repository.lboro.ac.uk/articles/Development_of_vacuum_technologies_for_the_preparation_of_high-purity_thin_films_in_simple_systems/9410687/files/17028365.pdf
4 Ives M. Fundamental studies of the PVD technique - ProQuest [Internet]. 1994 [cited 2025 Jun 10]. Available from: https://www.proquest.com/openview/bd7e75d3ebabc02b18bbbac8a44ec47a/1?pq-origsite=gscholar&cbl=2026366&diss=y
5 Safavi MS, Walsh FC, Surmeneva MA, Surmenev RA, Khalil-Allafi J. Electrodeposited hydroxyapatite-based biocoatings: Recent progress and future challenges. Coatings [Internet]. 2021 Jan 19 [cited 2025 Jun 9];11(1):110. Available from: https://www.mdpi.com/2079-6412/11/1/110/htm
6 Gudmundsson JT. Physics and technology of magnetron sputtering discharges. Plasma Sources Sci Technol [Internet]. [cited 2025 Jun 10]. Available from: https://iopscience.iop.org/article/10.1088/1361-6595/abb7bd/meta
7 Safi I. Recent aspects concerning DC reactive magnetron sputtering of thin films: A review. Surf Coat Technol [Internet]. 2000 May 22 [cited 2025 Jun 10];127(2–3): 203–18. Available from: https://www.sciencedirect.com/science/article/pii/S0257897200005661?casa_token=OP4XNYQphlIAAAAA:u4cCpGHAGIxSYlHXTkFb-6weeGMHIXmOj7JpAWNMKbKeWSPrpSIifpe5OnUKnZfwToYc389FEyI3
8 Stafe M, Marcu A, Puscas NN. Pulsed laser ablation of solids. Springer. 2014 [cited 2025 Jun 10];53. Available from: https://link.springer.com/10.1007/978-3-642-40978-3
9 Badiceanu M, Anghel S, Mihailescu N, Visan AI, Mihailescu CN, Mihailescu IN. Coatings functionalization via laser versus other deposition techniques for medical applications: A comparative review. Coatings [Internet]. 2022 Jan 7 [cited 2025 Jun 10];12(1):71. Available from: https://www.mdpi.com/2079-6412/12/1/71/htm
10 Uzakbaiuly B, Mukanova A, Zhang Y, Bakenov Z. Physical vapor deposition of cathode materials for all solid-state Li ion batteries: A review. Front Energy Res [Internet]. 2021 May 28 [cited 2025 Jun 8];9:625123. Available from: www.frontiersin.org
11 Gaikwad MU, Parihar A, and Khan R. Advanced materials: production, characterization and multidisciplinary applications - Google books [Internet]. CRC Press. 2024 [cited 2025 Jun 10].
12 Casati R. Aluminum matrix composites reinforced with alumina nanoparticles. Springer. 2016 [cited 2025 Jun 10]. Available from: http://link.springer.com/10.1007/978-3-319-27732-5
13 Wu Z, Fan X, Wang J, Li G, Gan Z, Zhang Z. Core loss reduction in Fe–6.5 wt.%Si/SiO_2 core–shell composites by ball milling coating and spark plasma sintering. J Alloys Compd [Internet]. 2014 Dec 25 [cited 2025 Jun 10]; 617:21–8. Available from: https://www.sciencedirect.com/science/article/

pii/S0925838814018507?casa_token=TYu0jOrjIugAAAAA:SqJLxq1EKCz pBnMKXvGvGShAesMXM0JS7nr9JSTQISsl7O4U4QMRdp4ML2isLDWOb6 cGGBeNuTY
14 Tsamos D, Krestou A, Papagiannaki M, Maropoulos S. An overview of the production of magnetic core-shell nanoparticles and their biomedical applications. Metals [Internet]. 2022 Mar 31 [cited 2025 Jun 8];12(4):605. Available from: https://www.mdpi.com/2075-4701/12/4/605/htm
15 Wintzheimer S, Luthardt L, Cao KLA, Imaz I, Maspoch D, Ogi T, et al. Multifunctional, hybrid materials design via spray-drying: Much more than just drying. Adv Mater [Internet]. 2023 Nov 23 [cited 2025 Jun 10];35(47):2306648. Available from: https://doi.org/10.1002/adma.202306648
16 Abdullah MF, Nuge T, Andriyana A, Ang BC, Muhamad F. Core–shell fibers: Design, roles, and controllable release strategies in tissue engineering and drug delivery. Polymers [Internet]. 2019 Dec 4 [cited 2025 Jun 10];11(12):2008. Available from: https://www.mdpi.com/2073-4360/11/12/2008/htm
17 Peng S, Li L, Kong Yoong Lee J, Tian L, Srinivasan M, Adams S, et al. Electrospun carbon nanofibers and their hybrid composites as advanced materials for energy conversion and storage. Nano Energy [Internet]. 2016 Apr 1 [cited 2025 Jun 10];22:361–95. Available from: https://www.sciencedirect.com/science/article/pii/S2211285516000525?casa_token=Q3gk6JVWJk0AAAAA:kqIP2OJIA1VhiD0X zhvCZJrokjy64sk9k27jD6HuVwGqa7gJ0LpbqyHXj2G5namaWC3BXplgD8w
18 Deshmukh R, Wagh P, Naik J. Solvent evaporation and spray drying technique for micro- and nanospheres/particles preparation: A review. Dry Technol [Internet]. 2016 Nov 17 [cited 2025 Jun 10];34(15):1758–72. Available from: https://www.tandfonline.com/doi/abs/10.1080/07373937.2016.1232271
19 Mohammed NK, Tan CP, Manap YA, Muhialdin BJ, Hussin ASM. Spray drying for the encapsulation of oils—A review. Molecules [Internet]. 2020 Aug 26 [cited 2025 Jun 8];25(17):3873. Available from: https://www.mdpi.com/1420-3049/25/17/3873/htm
20 Mochane MJ. Polymer encapsulated paraffin wax to be used as phase change material for energy storage [Internet]. University of the Free State (Qwaqwa Campus). 2011 [cited 2025 Jun 10]. Available from: http://hdl.handle.net/11660/2191
21 Pérez-Page M, Yu E, Li J, Rahman M, Dryden DM, Vidu R, et al. Template-based syntheses for shape controlled nanostructures. Adv Colloid Interface Sci [Internet]. 2016 Aug 1 [cited 2025 Jun 10];234:51–79. Available from: https://www.sciencedirect.com/science/article/pii/S0001868616301105?casa_token=D9HpFoXLA8cAAAAA:p9tm-QgpQzFwDIRkVKJHwGzkP2MUGUxklFZ erAsbPWKWh6Cn9YQ53i3lzPnHT6UWShJE1qEMF0w
22 Wei Q, Fu Y, Zhang G, Yang D, Meng G, Sun S. Rational design of novel nanostructured arrays based on porous AAO templates for electrochemical energy storage and conversion. Nano Energy [Internet]. 2019 Jan 1 [cited 2025 Jun 10];55:234–59. Available from: https://www.sciencedirect.com/science/article/pii/S221128551830795X?casa_token=A899aCX2nbIAAAAA:n0JKT1UTMTYrrDvNo NnnCyclDaEBF9U0_fMNFdlgZQacDIlsEBnHU33c9T46rksyhlxWlrCdhj4
23 Zhao T, Elzatahry A, Li X, Zhao D. Single-micelle-directed synthesis of mesoporous materials. Nat Rev Mater [Internet]. 2019 Oct 7 [cited 2025

Jun 10];4(12):775–91. Available from: https://www.nature.com/articles/s41578-019-0144-x

24 Zhang Q, Lee I, Joo JB, Zaera F, Yin Y. Core-shell nanostructured catalysts. Acc Chem Res [Internet]. 2013 Aug 20 [cited 2025 Jun 10];46(8):1816–24. Available from: https://doi.org/10.1021/ar300230s

25 Mitzi D. Solution processing of inorganic materials - Google books [Internet]. Wiley. 2008 [cited 2025 Jun 10].

26 Lee SH 1974. Synthesis and Characterization of Shell/core Nanoparticles and the Fabrication of Nanoparticle Thin films [Internet]. 2012 [cited 2025 Jun 10]. Available from: http://hdl.handle.net/10657/1184

27 Wu KJ, Tse ECM, Shang C, Guo Z. Nucleation and growth in solution synthesis of nanostructures – From fundamentals to advanced applications. Prog Mater Sci [Internet]. 2022 Jan 1 [cited 2025 Jun 10];123:100821. Available from: https://www.sciencedirect.com/science/article/pii/S0079642521000451?casa_token=7lAvGE2v-GUAAAAA:UwqOhaz44Ysz44VXVSHjYGPPfIsE0A675skY0ZArlG9OHMoZRWO97_74wIqB4HiqPVCTzo7OFo8

28 Navas D, Fuentes S, Castro-Alvarez A, Chavez-Angel E. Review on sol-gel synthesis of perovskite and oxide nanomaterials. Gels [Internet]. 2021 Dec 18 [cited 2025 Jun 9];7(4):275. Available from: https://www.mdpi.com/2310-2861/7/4/275/htm

29 Gawande MB, Goswami A, Asefa T, Guo H, Biradar AV, Peng DL, et al. Core–shell nanoparticles: Synthesis and applications in catalysis and electrocatalysis. Chem Soc Rev [Internet]. 2015 Oct 19 [cited 2025 Jun 10];44(21):7540–90. Available from: https://pubs.rsc.org/en/content/articlehtml/2015/cs/c5cs00343a

30 Tiwari SK, Bystrzejewski M, Zhu Y. Synthesis, properties, and applications of carbon-encapsulated metal nanoparticles. Coord Chem Rev [Internet]. 2024 Dec 1 [cited 2025 Jun 10];520:216125. Available from: https://www.sciencedirect.com/science/article/pii/S0010854524004715?casa_token=H_1u7SfqTXsAAAAA:LhV2igggSxAs05GG1Z1ULr02YWgbGJ1DXMcq-sLG2neyIHHf3asSoYOPrHx6C3Ef9bnPHUTfOyM

31 Patil S, Jagadale S. Co-precipitation methods for the synthesis of metal oxide nanostructures. In: Solution Methods for Metal Oxide Nanostructures [Internet]. 2023 Jan 1 [cited 2025 Jun 10];39–60. Available from: https://www.sciencedirect.com/science/article/abs/pii/B9780128243534000166

32 Borges J, Mano JF. Molecular interactions driving the layer-by-layer assembly of multilayers. Chem Rev [Internet]. 2014 Sep 24 [cited 2025 Jun 10];114(18):8883–942. Available from: https://doi.org/10.1021/cr400531v

33 Borges J, Rodrigues LC, Reis RL, Mano JF. Layer-by-layer assembly of light-responsive polymeric multilayer systems. Adv Funct Mater [Internet]. 2014 Sep 24 [cited 2025 Jun 10];24(36):5624–48. Available from: https://doi.org/10.1002/adfm.201401050

34 Choy KL. Chemical Vapour Deposition (CVD): advances, technology and applications - Google books [Internet]. CRC Press/Taylor & Francis Group. 2019 [cited 2025 Jun 10].

35 Sabzi M, Mousavi Anijdan SH, Shamsodin M, Farzam M, Hojjati-Najafabadi A, Feng P, et al. A Review on sustainable manufacturing of ceramic-based thin films by Chemical Vapor Deposition (CVD): Reactions kinetics and the deposition mechanisms. Coatings [Internet]. 2023 Jan 14 [cited 2025 Jun 9];13(1):188. Available from: https://www.mdpi.com/2079-6412/13/1/188/htm

36 Olowoyo JO, Gharahshiran VS, Zeng Y, Zhao Y, Zheng Y. Atomic/molecular layer deposition strategies for enhanced CO_2 capture, utilisation and storage materials. Chem Soc Rev [Internet]. 2024 Jun 4 [cited 2025 Jun 10];53(11):5428–88. Available from: https://pubs.rsc.org/en/content/articlehtml/2024/cs/d3cs00759f

37 Thanh NTK, Maclean N, Mahiddine S. Mechanisms of nucleation and growth of nanoparticles in solution. Chem Rev [Internet]. 2014 Aug 13 [cited 2025 Jun 10];114(15):7610–30. Available from: https://doi.org/10.1021/cr400544s

38 Spoială A, Ilie CI, Crăciun LN, Ficai D, Ficai A, Andronescu E. Magnetite-silica core/shell nanostructures: From surface functionalization towards biomedical applications—A review. Appl Sci [Internet]. 2021 Nov 22 [cited 2025 Jun 10];11(22):11075. Available from: https://www.mdpi.com/2076-3417/11/22/11075/htm

39 Aswathi VP, Meera S, Maria CGA, Nidhin M. Green synthesis of nanoparticles from biodegradable waste extracts and their applications: A critical review. Nanotechnol Environ Eng [Internet]. 2022 Aug 24 [cited 2025 Jun 10];8(2):377–97. Available from: https://link.springer.com/article/10.1007/s41204-022-00276-8

40 Jadhav V, Bhagare A, Ali IH, Dhayagude A, Lokhande D, Aher J, et al. Role of moringa oleifera on green synthesis of metal/metal oxide nanomaterials. J Nanomater [Internet]. 2022 Jan 1 [cited 2025 Jun 9];2022(1):2147393. Available from: https://doi.org/10.1155/2022/2147393

41 Lucas E, Martín AJ, Mitchell S, Nabera A, Santos LF, Pérez-Ramírez J, et al. The need to integrate mass- and energy-based metrics with life cycle impacts for sustainable chemicals manufacture. Green Chem [Internet]. 2024 Aug 27 [cited 2025 Jun 10];26(17):9300–9. Available from: https://pubs.rsc.org/en/content/articlehtml/2024/gc/d4gc00394b

42 Khatami M, Alijani HQ, Nejad MS, Varma RS. Core@shell nanoparticles: Greener synthesis using natural plant products. Appl Sci [Internet]. 2018 Mar 10 [cited 2025 Jun 10];8(3):411. Available from: https://www.mdpi.com/2076-3417/8/3/411/htm

43 Zou L, Zhu F, Long Z er, Huang Y. Bacterial extracellular electron transfer: A powerful route to the green biosynthesis of inorganic nanomaterials for multifunctional applications. J Nanobiotechnol [Internet]. 2021 Apr 27 [cited 2025 Jun 9];19(1):1–27. Available from: https://link.springer.com/articles/10.1186/s12951-021-00868-7

44 Abegunde SM, Afolayan BO, Ilesanmi TM. Ensuring sustainable plant-assisted nanoparticles synthesis through process standardization and reproducibility: Challenges and future directions – A review. Sus Chem One World [Internet]. 2024 Sep 1 [cited 2025 Jun 10];3:100014. Available from: https://www.sciencedirect.com/science/article/pii/S2950357424000143?casa_token=nTaYracMjPcAAAAA:AA0Qy0pxKHgEBnCyD04tPGSogaUHgwOb1TgCbMfZT2NuAw9ujs2mHwD57lc_kFhf-W06nvUu6B8

45 Chandler CD, Roger C, Hampden-Smith MJ. Chemical aspects of solution routes to perovskite-phase mixed-metal oxides from metal-organic precursors. Chem Rev [Internet]. 1993 [cited 2025 Jun 10];93(3):1205–41. Available from: /action/showCitFormats?doi=10.1021/cr00019a015&href=/doi/pdf/10.1021/cr00019a015

46 Wang X, Wang P, Li M, Li J. Advances in the preparation and biological applications of core@shell nanocrystals based on quantum dots and noble metal. RSC Adv [Internet]. 2024 Aug 20 [cited 2025 Jun 10];14(36):26308–24. Available from: https://pubs.rsc.org/en/content/articlehtml/2024/ra/d4ra05386a

47 Jayaraman V, Singh R, Anandnarayan A. Impact of sustainable manufacturing practices on consumer perception and revenue growth: An emerging economy perspective. Int J Prod Res [Internet]. 2012 Mar 1 [cited 2025 Jun 10];50(5): 1395–1410. Available from: https://scholar.google.com/scholar_url?url=https://www.tandfonline.com/doi/pdf/10.1080/00207543.2011.571939%3Fcasa_token%3D9rtFZEjqEhcAAAAA:uB4Az7djO_psUQhKv491aKyboUiU1R7ypvSp_Og8qo6VROIq_e_XTSWs-JcL2VVKPdZPc2RYMu8dfg&hl=en&sa=T&oi=ucasa&ct=ucasa&ei=VNNHaKv7B8Or6rQP277P8QY&scisig=AAZF9b-16ZulL6K7ERmpvtJCmrcs

48 Yadav M, Dutta M, Tanwar P, Jain R, Srivastava A, Sharma RK. Microwave-assisted C-C, C-O, C-N, C-S bond formation and multicomponent reactions using magnetic retrievable nanocatalysts. Curr Microw Chem [Internet]. 2021 Aug 1 [cited 2025 Jun 10];8(2):96–116. Available from: https://www.benthamdirect.com/content/journals/cmic/10.2174/2213335608666210804144559

49 Zou Y, Sun Z, Wang Q, Ju Y, Sun N, Yue Q, et al. Core-shell magnetic particles: Tailored synthesis and applications. Chem Rev [Internet]. 2025 Jan 22 [cited 2025 Jun 10];125(2):972–1048. Available from: https://doi.org/10.1021/acs.chemrev.4c00710

50 Niraula G, Shrivastava N, Akhtar K, Javed Y, Coaquira JAH, Sharma SK. Liquid-phase synthesis of multifunctional nanomaterials: A recent update. In: Magnetic Nanoheterostructures. Springer. 2020 [cited 2025 Jun 10];1–56. Available from: https://link.springer.com/chapter/10.1007/978-3-030-39923-8_1

51 Imran M, Affandi AM, Alam MM, Khan A, Khan AI. Advanced biomedical applications of iron oxide nanostructures based ferrofluids. Nanotechnology [Internet]. 2021 Jul 26 [cited 2025 Jun 10];32(42):422001. Available from: https://iopscience.iop.org/article/10.1088/1361-6528/ac137a

52 Liu X, Li Y, Zeng L, Li X, Chen N, Bai S, et al. A review on mechanochemistry: Approaching advanced energy materials with greener force. Adv Mater [Internet]. 2022 Nov 17 [cited 2025 Jun 10];34(46):2108327. Available from: https://doi.org/10.1002/adma.202108327

53 Leuenberger H. New trends in the production of pharmaceutical granules: The classical batch concept and the problem of scale-up. Eur J Pharm Biopharm [Internet]. 2001 Nov 1 [cited 2025 Jun 10];52(3):279–88. Available from: https://www.sciencedirect.com/science/article/pii/S0939641101002004?casa_token=f8xkhoKUkeEAAAAA:_cZiTKUESgwVlPO_ukX_0R4kxDTuGnZikRmm9noxiBS9pZLs1IlhGjE9mAivJMDvxsQYekhf7yY

54 Duflou JR, Sutherland JW, Dornfeld D, Herrmann C, Jeswiet J, Kara S, et al. Towards energy and resource efficient manufacturing: A processes and systems approach. CIRP Annals [Internet]. 2012 Jan 1 [cited 2025 Jun 10];61(2):587–609. Available from: https://www.sciencedirect.com/science/article/pii/S0007850612002016?casa_token=DLv8ZFa18pwAAAAA:ItdqbXfPVWcLPFFr7I8crajT1U2avSGBZfaU5B3Smam7Wq0tInVGnVVnDLC0T2By5ADoWGkaRKk

55 Mondal K, Sharma A. Recent advances in the synthesis and application of photocatalytic metal–metal oxide core–shell nanoparticles for environmental remediation and their recycling process. RSC Adv [Internet]. 2016 Sep 2 [cited 2025 Mar 7];6(87):83589–612. Available from: https://pubs.rsc.org/en/content/articlehtml/2016/ra/c6ra18102c

56 Shafiee A, Rabiee N, Ahmadi S, Baneshi M, Khatami M, Iravani S, et al. Core–shell nanophotocatalysts: Review of materials and applications. ACS Appl Nano Mater [Internet]. 2022 Jan 28 [cited 2025 Mar 7];5(1):55–86. Available from: https://pubs.acs.org/doi/full/10.1021/acsanm.1c03714

57 Fei X, Wang P, Zhang D, Wang H, Wu Z. Confined catalysts application in environmental catalysis: Current research progress and future prospects. ChemCatChem [Internet]. 2021 May 19 [cited 2025 Mar 7];13(10):2313–36. Available from: https://onlinelibrary.wiley.com/doi/full/10.1002/cctc.202001578

58 Jin S, Park BC, Ham WS, Pan L, Kim YK. Effect of the magnetic core size of amino-functionalized Fe_3O_4-mesoporous SiO_2 core-shell nanoparticles on the removal of heavy metal ions. Colloids Surf A: Physicochem Eng Asp. 2017 Oct 20;531:133–140.

59 Iftekhar S, Deb A, Heidari G, Sillanpää M, Lehto VP, Doshi B, et al. A review on the effectiveness of nanocomposites for the treatment and recovery of oil spill. Environ Sci Pollut Res [Internet]. 2023 Jan 7 [cited 2025 Mar 7];30(7):16947–83. Available from: https://link.springer.com/article/10.1007/s11356-022-25102-1

60 Bhattacharjee Y, Chatterjee D, Bose S. Core-multishell heterostructure with excellent heat dissipation for electromagnetic interference shielding. ACS Appl Mater Interfaces [Internet]. 2018 Sep 12 [cited 2025 Mar 7];10(36):30762–73. Available from: https://pubs.acs.org/doi/full/10.1021/acsami.8b10819

61 Li W, Elzatahry A, Aldhayan D, Zhao D. Core–shell structured titanium dioxide nanomaterials for solar energy utilization. Chem Soc Rev [Internet]. 2018 Nov 12 [cited 2025 Mar 7];47(22):8203–37. Available from: https://pubs.rsc.org/en/content/articlehtml/2018/cs/c8cs00443a

62 Bahadoran A, De Lile JR, Masudy-Panah S, Sadeghi B, Li J, Sabzalian MH, et al. Photocatalytic materials obtained from e-waste recycling: Review, techniques, critique, and update. J Manuf Mater Process [Internet]. 2022 Jun 26 [cited 2025 Mar 7];6(4):69. Available from: https://www.mdpi.com/2504-4494/6/4/69/htm

63 Ghosh Chaudhuri R, Paria S. Core/shell nanoparticles: Classes, properties, synthesis mechanisms, characterization, and applications. Chem Rev [Internet]. 2012 Apr 11 [cited 2025 Mar 7];112(4):2373–433. Available from: https://pubs.acs.org/doi/full/10.1021/cr100449n

64 Zou H, Luo Z, Yang X, Xie Q, Zhou Y. Toward emerging applications using core–shell nanostructured materials: A review. J Mater Sci. 2022 Jun 1;57(24):10912–42.

65 Gu Q, Zhu J, Weng G-J, Li J-J, Zhao JW. Core-satellite nanostructures and their biomedical applications. Microchimica Acta. 2022 Dec 1;189(12):470.

66 Ramadhan SA, Ali DS. Innovations in core–shell nanoparticles: Advancing drug delivery solutions and precision medicine. OMICS [Internet]. 2025 Mar 1 [cited 2025 Mar 7];29(3):73–86. Available from: https://www.liebertpub.com/doi/10.1089/omi.2024.0182

67 Albinali KE, Zagho MM, Deng Y, Elzatahry AA. A perspective on magnetic core–shell carriers for responsive and targeted drug delivery systems. Int J Nanomedicine [Internet]. 2019 [cited 2025 Mar 7];14:1707–23. Available from: https://www.tandfonline.com/action/journalInformation?journalCode=dijn20

68 Das M, Mohanty C, Sahoo SK. Ligand-based targeted therapy for cancer tissue. Expert Opin Drug Deliv [Internet]. 2009 Mar [cited 2025 Mar 7];6(3):285–304. Available from: https://www.tandfonline.com/doi/abs/10.1517/1742524090 2780166

69 Han X, Alu A, Liu H, Shi Y, Wei X, Cai L, et al. Biomaterial-assisted biotherapy: A brief review of biomaterials used in drug delivery, vaccine development, gene therapy, and stem cell therapy. Bioact Mater. 2022 Nov 1;17:29–48.

70 Shi Y, Zhang Y, Zhu L, Miao Y, Zhu Y, Yue B. Tailored drug delivery platforms: Stimulus-responsive core–shell structured nanocarriers. Adv Healthc Mater [Internet]. 2024 Jan 1 [cited 2025 Mar 7];13(1):2301726. Available from: https://onlinelibrary.wiley.com/doi/full/10.1002/adhm.202301726

71 Xu H, Li Z, Si J. Nanocarriers in gene therapy: A review. J Biomed Nanotechnol. 2014 Dec 1;10(12):3483–507.

72 Wei Hu S, Ding T, Tang H, Guo H, Cui W, Shu Y. Nanobiomaterial vectors for improving gene editing and gene therapy. Mater Today. 2023 Jun 1;66:114–36.

73 Khalilov R. A comprehensive review of advanced nano-biomaterials in regenerative medicine and drug delivery. Adv Biol Earth Sci. 2023;8(1):5–18.

74 Du Z, Cao G, Li K, Zhang R, Li X. Nanocomposites for the delivery of bioactive molecules in tissue repair: Vital structural features, application mechanisms, updated progress and future perspectives. J Mater Chem B [Internet]. 2020 Nov 25 [cited 2025 Mar 7];8(45):10271–89. Available from: https://pubs.rsc.org/en/content/articlehtml/2020/tb/d0tb01670e

75 Wang Z, Shen G. Flexible optoelectronic sensors: Status and prospects. Mater Chem Front [Internet]. 2023 Apr 11 [cited 2025 Mar 7];7(8):1496–519. Available from: https://pubs.rsc.org/en/content/articlehtml/2023/qm/d2qm01319c

76 Tehrani M. Advanced electrical conductors: An overview and prospects of metal nanocomposite and nanocarbon based conductors. Phys Status Solidi (a) [Internet]. 2021 Apr 1 [cited 2025 Mar 7];218(8):2000704. Available from: https://onlinelibrary.wiley.com/doi/full/10.1002/pssa.202000704

77 Zhuo MP, Wang XD, Liao LS. Construction and optoelectronic applications of organic core/shell micro/nanostructures. Mater Horiz [Internet]. 2020 Dec 7 [cited 2025 Mar 7];7(12):3161–75. Available from: https://pubs.rsc.org/en/content/articlehtml/2020/mh/d0mh00967a

78 Chen M, Lu L, Yu H, Li C, Zhao N. Integration of colloidal quantum dots with photonic structures for optoelectronic and optical devices. Adv Sci [Internet]. 2021 Sep 1 [cited 2025 Mar 7];8(18):2101560. Available from: https://onlinelibrary.wiley.com/doi/full/10.1002/advs.202101560

79 Koo JH, Kim DC, Shim HJ, Kim TH, Kim DH. Flexible and stretchable smart display: Materials, fabrication, device design, and system integration. Adv Funct Mater [Internet]. 2018 Aug 1 [cited 2025 Mar 7];28(35):1801834. Available from: https://onlinelibrary.wiley.com/doi/full/10.1002/adfm.201801834

80 Khan B, Riaz Z, Ahmad RuS, Khoo BL. Advancements in wearable sensors for cardiovascular disease detection for health monitoring. Mater Sci Eng R: Rep. 2024 Jun 1;159:100804.

81 Chen O, Wei H, Maurice A, Bawendi M, Reiss P. Pure colors from core–shell quantum dots. MRS Bull [Internet]. 2013 Sep [cited 2025 Mar 7];38(9):696–702. Available from: https://www.cambridge.org/core/journals/mrs-bulletin/article/abs/pure-colors-from-coreshell-quantum-dots/4B8AC6E7E8B1B554420EE8702AC724E5

82 Zhang X, Jiang C, Liang J, Wu W. Electrode materials and device architecture strategies for flexible supercapacitors in wearable energy storage. J Mater Chem A Mater [Internet]. 2021 Apr 6 [cited 2025 Mar 7];9(13):8099–128. Available from: https://pubs.rsc.org/en/content/articlehtml/2021/ta/d0ta12299h

4

Characterization Techniques for Core-shell Nanomaterials

The rapid evolution of core-shell nanomaterials in diverse fields such as catalysis, energy storage, drug delivery, and sensing demands a deep understanding of their physicochemical properties. Such understanding is only achievable through comprehensive and multi-modal characterization strategies. Unlike simple nanostructures, core-shell systems possess inherent complexities, including multilayered architectures, diverse material compositions, interface dynamics, and tunable surface functionalities. Hence, their effective characterization calls for advanced techniques capable of probing not just the morphology or composition but also the interfacial properties, internal structure, and dynamic behavior under various conditions. Characterization of core-shell nanomaterials is essential for verifying the successful formation of the core-shell architecture, measuring the shell thickness, evaluating uniformity and crystallinity, assessing elemental distribution, and confirming functional properties such as thermal stability or optical activity. Moreover, the specific applications of these materials often dictate the choice of techniques—for instance, electron microscopy for structural insights, spectroscopy for chemical and electronic information, or thermal analysis for stability evaluation. Recent advances in real time and in situ techniques have further enabled researchers to track the growth mechanisms and operational performance of core-shell nanostructures under practical conditions, offering insights previously unattainable through conventional post-synthesis methods. In parallel, the integration of machine learning and artificial intelligence (AI) into characterization workflows has begun transforming the field by automating data interpretation, enhancing predictive capabilities, and accelerating the discovery of structure–property relationships. As we move toward high-throughput material development and smart manufacturing, characterization tools are becoming more data-rich, precise, and application-focused. The chapter provides a structured overview of the most prominent techniques used to characterize core-shell nanomaterials, categorized under microscopy, spectroscopy, thermal and mechanical analysis, in situ and real-time monitoring, and emerging AI-powered approaches. The goal is to offer a comprehensive guide for selecting and implementing appropriate characterization methodologies tailored to specific core-shell architectures and their intended applications.

Core-Shell Nanomaterials: From Fundamentals to Applications, First Edition. Shreya Sharma et al.

4.1 Microscopy Techniques

Microscopy techniques represent a cornerstone in the structural and morphological analysis of core-shell nanomaterials. These methods enable direct visualization of nanoscale features, allowing researchers to assess critical aspects such as particle size, shape, surface morphology, shell thickness, and the integrity of the core-shell architecture. Due to the layered and often compositionally distinct nature of core-shell systems, high-resolution imaging becomes essential to confirm the successful formation and uniformity of both the core and shell components. Unlike bulk characterization techniques, microscopy provides spatially resolved information that is vital for understanding how the core and shell materials interact, how uniformly the shell encapsulates the core, and whether any defects, voids, or asymmetries exist within the structure. These attributes significantly influence the functional performance of the material, especially in applications requiring precise control over surface area, diffusion barriers, or interfacial interactions. Advanced microscopy techniques extend beyond simple imaging to include capabilities such as elemental mapping, crystallographic orientation analysis, three-dimensional (3D) reconstruction, and real-time observation under varying environmental or operational conditions [1]. The ability to couple imaging with spectroscopic modalities—such as energy-dispersive X-ray spectroscopy (EDS) or electron energy loss spectroscopy (EELS)—further enhances the understanding of compositional gradients and core-shell boundaries at the nanoscale. Given the broad range of available microscopy methods, the choice of technique often depends on the desired resolution, sample preparation constraints, and whether surface or internal structure needs to be probed. In this section, we explore the most widely used microscopy techniques for characterizing core-shell nanomaterials, including scanning electron microscopy (SEM), transmission electron microscopy (TEM), atomic force microscopy (AFM), and focused ion beam (FIB) imaging, each offering unique advantages for different characterization objectives.

A comparative overview of commonly used microscopy techniques for core-shell nanomaterials is presented in Figure 4.1, which categorizes each method based on its ability to probe surface features, internal structure, or both. This flowchart aids in selecting the most appropriate technique depending on the specific analytical needs of the study.

4.1.1 SEM for Morphology Analysis

SEM is one of the most widely employed techniques for the morphological characterization of core-shell nanomaterials. SEM operates by scanning a focused beam of high-energy electrons across the surface of a sample. These electrons interact with the atoms at or near the surface, producing secondary and backscattered electrons (SE and BSE) that are detected to form high-resolution images. The resulting micrographs provide valuable information about surface topography, particle size distribution, and agglomeration behavior. In the context of core-shell nanomaterials, SEM serves as a powerful tool for confirming the uniformity and consistency of shell coverage across core particles [2]. Though SEM cannot directly reveal

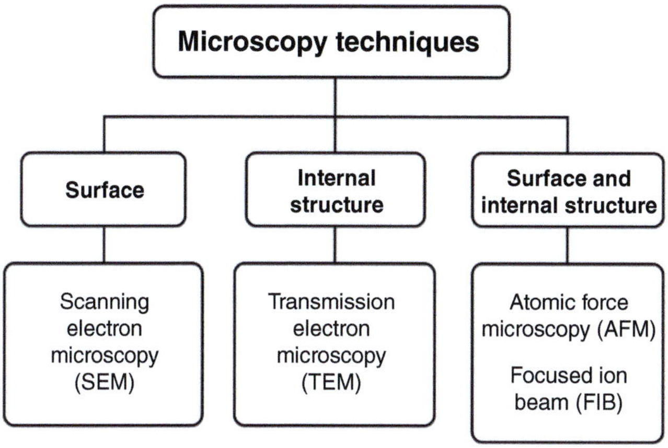

Figure 4.1 Categorization of microscopy techniques based on structural target. Generated with AI using ChatGPT (version GPT-4).

the internal structure of core-shell systems (unless the shell is partially broken or etched), it offers critical insights into surface morphology and texture. For instance, the smoothness or roughness of the shell, presence of pores, and even nanostructured features such as ridges or coatings can be effectively visualized. Such morphological features often correlate strongly with functional properties such as surface reactivity, catalytic activity, and interfacial interactions. A simplified schematic of the SEM setup, illustrating the major components and the pathway of the electron beam, is shown in Figure 4.2.

High-resolution field-emission SEM (FE-SEM) offers superior imaging capabilities, often with resolution down to a few nanometers, making it suitable for

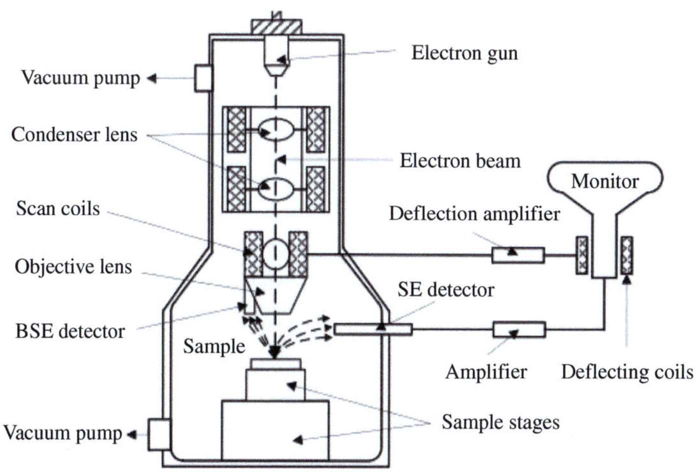

Figure 4.2 Schematic of a scanning electron microscope (SEM) setup: The diagram illustrates key components of an SEM, including the electron gun, condenser lens, scan coils, objective lens, detectors (secondary electrons [SE] and backscattered electrons [BSE]), and vacuum chamber. The system generates and focuses an electron beam to scan the sample surface, collecting interaction signals to produce high-resolution digital images. Reproduced with permission from [4]/IEEE.

nanoscale analysis of even very small core-shell particles. Moreover, SEM can be coupled with complementary analytical techniques such as Energy Dispersive X-ray Spectroscopy (EDS), which enables elemental mapping of the sample surface [3]. This combination is especially useful in confirming the spatial distribution of elements and the integrity of the core-shell configuration. Sample preparation for SEM is relatively straightforward but must be done carefully to avoid distortion or charging effects, especially for insulating or organic shell materials. Gold or carbon coatings are often applied to enhance conductivity, although such treatments can sometimes mask fine surface details. Recent advances in low-voltage and environmental SEM (ESEM) have allowed imaging of non-conductive and even hydrated samples with minimal preparation, expanding the applicability of SEM to more diverse core-shell systems. Overall, SEM remains an indispensable technique in the characterization toolkit for core-shell nanomaterials, offering rapid, high-resolution morphological assessment that guides synthesis optimization and quality control in both research and industrial settings.

4.1.2 TEM and High-resolution TEM

TEM is a cornerstone technique for the structural and compositional characterization of core-shell nanomaterials. Unlike SEM, which provides surface images, TEM offers insight into the internal structure of materials by transmitting a beam of electrons through an ultrathin sample. This unique capability makes TEM particularly valuable for visualizing the core and shell architecture, determining shell thickness, and observing interfaces between the core and shell. One of the primary advantages of TEM in the study of core-shell nanostructures is its ability to deliver direct imaging of both the core and the shell, enabling clear confirmation of successful encapsulation and symmetry. In core-shell systems with well-defined crystalline domains, TEM can also be used to detect lattice fringes, providing information about crystallinity and orientation. The contrast in TEM images often arises from differences in atomic number (Z-contrast), thickness, and density between the core and shell materials, thus offering a natural means to distinguish between the two layers.

High-resolution TEM (HRTEM) takes these capabilities further by allowing atomic-level resolution imaging, often below 1 Å. HRTEM is particularly useful for elucidating the crystallographic structure of core-shell nanomaterials, visualizing atomic arrangements, dislocations, and defects at interfaces. It is especially beneficial in the study of epitaxial growth of shell materials on crystalline cores and in examining strain or mismatch-induced phenomena at the core-shell boundary. In addition to imaging, TEM can be integrated with a suite of analytical tools such as Selected Area Electron Diffraction (SAED), EELS, and EDS [5]. These techniques enable phase identification, chemical composition analysis, and elemental mapping at nanometer resolution. For example, EDS mapping in TEM can confirm whether elements are uniformly distributed within the shell or whether there are gradients or segregations.

Sample preparation for TEM is more challenging compared to SEM. Samples must be extremely thin (typically <100 nm) to allow sufficient electron transmission, which often involves careful drop-casting of dilute nanoparticle suspensions on TEM grids or FIB thinning for bulk specimens. Care must also be taken to avoid beam-induced damage, particularly for materials with organic or polymeric shells. Overall, TEM and HRTEM are indispensable tools in the comprehensive analysis of core-shell nanomaterials, offering unparalleled insight into their internal structure, composition, and interface properties. These capabilities are critical for understanding performance in applications ranging from catalysis and photonics to drug delivery and energy storage.

4.1.3 AFM for Surface Analysis

AFM is a powerful tool used to investigate the surface morphology, roughness, and mechanical properties of core-shell nanomaterials at nanometer to atomic scale. Unlike electron-based imaging techniques such as SEM and TEM, AFM relies on the interaction forces between a sharp probe and the sample surface, making it especially useful for non-conductive, soft, or biological materials that are otherwise difficult to image using electron microscopy. AFM operates by scanning a cantilever with a nanometer-scale tip across the sample surface. As the tip moves over the material, interactions such as van der Waals forces or electrostatic forces cause the cantilever to deflect [6]. These deflections are measured with a laser and photodiode system, providing a 3D topographical map of the surface with exceptional vertical resolution, often down to sub-nanometer levels. A schematic illustration of an in situ AFM setup is shown in Figure 4.3, where the AFM tip scans continuously while molecular or ionic species self-assemble on the substrate in real time.

In the context of core-shell nanomaterials, AFM is highly valuable for characterizing surface smoothness, uniformity of shell coverage, and particle aggregation

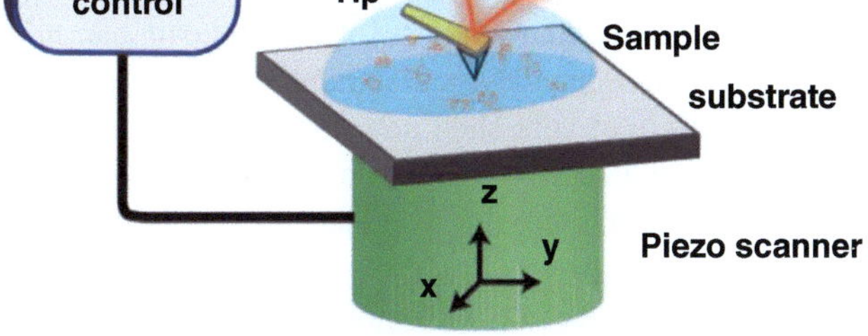

Figure 4.3 Schematic illustration of in situ Atomic Force Microscopy (AFM) setup. The AFM tip continuously scans the sample surface to capture real-time topographical images, while ions or molecules in the surrounding solution undergo self-assembly into ordered nanostructures. [8]/MDPI/CC BY 4.0.

behavior. For instance, when shell materials are formed via surface functionalization or coating, AFM can be used to verify the homogeneity and thickness of the shell layer, especially when working with soft or polymeric shells. Additionally, the force spectroscopy mode of AFM allows for mechanical property analysis such as stiffness, elasticity, and adhesion forces, which are crucial for applications in drug delivery, sensors, or stimuli-responsive systems. AFM also enables imaging in a variety of environments, including air, liquid, and vacuum, making it ideal for studying biofunctionalized or environmentally sensitive core-shell nanomaterials under realistic conditions [7]. This capability is particularly useful for nanocarriers or biosensors where surface interaction with fluids or biomolecules is a key functionality. Moreover, modes such as tapping mode and noncontact mode reduce the risk of damaging delicate samples, which is critical when analyzing nanomaterials with fragile organic shells.

Advanced AFM techniques, including Conductive AFM (C-AFM), Magnetic Force Microscopy (MFM), and Kelvin Probe Force Microscopy (KPFM), provide further insights into electrical, magnetic, and surface potential characteristics of core-shell nanostructures. These advanced modes enable the correlation of topographical data with functional properties, offering a comprehensive view of performance-relevant features at the nanoscale. Despite its versatility, AFM is generally slower in scanning speed compared to electron microscopy, and the analysis is limited to relatively small surface areas. However, for applications where surface properties are paramount, AFM remains an indispensable characterization method, complementing other techniques by providing high-resolution, label-free, and non-destructive analysis.

4.1.4 FIB and 3D Imaging Techniques

FIB technology has become an essential tool for nanoscale characterization and manipulation, especially in the study of complex core-shell nanostructures. Originally developed for microfabrication and semiconductor analysis, FIB has evolved into a versatile platform capable of site-specific material removal, deposition, and 3D reconstruction of nanomaterials. In core-shell systems, where internal structural integrity, layer thickness, and interfaces are critically important, FIB provides unparalleled capabilities for high-resolution cross-sectional imaging and in-depth compositional analysis. FIB operates by directing a beam of high-energy ions, typically gallium (Ga^+) onto the sample surface [9]. This ion beam can mill away material with nanometer precision, allowing researchers to create cross sections that reveal the inner architecture of core-shell nanostructures. When combined with SEM in a dual-beam system, FIB-SEM allows simultaneous imaging and milling, enabling real-time observation of internal features such as core-shell interfaces, voids, and multilayer arrangements.

One of the key strengths of FIB is its role in 3D tomography. By sequentially milling thin slices of a sample and imaging each layer, FIB-SEM tomography can reconstruct a high-resolution 3D model of the material. This is particularly useful for core-shell nanoparticles, nanorods, and multilayered coatings, where understanding the spatial distribution and continuity of shells is vital. The resulting 3D

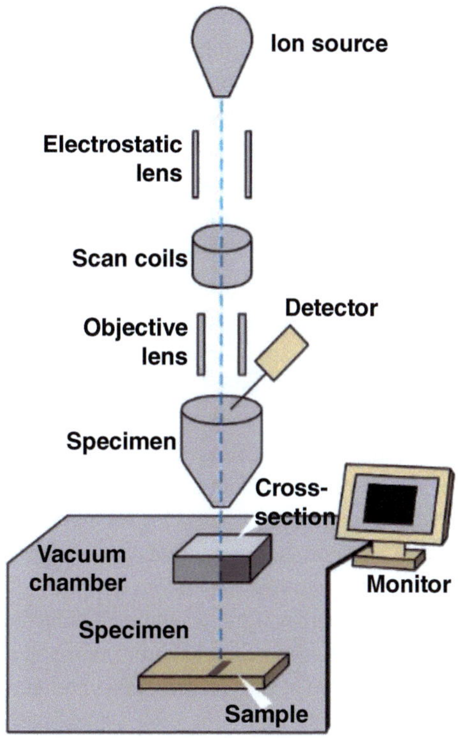

Figure 4.4 Schematic diagram illustrating the setup of a focused ion beam (FIB) system. The diagram highlights key components such as the ion source, electron column (in dual-beam systems), sample stage, detectors, and the milling/imaging process, emphasizing its use for cross-sectional imaging and nanoscale fabrication. Generated with AI using ChatGPT (version GPT-4).

datasets allow for quantitative analysis of shell thickness, uniformity, and porosity, which are critical for performance in applications like catalysis, drug delivery, and photonics. A schematic of the FIB system is shown in Figure 4.4, illustrating the key components such as the ion column, electron beam (in dual-beam configurations), and detectors that enable both milling and high-resolution imaging [10]. Beyond imaging, FIB also enables site-specific sample preparation for advanced TEM. It allows the extraction of thin lamellae from specific regions of interest in a nanostructure, preserving core-shell interfaces for high-resolution TEM or STEM analysis. This precision is particularly beneficial when dealing with heterogeneous or asymmetric core-shell architectures. However, it is important to consider some limitations of FIB. The ion beam can potentially induce damage or modify surface chemistry due to the high energy of ion bombardment. For sensitive materials, alternative ion sources such as helium or neon in Helium Ion Microscopy (HIM) systems, or cryogenic FIB techniques, may be employed to mitigate these effects.

In addition to FIB, 3D imaging techniques such as electron tomography (in TEM), X-ray computed tomography (XCT), and serial block-face SEM (SBF-SEM) offer complementary capabilities. Electron tomography enables nanometer-scale 3D reconstructions from tilt-series images, especially suited for smaller nanoparticles. XCT, on the other hand, provides non-destructive 3D imaging of larger structures and is particularly useful for assessing macroscopic distributions of core-shell materials in composites or coatings [11]. Altogether, FIB and 3D imaging techniques provide a comprehensive toolbox for structural elucidation of core-shell nanomaterials, bridging the gap between surface features and internal architecture. These methods are pivotal for correlating structure–function relationships and guiding the rational design of next-generation core-shell materials.

4.2 Spectroscopy Techniques

Spectroscopy plays a fundamental role in the comprehensive characterization of core-shell nanomaterials, providing valuable information about their optical, vibrational, chemical, and electronic properties. Unlike microscopy techniques that focus primarily on morphology and topography, spectroscopy delves into the intrinsic nature of materials—revealing how atoms and molecules interact with electromagnetic radiation. These interactions serve as fingerprints, enabling the identification of material composition, chemical bonding, structural integrity, and electronic states, all of which are crucial in assessing the performance and functionality of core-shell systems [12]. Core-shell nanomaterials often exhibit enhanced or altered spectroscopic signatures due to size-dependent quantum effects, shell-induced strain, or interfacial interactions between the core and shell layers. Understanding these spectral features helps researchers evaluate whether the shell uniformly covers the core, whether there is interdiffusion at the interface, or whether specific surface modifications have been successfully achieved. This makes spectroscopy a powerful tool for both qualitative and quantitative evaluation across various synthesis and application contexts.

The diversity of spectroscopic methods, ranging from ultraviolet-visible (UV-Vis) and photoluminescence (PL) spectroscopy to more advanced techniques like Raman, Fourier Transform Infrared (FTIR), X-ray Photoelectron Spectroscopy (XPS), and EDS allows researchers to probe different aspects of core-shell nanostructures. For instance, UV-Vis and PL can assess optical properties and bandgap behavior, while FTIR and Raman spectroscopy provide insights into bonding characteristics and surface functionalities [13]. XPS, on the other hand, is invaluable for surface elemental analysis and oxidation state determination, offering atomic-level insights into the shell composition and interfacial chemistry. In many applications, such as catalysis, sensing, biomedical imaging, and optoelectronics, the surface properties and electronic configuration of core-shell materials dictate their performance. Hence, spectroscopy techniques serve as indispensable tools for tailoring material properties and validating functional performance. Moreover, when coupled with microscopy or thermal analysis methods, spectroscopy provides a more holistic understanding of material behavior under various operational environments [14]. As spectroscopic instrumentation becomes increasingly sensitive and versatile, modern approaches now allow for nanoscale resolution, in situ monitoring, and even time-resolved measurements. These advancements are especially relevant for dynamic or reactive core-shell systems where property changes over time or under external stimuli need to be observed in real time. In the following subsections, various spectroscopy techniques commonly employed in core-shell nanomaterial characterization will be explored, emphasizing their principles, advantages, and application-specific relevance.

4.2.1 UV-Vis and PL Spectroscopy

Ultraviolet-Visible (UV-Vis) spectroscopy (Figure 4.5a) is a fundamental and versatile characterization technique extensively employed to investigate the optical properties of core-shell nanomaterials. This method measures the absorption of light across the ultraviolet and visible spectrum, providing valuable information about electronic transitions, bandgap energies, and surface plasmon resonance (SPR) effects. In the context of core-shell nanoparticles, UV-Vis spectroscopy is particularly insightful because the presence of a shell around a core can drastically influence these optical properties. For semiconductor core-shell nanostructures, UV-Vis spectra often show distinctive shifts in absorption edges compared to bare cores. These shifts are attributable to quantum confinement effects, shell-induced strain, or changes in electronic structure due to interfacial interactions between the core and shell materials [15]. For example, when a shell material with a larger bandgap encapsulates a semiconductor core, the absorption onset may shift to shorter wavelengths (blueshift), reflecting enhanced confinement. Conversely, in some cases, shell growth can lead to redshifts due to changes in core electronic states or the formation of interface states. This sensitivity makes UV-Vis spectroscopy a powerful tool for tracking shell formation, thickness variation, and compositional changes during synthesis. In metallic core-shell nanoparticles, UV-Vis spectroscopy primarily probes SPR, the collective oscillation of conduction electrons stimulated by incident light. The position, intensity, and shape of SPR peaks are strongly influenced

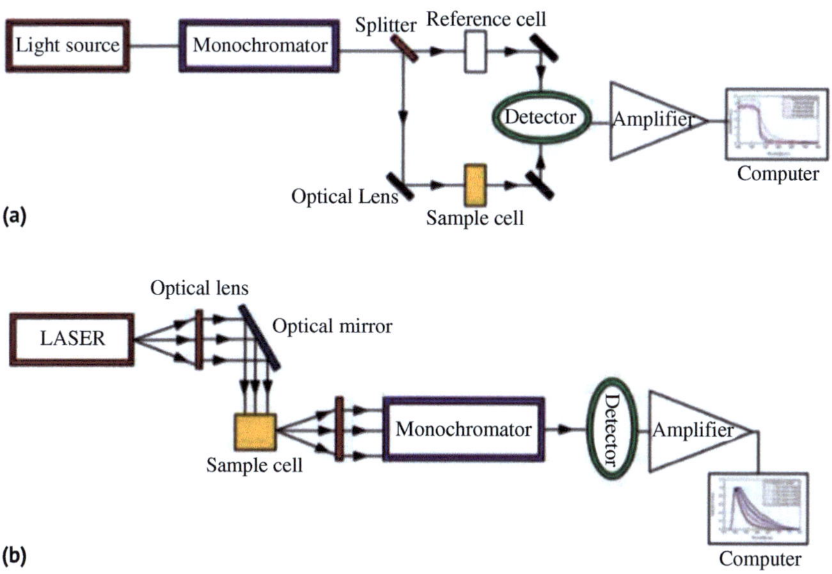

(a) (b)

Figure 4.5 Schematic representations of (a) a conventional UV-Vis absorption spectroscopy setup and (b) a laser-based UV-Vis spectroscopy configuration. Both systems include key optical components such as light sources, monochromators, detectors, and sample/reference cells, but the laser-based setup offers enhanced precision and intensity for high-resolution spectral analysis. [19]/MDPI/CC BY 4.0.

by the core size, shell thickness, shell composition, and the surrounding dielectric environment. For example, a gold core coated with a silica shell will exhibit a characteristic SPR peak that shifts depending on the shell thickness [16]. This tunability of plasmonic features enables UV-Vis spectroscopy to monitor the growth of metal shells with high precision and to tailor optical responses for applications in sensing, photothermal therapy, and plasmon-enhanced catalysis.

PL spectroscopy (Figure 4.5b) serves as a complementary technique to UV-Vis by providing insights into the emission behavior of core-shell nanomaterials upon photoexcitation. PL measures the wavelength and intensity of light emitted when excited electrons in the material relax back to their ground state, revealing information about electronic structure, defect states, and energy transfer mechanisms within the material. Core-shell quantum dots and semiconductor nanoparticles often demonstrate significantly enhanced PL intensity and stability relative to their core-only counterparts, mainly due to surface passivation effects imparted by the shell. The shell layer effectively reduces non-radiative recombination by isolating the core from surface defects and environmental quenchers, resulting in brighter and more stable luminescence. Additionally, shifts in PL emission wavelengths can indicate changes in core-shell interface quality, shell thickness, or doping levels. For example, increasing shell thickness often leads to redshifts in emission peaks due to strain or altered electronic coupling. Time-resolved photoluminescence (TRPL) techniques further provide dynamic insights by measuring carrier lifetimes and recombination dynamics, critical for understanding processes like charge separation, exciton migration, and non-radiative decay pathways [17]. These insights are particularly important in optimizing core-shell materials for optoelectronic devices such as LEDs, lasers, and solar cells.

One of the significant advantages of UV-Vis and PL spectroscopy lies in their non-destructive nature, rapid data acquisition, and minimal sample preparation requirements. These attributes allow for routine monitoring of synthesis progress and quality control, enabling researchers to fine-tune synthesis parameters in real time. However, it is important to consider potential confounding factors such as particle aggregation, polydispersity, and environmental effects (e.g., solvent polarity, temperature), which can affect optical measurements and complicate data interpretation. Together, UV-Vis and PL spectroscopy provide a comprehensive picture of both ground-state absorption and excited-state emission phenomena in core-shell nanomaterials [18]. Their combined application supports the rational design and engineering of core-shell architectures with tailored optical properties for a broad spectrum of applications, including photocatalysis, biological imaging, photovoltaics, and sensing technologies. By providing detailed insights into the interplay between core and shell components, these techniques enable optimization of shell thickness, composition, and interfacial quality to achieve enhanced performance in desired applications.

4.2.2 Raman Spectroscopy and Surface-enhanced Raman Spectroscopy

Raman spectroscopy is a powerful vibrational spectroscopic technique widely employed for the detailed structural and chemical characterization of core-shell nanomaterials. It relies on the inelastic scattering of monochromatic light, usually from a laser source, where incident photons interact with molecular vibrations or phonons in a material, resulting in energy shifts that are characteristic of specific chemical bonds and lattice structures. This technique is highly sensitive to molecular and crystallographic changes, making it ideal for probing the composition, crystallinity, phase, and strain effects within both the core and shell components of nanostructures. For core-shell materials, Raman spectroscopy provides crucial insights into the interfacial interactions between the core and shell, revealing changes in vibrational modes that arise from lattice mismatches, strain, or chemical bonding at the interface [20]. For instance, in semiconductor core-shell nanoparticles, the presence of a shell often modifies the phonon modes of the core material, resulting in peak shifts, broadening, or the emergence of new peaks in the Raman spectra. These spectral changes serve as fingerprints for the successful coating of the core and can help estimate shell thickness or detect alloying effects. Similarly, Raman spectroscopy can detect phase transitions or polymorphic forms in complex core-shell systems, providing guidance for tuning their physical and chemical properties. In metallic core-shell nanomaterials, Raman spectroscopy is equally valuable but often used in conjunction with Surface-Enhanced Raman Spectroscopy (SERS) to amplify weak vibrational signals. SERS is a specialized form of Raman spectroscopy that exploits the plasmonic properties of noble metal nanostructures such as gold or silver to enhance Raman signals by several orders of magnitude [21]. This enhancement arises from localized SPRs that increase the electromagnetic field near the nanoparticle surface, dramatically boosting the Raman scattering intensity of molecules adsorbed on or near the metallic shell.

The synergy between core-shell architectures and SERS is particularly notable. Core-shell nanoparticles with a plasmonic metal shell, such as gold-coated silica or silver-coated semiconductor cores, serve as highly effective SERS substrates. As shown in Figure 4.6, the schematic representation of a Raman spectrometer setup used for SERS includes a 785-nm laser, dichroic optical elements for directing the laser beam, and a sample substrate embedded with silver nanoparticles. These nanostructures can be engineered to optimize plasmonic resonance conditions by tuning shell thickness, morphology, and composition, thereby maximizing signal enhancement [22]. This capability enables ultrasensitive detection of trace amounts of analytes, including environmental pollutants, biomolecules, and chemical warfare agents, down to single-molecule levels in some cases. Additionally, the shell can provide chemical stability and biocompatibility to the plasmonic core, expanding the applicability of SERS substrates in biological and medical sensing. For example, biocompatible shells prevent nanoparticle aggregation and nonspecific binding, enhancing reproducibility and selectivity in complex biological environments. The tunability of core-shell structures also allows the design of multiplexed SERS sensors for simultaneous detection of multiple targets by using different Raman reporter molecules.

Raman and SERS techniques complement other characterization methods by offering non-destructive, label-free chemical specificity and molecular fingerprinting. They are invaluable for investigating not only the structural integrity and chemical composition of core-shell nanomaterials but also the dynamics of chemical reactions occurring at their surfaces or interfaces. In situ and time-resolved Raman spectroscopy can monitor catalytic reactions or phase changes, providing mechanistic insights crucial for optimizing core-shell catalysts and functional materials [23].

Despite their strengths, challenges in Raman and SERS characterization include fluorescence background interference, which can obscure weak Raman signals,

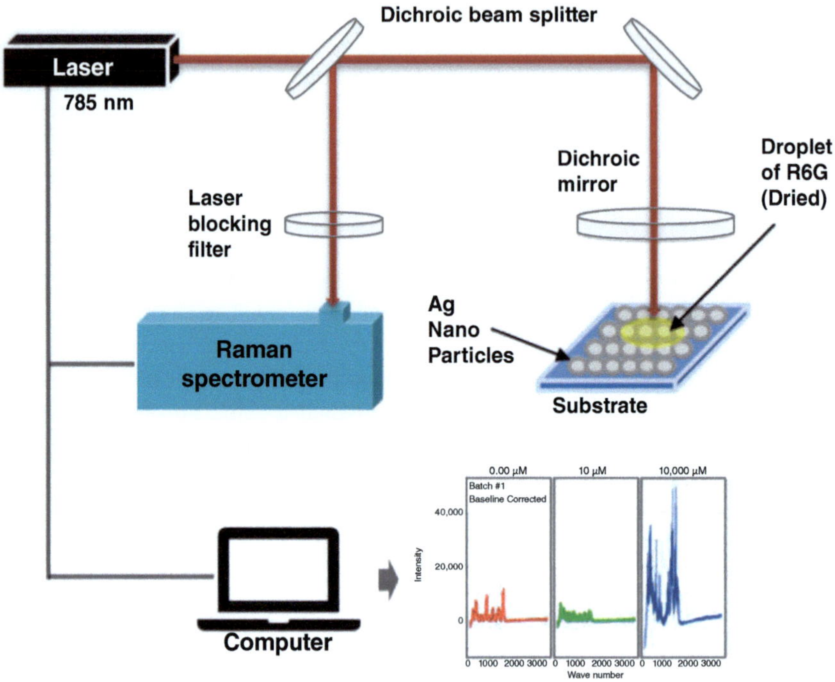

Figure 4.6 Schematic representation of a Raman spectrometer setup for Surface-enhanced Raman Scattering (SERS) measurements. The system includes a 785-nm laser source, dichroic beam splitter and mirror, a substrate containing silver nanoparticles with a dried droplet of R6G analyte, and a Raman spectrometer connected to a computer for data acquisition and spectral analysis. [24]/MDPI/CC BY 4.0.

and the need for careful substrate design to achieve reproducible enhancement. Advanced approaches such as tip-enhanced Raman spectroscopy (TERS) and combining Raman with complementary techniques like electron microscopy have been developed to overcome these limitations and provide nanoscale spatial resolution. Raman spectroscopy and SERS are indispensable tools in the characterization toolkit for core-shell nanomaterials. They provide detailed molecular and structural information that guides the design, synthesis, and application of core-shell systems in fields ranging from catalysis and energy conversion to biosensing and environmental monitoring. The ability to enhance and detect Raman signals with high sensitivity through tailored core-shell architectures opens new avenues for developing advanced functional nanomaterials with precise chemical control.

4.2.3 Fourier Transform Infrared Spectroscopy

Fourier Transform Infrared Spectroscopy (FTIR) is an essential analytical technique used to investigate the chemical composition and molecular structure of core-shell nanomaterials. It works by measuring the absorption of infrared radiation by the sample as a function of wavelength or frequency, revealing characteristic vibrational modes of chemical bonds within the material. Each functional

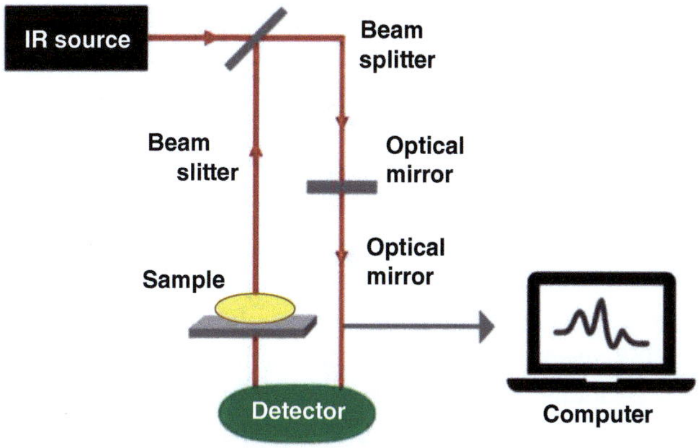

Figure 4.7 Schematic diagram of the Fourier Transform Infrared (FTIR) Spectroscopy setup, illustrating the path of infrared light through the interferometer, sample compartment, and detector for molecular vibration analysis. Generated with AI using ChatGPT (version GPT-4).

group and molecular bond absorbs IR radiation at specific frequencies, producing a unique spectral fingerprint that can be used to identify and quantify components in complex nanostructures [25], shown in Figure 4.7. In the context of core-shell nanomaterials, FTIR spectroscopy plays a crucial role in confirming the presence and chemical nature of both the core and shell materials. It helps identify surface functional groups, ligand molecules, and chemical bonds at the interface, which are often key to the material's stability, dispersion, and functionality. For instance, in polymer-coated metal nanoparticles, FTIR can verify the successful attachment of polymer chains by detecting characteristic peaks corresponding to C—H, C=O, or N—H stretching vibrations. Similarly, in oxide shell coatings on metal or semiconductor cores, FTIR can detect metal–oxygen bond vibrations that confirm shell formation and crystallinity.

FTIR is especially valuable for monitoring surface modifications and post-synthesis functionalization of core-shell nanomaterials. Surface chemistry plays a decisive role in determining the interaction of these materials with their environment, including biological systems, catalytic reactants, or adsorbates. FTIR enables the identification of surface-bound molecules, such as organic ligands, stabilizers, or biomolecules, which influence solubility, biocompatibility, and catalytic activity [26]. It also facilitates the detection of residual precursors or contaminants that may affect performance. Another key advantage of FTIR lies in its non-destructive nature and minimal sample preparation requirements, allowing the analysis of powders, thin films, and colloidal suspensions. When coupled with techniques such as attenuated total reflectance (ATR), FTIR can probe surface and near-surface regions with high sensitivity, making it ideal for investigating shell layers, thin coatings, or surface functional groups on core-shell nanomaterials. This capability is essential for assessing shell uniformity, thickness, and chemical integrity, which directly impact the material's optical, electronic, and catalytic properties. FTIR also supports the study of dynamic processes such as chemical reactions, adsorption/desorption phenomena, and environmental stability under various conditions. In situ FTIR measurements can monitor changes in surface chemistry during catalytic reactions or exposure to gases and liquids, providing real-time insights into reaction mechanisms and active site transformations [27]. This makes FTIR an indispensable tool for optimizing the design and performance of core-shell catalysts, sensors, and biomedical agents.

Despite its strengths, FTIR has limitations in spatial resolution compared to electron microscopy, which may hinder detailed mapping of core-shell heterogeneity at the nanoscale. However, recent advances like FTIR microscopy and nano-FTIR combine infrared spectroscopy with high-resolution imaging, enabling chemical mapping with nanoscale precision and expanding FTIR's applicability in nanomaterials research. Overall, FTIR spectroscopy offers comprehensive molecular-level insights into the chemical structure, surface chemistry, and functionalization of core-shell nanomaterials [28]. It complements other structural and elemental characterization techniques by providing detailed vibrational fingerprints critical for understanding and tailoring the interface and shell properties. Its versatility, sensitivity, and adaptability make FTIR an essential technique for advancing the synthesis, characterization, and application of core-shell nanomaterials across energy, environmental, and biomedical fields.

4.2.4 XPS for Surface Composition Analysis

X-ray Photoelectron Spectroscopy (XPS), also known as Electron Spectroscopy for Chemical Analysis (ESCA), is a highly sensitive surface analytical technique widely used to probe the elemental composition, chemical states, and electronic environment of atoms within the outermost 5–10 nm of core-shell nanomaterials. XPS provides quantitative information on the elemental makeup and the oxidation states of the elements present, making it an indispensable tool for understanding surface chemistry, especially at the core-shell interface [29]. In core-shell nanomaterials, surface and interface properties critically influence the overall behavior and performance, including catalytic activity, optical responses, stability, and biocompatibility. XPS helps elucidate these characteristics by detecting photoelectrons ejected from the sample surface upon irradiation with monochromatic X-rays. The kinetic energy of these emitted electrons is measured, providing binding energy values characteristic of specific elements and their chemical states. One of the key advantages of XPS is its ability to distinguish between different oxidation states and chemical environments of the same element. For instance, in a core-shell system comprising metal oxide shells on metallic cores, XPS can differentiate between metallic states, oxide forms, and hydroxide species by analyzing peak shifts and satellite features [2]. This capability enables precise determination of surface oxidation, shell composition, and the presence of contaminants or adsorbed species.

XPS also enables depth profiling through ion sputtering, where successive layers of the sample are etched away while spectra are recorded. This provides insights into the thickness, uniformity, and compositional gradients of the shell layer and the interface with the core. Depth profiling is particularly valuable in complex core-shell structures where chemical gradients or interdiffusion affect material properties. Moreover, XPS can characterize functional groups on the nanomaterial surface, including organic ligands, polymers, and biomolecules used for surface functionalization or stabilization [30]. This helps confirm successful surface modifications and evaluate the nature of chemical bonding, which is essential for tailoring surface reactivity and interaction with external environments, such as biological media or catalytic reactants. XPS measurements are generally non-destructive and require minimal sample preparation, allowing the analysis of powders, films, and even nanoparticles deposited on substrates. The technique is highly complementary to microscopy and other spectroscopic methods, providing a chemical context to structural and morphological information.

However, XPS also has some limitations. It requires high vacuum conditions, which may restrict the analysis of volatile or hydrated samples. The spatial resolution, typically in the micrometer range, is lower than that of electron microscopy, though recent developments in imaging XPS allow chemical mapping at sub-micron resolution. In recent years, advancements such as ambient pressure XPS (AP-XPS) have enabled the study of core-shell nanomaterials under near-realistic environmental conditions, providing insights into surface reactions, catalytic mechanisms, and material stability in operational environments. XPS serves as a powerful technique for detailed surface composition and chemical state analysis of core-shell nanomaterials. Its ability to probe elemental identity, oxidation state, and chemical bonding at the surface and interfaces makes it critical for understanding and optimizing the functionality of these advanced materials across applications in catalysis, energy storage, sensing, and biomedicine.

4.2.5 EDS for Elemental Mapping

EDS or EDX is a widely used analytical technique that provides qualitative and semi-quantitative elemental analysis of materials by detecting characteristic X-rays emitted from a specimen when it is bombarded with a high-energy electron beam. EDS is often integrated with electron microscopy techniques, such as SEM and TEM, allowing for localized chemical characterization and elemental mapping at micro to nanoscale resolutions. For core-shell nanomaterials, EDS is invaluable in determining the elemental distribution within the core and shell regions, confirming the presence and uniformity of each component, and revealing interdiffusion or contamination [31]. Since the physical and chemical properties of core-shell structures are closely linked to their compositional heterogeneity, EDS mapping provides crucial spatial information that supports understanding of structure–property relationships. The working principle of EDS involves the generation of characteristic X-rays unique to each element when incident electrons excite inner-shell electrons in atoms of the sample. These characteristic X-rays are collected by a semiconductor detector and analyzed to generate elemental spectra. The intensity of each peak corresponds to the concentration of that element, allowing for elemental identification and approximate quantification. One of the significant advantages of EDS lies in its ability to perform simultaneous multi-elemental analysis within localized regions, enabling rapid and efficient characterization of complex core-shell morphologies [32]. Elemental maps created from EDS data visually depict the spatial distribution of elements, highlighting whether the shell material fully encapsulates the core and revealing any compositional gradients or segregation phenomena.

When combined with high-resolution electron microscopy, EDS can achieve nanoscale elemental mapping, providing detailed insight into the shell thickness, uniformity, and the presence of dopants or secondary phases. This level of analysis is essential for optimizing synthesis protocols and ensuring reproducible nanomaterial performance. Moreover, EDS analysis is relatively straightforward and requires minimal sample preparation. It can be performed on powders, thin films, and nanoparticles mounted on substrates, facilitating broad applicability across research and industrial settings.

However, EDS has some limitations. The technique is generally less sensitive to light elements such as hydrogen, helium, and lithium due to their low X-ray emission energies. Additionally, the spatial resolution of EDS depends on the electron beam interaction volume, which can be larger than the size of individual nanoparticles, potentially causing signal overlap or averaging effects in densely packed samples. Recent advancements, including Silicon Drift Detectors (SDDs) and improved software algorithms, have enhanced sensitivity, resolution, and data analysis capabilities [33]. In addition to elemental mapping, EDS can be used in line scan mode to track compositional changes across the core-shell interface and in point analysis mode to determine elemental ratios at specific locations. These complementary approaches provide comprehensive chemical characterization critical for elucidating material behavior.

EDS stands as a vital tool for elemental identification and spatial distribution analysis in core-shell nanomaterials. Its integration with electron microscopy techniques offers a powerful combination for correlating structural and chemical information, thereby advancing the rational design and application of core-shell nanostructures in catalysis, energy storage, sensing, and biomedical fields.

4.3 Thermal and Mechanical Analysis

Thermal and mechanical properties are critical aspects that influence the performance, stability, and durability of core-shell nanomaterials across diverse applications. Understanding how these nanostructures respond to temperature variations, mechanical stress, and environmental conditions is essential for tailoring materials to meet specific functional requirements, especially in fields such as energy storage, catalysis, flexible electronics, and biomedicine [34]. Thermal analysis techniques provide insight into the stability, phase transitions, decomposition behavior, and thermal conductivity of core-shell materials. These properties not only impact the synthesis and processing routes but also affect the operational lifespan and safety of nanomaterials under real-world conditions. For instance, knowledge of thermal degradation temperatures helps in optimizing annealing protocols, while phase transition studies reveal structural transformations that can alter functionality.

Mechanical characterization is equally important, as the core-shell architecture often enhances mechanical strength, flexibility, or hardness compared to individual components. Evaluating parameters such as hardness, elastic modulus, and viscoelastic behavior allows researchers to understand the load-bearing capacity and deformation mechanisms of these nanomaterials. Such information is critical when core-shell structures are incorporated into composites, coatings, or devices subjected to mechanical stress [34]. The interplay between thermal and mechanical responses can further elucidate material performance under cyclic thermal loading, impact resistance, and fatigue behavior. These factors ultimately dictate reliability and endurance in practical applications, particularly in harsh or dynamic environments.

This section will explore the primary thermal and mechanical characterization techniques employed for core-shell nanomaterials. These methods help bridge the gap between nanoscale structural features and macroscale functional properties, thereby guiding the rational design and optimization of next-generation materials.

4.3.1 Thermogravimetric Analysis (TGA) for Stability Testing

TGA is a fundamental technique widely used to evaluate the thermal stability and compositional integrity of core-shell nanomaterials. By measuring the change in sample mass as a function of temperature or time under controlled atmospheres (such as nitrogen, air, or oxygen), TGA provides valuable insights into decomposition temperatures, oxidation behavior, moisture content, and the presence of volatile components. In core-shell structures, TGA is particularly useful for assessing the stability of both core and shell materials, as well as their interactions. For example, TGA can reveal the temperature range over which the shell material protects the core from oxidation or degradation, indicating the effectiveness of the core-shell encapsulation. It also helps determine the thermal decomposition pathways of organic or polymeric shells, which is critical for applications requiring thermal endurance, such as catalysis or electronics [35]. Typical TGA experiments involve heating the sample at a controlled rate (commonly 5–20 °C/min) while continuously recording the mass loss. Distinct mass loss steps can be correlated with specific processes, such as desorption of surface adsorbates, degradation of polymer shells, or oxidation of metallic cores. Coupling TGA with derivative thermogravimetry (DTG) further enhances the ability to pinpoint exact temperatures where major weight changes occur, enabling precise analysis of complex multi-component core-shell systems. Moreover, TGA can be combined with complementary techniques like Differential Scanning Calorimetry (DSC) or mass spectrometry (TGA-MS) to gain deeper understanding of thermal events and evolved gases during decomposition. Such multi-modal analysis is invaluable for designing core-shell materials with tailored thermal properties and improved long-term stability. Overall, TGA serves as a rapid, reliable, and quantitative tool to characterize the thermal robustness of core-shell nanomaterials, guiding synthesis optimization and ensuring suitability for intended applications that involve thermal cycling or exposure to elevated temperatures.

4.3.2 DSC for Phase Transition Studies

DSC is a powerful thermal analysis technique extensively employed to investigate phase transitions, crystallinity, and heat capacity changes in core-shell nanomaterials. By measuring the heat flow into or out of a sample relative to an inert reference as a function of temperature or time, DSC reveals critical information about melting points, crystallization, glass transitions, and other thermal events intrinsic to both core and shell components. In the context of core-shell nanostructures, DSC plays a pivotal role in understanding how the shell influences the thermal behavior and phase stability of the core material, and vice versa. For example, the presence of a shell can alter the melting or crystallization temperatures of the core due to strain effects, interfacial interactions, or nanoscale confinement [36]. Such changes are critical for tailoring material properties in applications ranging from drug delivery, where controlled release depends on thermal transitions, to catalysis, where phase stability affects activity. DSC analysis typically involves heating and cooling cycles at controlled rates (e.g., 5–20 °C/min), which allow for the detection of endothermic and exothermic events. These events correspond to phase changes such as melting, solid-solid transitions, or polymer relaxation phenomena. Figure 4.8, DSC setup highlighting the positions of the reference and sample pans used to measure thermal transitions and heat flow differences.

The peak temperatures, enthalpy changes, and onset points derived from DSC thermograms provide quantitative measures of these transitions and their reversibility. Moreover, DSC helps evaluate the degree of crystallinity of shell materials, particularly in polymer-coated core-shell nanoparticles, which directly influences mechanical strength, permeability, and thermal resistance. Understanding these thermal transitions through DSC enables optimization of synthesis parameters and selection of shell materials to achieve desired thermal and functional properties. When combined with other characterization methods such as TGA or X-ray diffraction (XRD), DSC offers comprehensive insights into the thermal stability and structural evolution of core-shell nanomaterials under varying conditions [2]. This integrative approach is essential for developing robust nanostructures with enhanced performance in diverse fields, including electronics, energy storage, and biomedicine.

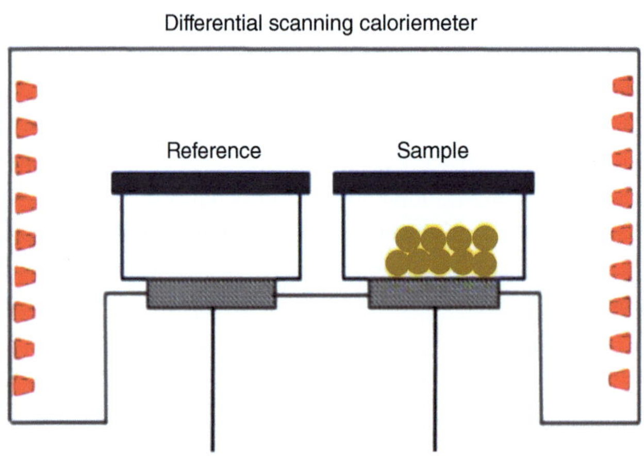

Figure 4.8 Differential Scanning Calorimetry (DSC) setup highlighting the positions of the reference and sample pans used to measure thermal transitions and heat flow differences. Generated with AI using ChatGPT (version GPT-4).

4.3.3 Nanoindentation for Mechanical Property Evaluation

Nanoindentation is a precision technique used to evaluate the mechanical properties of core-shell nanomaterials at the nanoscale, including hardness, elastic modulus, and viscoelastic behavior. This method involves pressing a sharp indenter—typically made of diamond—into the surface of a material with controlled force while continuously measuring the depth of penetration. The resulting load-displacement data provide direct insight into the material's resistance to deformation and its elastic recovery. For core-shell nanostructures, nanoindentation is particularly valuable because it enables localized probing of mechanical characteristics of both the core and shell components independently or as a composite system [37]. The shell, often designed to provide mechanical protection or functional robustness, can be characterized for its ability to enhance hardness or stiffness relative to the core. Likewise, the technique can reveal how interfacial properties and core-shell bonding affect the overall mechanical performance.

One of the key advantages of nanoindentation is its ability to assess thin films and coatings on substrates, which is directly relevant for many core-shell systems where the shell layer thickness may be just a few nanometers to micrometers. By varying indentation depth, researchers can isolate the shell's mechanical response or probe deeper to include the core's influence, allowing detailed mapping of mechanical gradients within the nanostructure [38]. Beyond hardness and modulus, advanced nanoindentation techniques enable measurement of time-dependent mechanical properties such as creep and stress relaxation, critical for understanding durability and performance under cyclic or prolonged loading. This is essential for applications in flexible electronics, biomedical implants, and protective coatings, where mechanical integrity over time is paramount. Nanoindentation data often complement other characterization tools such as electron microscopy, which provides morphological context to mechanical measurements. Together, these techniques offer a comprehensive view of how synthesis methods, material composition, and nanoscale architecture influence mechanical behavior in core-shell nanomaterials.

The nanoindentation stands as an indispensable technique for the quantitative evaluation of mechanical properties in core-shell nanostructures, guiding the design of materials with tailored mechanical resilience suited for advanced technological applications.

4.3.4 Dynamic Mechanical Analysis and Stress–Strain Behavior

Dynamic Mechanical Analysis (DMA) is a powerful technique used to probe the viscoelastic properties of core-shell nanomaterials by applying an oscillatory force and measuring the resulting material response. DMA provides valuable information on the storage modulus (elastic behavior), loss modulus (viscous behavior), and damping characteristics (tan delta), which reflect how a material stores and dissipates energy under dynamic loading conditions. This is particularly crucial for core-shell materials that often experience varying mechanical stresses in real-world applications such as flexible electronics, sensors, and biomedical devices. In DMA, samples are subjected to oscillatory deformation over a range of frequencies and temperatures, allowing the study of phase transitions, relaxation processes, and mechanical stability. For core-shell nanostructures, this technique can distinguish the contributions of the core and shell materials to the overall mechanical response [39]. For example, a stiff shell surrounding a softer core might increase the composite's storage modulus while affecting energy dissipation differently depending on temperature or strain rate.

Stress–strain behavior, typically evaluated through tensile or compressive testing, complements DMA by providing direct measurements of material strength, elasticity, and failure modes. In core-shell systems, stress–strain curves reveal important characteristics such as yield strength, ultimate tensile strength, elongation at break, and toughness. These parameters help in understanding how the shell enhances mechanical reinforcement, mitigates crack propagation, or accommodates deformation through mechanisms like strain hardening or energy absorption. The integration of DMA and stress–strain analysis is essential for designing core-shell nanomaterials with optimized mechanical performance. While DMA offers insights into the time-dependent and temperature-dependent mechanical properties, stress–strain testing highlights the static mechanical limits and deformation mechanisms [40]. Together, these approaches inform the selection of material combinations, shell thicknesses, and synthesis conditions to tailor mechanical behavior for specific applications.

Additionally, DMA can be employed to investigate the interface quality between the core and shell, as poor interfacial bonding often manifests as reduced mechanical integrity or abnormal damping behavior. By tuning the core-shell interface, materials scientists can engineer nanostructures that better withstand mechanical fatigue and exhibit enhanced durability. DMA and stress–strain testing are complementary techniques that provide a comprehensive understanding of the mechanical performance of core-shell nanomaterials, supporting their advancement in applications demanding both resilience and functional adaptability.

4.4 Real-time Monitoring and In Situ Characterization

Real-time monitoring and in situ characterization techniques have revolutionized the study of core-shell nanomaterials by enabling direct observation of their dynamic behavior under operational or environmental conditions. Unlike traditional ex situ methods, which analyze samples before or after a process, in situ techniques provide continuous, time-resolved data that capture structural, chemical, and morphological changes as they happen. This capability is crucial for understanding reaction mechanisms, phase transformations, interfacial phenomena, and degradation pathways that govern the performance and stability of core-shell nanostructures. In situ characterization methods can simulate realistic environments such as varying temperatures, pressures, chemical atmospheres, or applied fields (electrical, mechanical, magnetic), allowing researchers to study how core-shell materials respond in real-world settings [41]. These techniques help unveil transient states and intermediate species that are often elusive with conventional approaches, thus offering deeper insight into nucleation, growth, shell formation, and functional responses.

The integration of real-time monitoring tools with advanced microscopy and spectroscopy enables a multi-modal approach that combines structural imaging with chemical and electronic state analysis. This holistic understanding is vital for optimizing synthesis protocols, enhancing material properties, and tailoring core-shell architectures for targeted applications in catalysis, energy storage, sensing, and biomedicine. Moreover, real-time and in situ characterization facilitate feedback-controlled synthesis and processing, paving the way for adaptive manufacturing methods that improve reproducibility and quality. As instrumentation and analytical software advance, these approaches increasingly incorporate automation and data analytics, further accelerating material discovery and deployment.

In the following subsections, key in situ and real-time techniques such as in situ TEM, operando microscopy, spectroscopy methods, surface and interface monitoring, and environmental/live-cell imaging are discussed to highlight their applications, advantages, and challenges in the field of core-shell nanomaterials.

4.4.1 In Situ TEM and Operando Microscopy for Dynamic Analysis

TEM has long been a cornerstone technique for high-resolution imaging of nanoscale materials. The advent of in situ TEM has significantly expanded its capabilities by enabling direct observation of dynamic processes within core-shell nanomaterials under controlled environmental conditions. In situ TEM allows researchers to monitor real-time morphological, structural, and compositional changes with atomic or near-atomic resolution while applying stimuli such as heating, electrical biasing, mechanical stress, or gas/liquid environments [42]. This dynamic imaging capability is particularly valuable for studying the growth mechanisms of the shell over the core, phase transformations, crystallization processes, and defect evolution in core-shell architectures. For example, in situ heating stages within TEM enable the visualization of thermal stability, sintering, or shell crystallization kinetics, revealing insights that static images cannot capture [43]. Similarly, applying electrical bias inside the TEM allows exploration of electrochemical reactions, such as those occurring in battery electrode materials or catalytic systems.

Complementing in situ TEM, operando microscopy takes the analysis further by enabling observation during actual device operation or chemical reactions, bridging the gap between model systems and functional applications. Operando approaches capture the behavior of core-shell nanomaterials under realistic working conditions, such as gas exposure in catalytic converters or charge/discharge cycles in energy storage devices. This facilitates correlation between nanoscale structural changes and macroscopic performance metrics. Recent advances in detector technology, environmental cells, and sample holders have broadened the scope of in situ and operando TEM to include liquid-phase TEM, gas environments, and mechanical testing inside the microscope. This versatility enables investigation of a wide range of phenomena relevant to core-shell systems, including corrosion, ion diffusion, and shell degradation.

Despite challenges such as beam damage and complex data interpretation, in situ TEM and operando microscopy remain indispensable for elucidating the fundamental mechanisms that govern the synthesis, functionality, and durability of core-shell nanomaterials. Integration with complementary techniques, such as spectroscopy, further enriches the insights gained from these powerful real-time characterization tools.

4.4.2 In Situ Spectroscopy for Reaction Mechanism Studies

In situ spectroscopy techniques have emerged as powerful tools for probing the reaction mechanisms and dynamic chemical processes occurring in core-shell nanomaterials under operational conditions. Unlike ex situ methods, which analyze samples before or after reactions, in situ spectroscopy allows continuous monitoring of structural, electronic, and chemical changes as they happen, providing real-time insights into reaction pathways, intermediate species, and kinetics. Several spectroscopic methods are widely employed for in situ studies of core-shell materials. For instance, in situ Raman spectroscopy enables tracking of vibrational modes related to chemical bonds, phase changes, or surface adsorbates during catalytic reactions or material transformations [44]. The high sensitivity of Raman to molecular structure and its compatibility with various environments (gas, liquid, solid) make it ideal for monitoring surface chemistry on core-shell catalysts and sensors.

Similarly, in situ FTIR spectroscopy provides detailed information on surface species and reaction intermediates by measuring characteristic infrared absorption bands. This technique is instrumental in elucidating mechanisms in processes such as photocatalysis, hydrogenation, and oxidation reactions facilitated by core-shell nanostructures [45]. X-ray-based in situ techniques, including X-ray Absorption Spectroscopy (XAS) and XPS, offer valuable insights into oxidation states, electronic structures, and elemental compositions during reactions. When combined with synchrotron radiation sources, these methods achieve high temporal and spatial resolution, enabling detailed mapping of changes in core and shell components.

The application of in situ spectroscopy extends beyond catalysis to energy storage, sensing, and environmental remediation, where understanding real-time material behavior under working conditions is critical. These techniques help decipher how shell modifications influence reaction selectivity, stability, and overall performance of core-shell materials. Challenges such as maintaining sample integrity, controlling environmental variables, and interpreting complex spectral data remain, but advancements in instrumentation and data analysis methods are continuously improving the reliability and applicability of in situ spectroscopy. Overall, in situ spectroscopic methods are indispensable for unraveling the intricate reaction mechanisms in core-shell nanomaterials, enabling rational design and optimization for enhanced functionality in diverse applications.

4.4.3 Real-time Surface and Interface Monitoring Techniques

The surface and interface properties of core-shell nanomaterials play a pivotal role in determining their overall performance across applications such as catalysis, sensing, energy storage, and biomedicine. Real-time monitoring of these surfaces and interfaces during synthesis, processing, or operational use provides critical insights into dynamic changes that affect stability, reactivity, and functional behavior. Various advanced techniques have been developed to observe these phenomena with high spatial and temporal resolution. One key approach involves real-time monitoring through surface-sensitive spectroscopies, such as in situ XPS and sum-frequency generation (SFG) spectroscopy [46]. These techniques provide detailed chemical and electronic information at interfaces, enabling the observation of adsorption/desorption processes, surface reconstruction, and chemical modifications as they occur. Another powerful technique is real-time ellipsometry, which measures changes in film thickness and refractive index at the surface, often applied during layer-by-layer growth or coating processes in core-shell nanomaterial fabrication. This optical method provides continuous feedback on shell formation dynamics and interface quality, essential for precise control over shell thickness and uniformity.

Quartz crystal microbalance (QCM) technology is widely employed for sensitive mass measurement of adsorbed species on surfaces, enabling detection of nanoscale deposition or etching events in real time [47]. QCM studies help unravel interaction kinetics at the core-shell interface, which are crucial for optimizing catalytic or sensor functionalities. Additionally, advanced microscopy techniques such as real-time scanning probe microscopy (SPM) enable nanoscale imaging of surface morphology and topography changes during reaction or processing. When combined with conductive atomic force microscopy (c-AFM) or KPFM, these methods reveal local electrical properties and charge distribution variations at the core-shell interface [48]. Environmental and liquid cell TEM also facilitates direct visualization of surface and interface phenomena in near-native conditions, capturing growth dynamics, morphological transformations, and interfacial reactions with atomic resolution.

Despite the immense potential, challenges remain in achieving simultaneous high temporal and spatial resolution, minimizing beam-induced damage, and interpreting complex multi-modal data. Integration of real-time monitoring tools with machine learning algorithms is an emerging trend to address these limitations, enabling automated pattern recognition and predictive analysis of interface dynamics.

4.4.4 Environmental and Live-cell Imaging for Bioapplications

The use of core-shell nanomaterials in biomedical and environmental applications demands thorough understanding of their interactions with living systems and their behavior under physiological or environmental conditions. Environmental and live-cell imaging techniques have emerged as essential tools to visualize and analyze these interactions in real time, providing crucial information about biocompatibility, cellular uptake, intracellular trafficking, and nanomaterial stability in complex biological milieus. Live-cell imaging employs advanced fluorescence microscopy methods such as confocal laser scanning microscopy (CLSM), two-photon microscopy, and spinning disk microscopy to monitor core-shell nanoparticles within living cells with high spatial and temporal resolution [49]. These techniques allow researchers to track nanoparticle localization, distribution, and dynamics, revealing insights into cellular internalization pathways, endosomal escape, and intracellular fate. Fluorescent tagging of core-shell materials, combined with organelle-specific dyes or genetically encoded markers, further enhances the ability to decipher intricate cell-nanomaterial interactions. Environmental imaging techniques are tailored to study core-shell nanomaterial behavior in natural and engineered ecosystems. Techniques such as hyperspectral imaging and SEM equipped with environmental chambers enable observation of nanoparticles in aqueous environments, soil matrices, or atmospheric conditions without significant sample alteration. These approaches are vital for assessing nanomaterial stability, aggregation, and potential toxicity in real-world settings.

Cryogenic electron microscopy (cryo-EM) and super-resolution fluorescence microscopy provide nanoscale structural details of core-shell nanoparticles within biological systems, preserving their native hydrated state and avoiding artifacts associated with conventional sample preparation. This high-resolution imaging is crucial for understanding the interactions at the bio-nano interface and for guiding the design of safer and more effective nanomaterials. Additionally, integrating live-cell imaging with functional assays such as reactive oxygen species (ROS) detection, mitochondrial activity measurement, and apoptosis markers enables a comprehensive evaluation of the biological impact of core-shell nanomaterials [50]. Such multi-modal approaches help optimize nanomaterial properties for applications like targeted drug delivery, imaging contrast enhancement, biosensing, and environmental remediation.

However, challenges persist, including phototoxicity, limited penetration depth, and potential interference of imaging probes with biological functions. Continuous advancements in imaging probes, instrumentation, and computational analysis are addressing these limitations, paving the way for more accurate and informative live-cell and environmental imaging studies. The environmental and live-cell imaging techniques are indispensable for elucidating the complex behavior of core-shell nanomaterials in biological and ecological contexts. These insights accelerate the development of nanomaterials with enhanced efficacy, safety, and environmental compatibility for diverse bioapplications.

4.5 Advanced Characterization Techniques

As core-shell nanomaterials continue to evolve in complexity and functionality, conventional characterization methods sometimes fall short in providing the detailed insights required for understanding their intricate structures, compositions, and dynamic behaviors. Advanced characterization techniques have thus become indispensable tools in the study of these materials, offering unparalleled resolution, sensitivity, and multidimensional analysis capabilities. These cutting-edge methods enable researchers to probe the nanoscale architecture of core-shell structures with atomic or near-atomic precision, revealing not only the morphology and elemental distribution but also electronic states, magnetic properties, and chemical bonding environments. The ability to analyze materials under operando or in situ conditions further provides critical understanding of how core-shell nanomaterials respond to external stimuli such as temperature, pressure, electromagnetic fields, or chemical environments. Additionally, advanced techniques harnessing synchrotron radiation, neutron scattering, and cryo-EM have opened new avenues for investigating materials in their native states or during real-time transformations, thereby offering insights into stability, phase transitions, and reaction mechanisms that are otherwise inaccessible. With the growing demand for multifunctional and high-performance nanomaterials in energy, catalysis, medicine, and environmental applications, these advanced characterization tools are essential for correlating structure–property relationships and guiding rational design and synthesis. They also play a pivotal role in quality control and standardization, crucial for scaling up production and ensuring reproducibility. The following subsections will explore several key advanced characterization methods that have significantly advanced the frontiers of core-shell nanomaterial research, highlighting their principles, capabilities, and specific applications.

4.5.1 Synchrotron-based X-ray Techniques for Nanoscale Analysis

Synchrotron-based X-ray techniques have revolutionized the characterization of core-shell nanomaterials by providing powerful, high-resolution analytical capabilities that are difficult to achieve with conventional laboratory equipment. These methods exploit the intense, tunable, and highly collimated X-ray beams generated by synchrotron radiation facilities, enabling detailed investigation of structural, chemical, and electronic properties at the nanoscale. One of the primary advantages of synchrotron X-rays is their ability to penetrate deeply into materials, allowing non-destructive probing of core-shell architectures in three dimensions. Techniques such as XAS, including X-ray absorption near edge structure (XANES) and extended X-ray absorption fine structure (EXAFS), offer element-specific insights into oxidation states, coordination environments, and local atomic structures within both core and shell components [51]. These details are critical for understanding catalytic activity, charge transfer mechanisms, and other functional properties. Synchrotron-based XRD methods provide high angular resolution, enabling the detection of subtle phase changes, crystallinity, and strain within core-shell nanoparticles. This is particularly useful for assessing how shell growth affects the crystal structure of the core and vice versa. Additionally, advanced imaging modalities such as X-ray microscopy and tomography leverage synchrotron sources to generate 3D reconstructions of nanomaterial morphology and elemental distributions with nanoscale resolution. These capabilities facilitate direct visualization of shell thickness uniformity, porosity, and interfacial characteristics.

Time-resolved synchrotron techniques further enable operando studies, where the evolution of core-shell structures can be monitored in real-time under working conditions such as temperature changes, reactive gas environments, or applied electric fields. This dynamic insight is invaluable for optimizing synthesis protocols and enhancing material performance in practical applications. While access to synchrotron facilities is limited and requires careful experimental planning, the unique and comprehensive data they provide make them indispensable for advancing core-shell nanomaterial research. Their integration with complementary characterization methods continues to push the boundaries of nanoscale analysis and deepen our understanding of these complex materials. Figure 4.9 shows the flowchart illustrating advanced characterization techniques employed for core-shell nanomaterials.

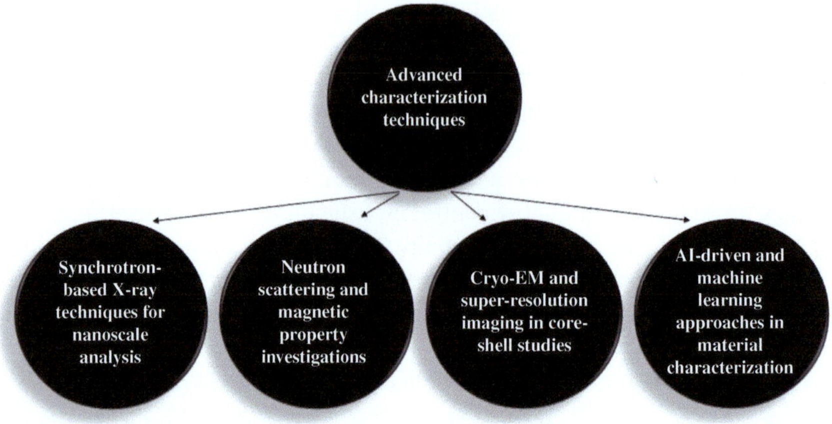

Figure 4.9 Flowchart illustrating advanced characterization techniques employed for core-shell nanomaterials. The chart categorizes key methods into four main domains. Generated with AI using ChatGPT (version GPT-4).

4.5.2 Neutron Scattering and Magnetic Property Investigations

Neutron scattering techniques provide a complementary and powerful approach to characterizing core-shell nanomaterials, particularly for probing atomic arrangements, magnetic structures, and dynamics that are often inaccessible by other methods. Unlike X-rays, neutrons interact primarily with atomic nuclei and are sensitive to light elements such as hydrogen, making them especially valuable for studying organic or polymer-based shells and interfaces in core-shell systems. Small-angle neutron scattering (SANS) is extensively utilized to investigate size, shape, and spatial distribution of nanoparticles within a matrix, as well as to resolve core-shell morphologies with high precision [52]. Its sensitivity to isotopic substitution also allows for contrast variation, enabling selective highlighting of the core or shell components by deuteration or isotope labeling. This technique is critical for understanding how shell thickness and uniformity influence the overall nanostructure. Neutron diffraction provides insight into crystallographic phases and magnetic ordering within core-shell materials, facilitating studies of complex magnetic interactions at the core-shell interface. This is particularly important for magnetic nanoparticles used in data storage, spintronics, or biomedical applications such as targeted drug delivery and magnetic hyperthermia.

Complementing scattering methods, neutron reflectometry offers detailed profiling of thin films and layered structures, yielding depth-resolved information on composition, density, and magnetic properties at interfaces, which is vital for engineering tailored core-shell architectures with optimized magnetic behavior. Magnetic property investigations commonly incorporate techniques such as vibrating sample magnetometry (VSM), superconducting quantum interference device (SQUID) magnetometry, and Mössbauer spectroscopy to analyze magnetization, coercivity, and magnetic anisotropy in core-shell nanoparticles [53]. These studies elucidate how shell materials influence magnetic coupling, surface spin disorder, and overall magnetic performance. The combined application of neutron scattering and magnetic measurements provides a comprehensive understanding of the interplay between structure and magnetism in core-shell nanomaterials. This insight drives the design of novel multifunctional nanostructures with controlled magnetic responses tailored for advanced technological applications.

4.5.3 Cryo-EM and Super-resolution Imaging in Core-shell Studies

Cryo-EM has emerged as a revolutionary imaging technique that enables the visualization of nanostructures in their native hydrated state without the need for staining or drying. This feature makes cryo-EM particularly valuable for studying soft or hybrid core-shell nanomaterials, including those composed of organic–inorganic interfaces, biomolecules, or polymeric shells [54]. By rapidly freezing the sample in vitreous ice, cryo-EM preserves the morphology and fine structural details of nanomaterials, allowing for near-atomic resolution imaging under cryogenic conditions. In core-shell nanomaterials, cryo-EM facilitates the direct visualization of shell integrity, thickness uniformity, and structural defects, especially in delicate or labile systems that are prone to damage under conventional TEM conditions. Moreover, with the advent of single-particle analysis and tomographic reconstruction techniques, cryo-EM enables 3D imaging of individual core-shell nanoparticles, revealing internal structural arrangements and heterogeneities that are crucial for understanding material performance. In parallel, super-resolution imaging techniques such as stimulated emission depletion (STED) microscopy, photoactivated localization microscopy (PALM), and stochastic optical reconstruction microscopy (STORM) have significantly surpassed the diffraction limit of traditional optical microscopy [55]. These methods allow for nanometer-scale visualization of surface features, functional groups, and interactions at the shell surface, particularly in biological and environmental systems where labeling is possible.

Super-resolution fluorescence imaging is increasingly employed in the study of core-shell nanoparticles for biomedical applications, such as drug delivery and diagnostics. These techniques offer spatial resolution down to ~10–20 nm, which is sufficient to resolve shell coatings, functional moieties, and receptor interactions in living cells and tissues. Moreover, real-time dynamic studies of labeled core-shell structures are feasible, providing insight into cellular uptake, trafficking, and interaction mechanisms in biological environments. Together, cryo-EM and super-resolution imaging bridge the gap between high-resolution structural imaging and real-time functional analysis. They enable researchers to study core-shell nanomaterials with unprecedented detail and contextual relevance, offering valuable perspectives for rational material design and application in complex systems such as biomedicine, catalysis, and soft-matter nanotechnology.

4.5.4 AI-driven and Machine Learning Approaches in Material Characterization

The integration of AI and machine learning (ML) into material characterization represents a transformative advancement in the field of nanomaterials, particularly in the analysis of complex core-shell structures. These computational techniques offer powerful tools to automate data interpretation, uncover hidden patterns, and accelerate the discovery of structure–property relationships that are often inaccessible through conventional analytical methods. In microscopy and spectroscopy, AI and ML algorithms are being increasingly applied to enhance image resolution, reduce noise, and segment complex features in high-throughput datasets. For example, convolutional neural networks (CNNs) can identify and classify core-shell morphologies, detect shell thickness variations, and quantify defects with high precision from TEM, SEM, or AFM images. Similarly, spectral data from techniques such as Raman, FTIR, or XPS can be rapidly deconvoluted using unsupervised learning methods like principal component analysis (PCA) or clustering algorithms, revealing compositional gradients, bonding variations, or phase transitions. ML is also driving the development of predictive models that correlate synthesis parameters with final material properties. By training algorithms on large experimental or simulated datasets, researchers can predict the outcomes of synthesis routes, optimize processing conditions, and guide the design of new core-shell architectures with tailored functionalities. Reinforcement learning and Bayesian optimization techniques are being used to automate experimental workflows, enabling real-time feedback and adaptive experimentation, thereby minimizing trial-and-error efforts.

Moreover, AI models are proving instrumental in in situ and operando studies, where dynamic datasets are often complex and multidimensional. ML tools can analyze time-resolved spectra or imaging sequences to identify transient phases, reaction intermediates, or structural transformations in core-shell systems under varying environmental conditions. These insights are invaluable for catalytic, biomedical, and energy-related applications, where understanding material behavior in real time is critical. The synergy between AI and materials characterization also extends to data management and knowledge discovery. AI-powered platforms can curate and mine large repositories of experimental data and publications, extract relevant insights, and assist in literature-based hypothesis generation, thus facilitating informed experimental design and accelerating innovation. As the field evolves, the incorporation of explainable AI (XAI) and physics-informed ML models will further enhance the interpretability and reliability of computational predictions. Together, AI and ML are reshaping how core-shell nanomaterials are characterized, understood, and engineered—ushering in a new era of intelligent materials research.

4.6 Conclusion

The characterization of core-shell nanomaterials is an essential and multifaceted process that underpins their successful design, development, and application. As this chapter has demonstrated, a comprehensive suite of techniques—ranging from classical microscopy and spectroscopy to cutting-edge in situ and AI-driven approaches—plays a pivotal role in unveiling the structural, compositional, thermal, mechanical, and dynamic attributes of these complex nanostructures. Each method offers unique insights, and when used synergistically, they provide a holistic understanding of the core-shell architecture from nanoscale morphology to interfacial behavior and functional performance. With the increasing sophistication of core-shell nanomaterials tailored for energy, environmental, and biomedical applications, the demand for more precise, real-time, and high-throughput characterization tools continues to grow. Emerging trends such as ML-guided analysis, operando techniques, and integration of multimodal data are not only enhancing the depth of characterization but also accelerating materials discovery and optimization. Looking forward, the convergence of experimental techniques with computational intelligence will drive the next phase of innovation in nanomaterials characterization. This integrated approach will be instrumental in addressing challenges related to reproducibility, scalability, and predictive design, ultimately enabling the rational engineering of core-shell systems with unprecedented precision and efficiency.

References

1 Das S, Biswas J, Siddique IM. Mechanical characterization of materials using advanced microscopy techniques. World J Adv Res Rev [Internet]. 2024 Mar 30 [cited 2025 Jun 10];21(3):274–83. Available from: https://wjarr.co.in/wjarr-2024-0742

2 Ghosh Chaudhuri R, Paria S. Core/shell nanoparticles: Classes, properties, synthesis mechanisms, characterization, and applications. Chem Rev [Internet]. 2012 Apr 11 [cited 2025 Jun 10];112(4):2373–433. Available from: https://doi.org/10.1021/cr100449n

3 Girão AV, Caputo G, Ferro MC. Application of scanning electron microscopy–energy dispersive X-ray spectroscopy (SEM-EDS). Compr Anal Chem [Internet]. 2017 Jan 1 [cited 2025 Jun 10];75:153–68. Available from: https://www.sciencedirect.com/science/article/abs/pii/S0166526X16301519

4 Yao S, Li H, Pang S, Zhu B, Zhang X, Fatikow S. A review of computer microvision-based precision motion measurement: Principles, characteristics, and applications. IEEE Trans Instrum Meas. 2021;70:1–28.

5 Lin Y, Zhou M, Tai X, Li H, Han X, Yu J. Analytical transmission electron microscopy for emerging advanced materials. Matter [Internet]. 2021 Jul 7 [cited 2025 Jun 10];4(7):2309–39. Available from: https://www.cell.com/action/showFullText?pii=S2590238521002241

6 Palaci I. Atomic force microscopy studies of nanotribology and nanomechanics [Doctoral thesis]. 2007 [cited 2025 Jun 10]. Available from: https://infoscience.epfl.ch/handle/20.500.14299/10050

7 Chatterjee K, Sarkar S, Jagajjanani Rao K, Paria S. Core/shell nanoparticles in biomedical applications. Adv Colloid Interface Sci [Internet]. 2014 Jul 1 [cited 2025 Jun 10];209:8–39. Available from: https://www.sciencedirect.com/science/article/pii/S0001868613001899?casa_token=T3mCmCVkFDkAAAAA:9y_-GmS1Bc0jR5cjtRTGGqWuWnb9X_wdJyEmaEmAS4YhWd6CwuBbVJChLlt960rBSAgmya2bmuw

8 Zeng C, Vitale-Sullivan C, Ma X. In situ atomic force microscopy studies on nucleation and self-assembly of biogenic and bio-inspired materials. Minerals

[Internet]. 2017 Aug 31 [cited 2025 Jun 6];7(9):158. Available from: https://www.mdpi.com/2075-163X/7/9/158/htm

9 Li P, Chen S, Dai H, Yang Z, Chen Z, Wang Y, et al. Recent advances in focused ion beam nanofabrication for nanostructures and devices: Fundamentals and applications. Nanoscale [Internet]. 2021 Jan 28 [cited 2025 Jun 10];13(3):1529–65. Available from: https://pubs.rsc.org/en/content/articlehtml/2021/nr/d0nr07539f

10 Raffa V, Castrataro P, Menciassi A, Dario P. Focused ion beam as a scanning probe: Methods and applications. In: Applied Scanning Probe Methods II. Springer. 2006 [cited 2025 Jun 10];361–412. Available from: https://link.springer.com/chapter/10.1007/3-540-27453-7_11

11 She Y, Tang J, Wang C, Wang Z, Huang Z, Yang Y. Nano-additive manufacturing and non-destructive testing of nanocomposites. Nanomaterials [Internet]. 2023 Oct 1 [cited 2025 Jun 10];13(20):2741. Available from: https://pmc.ncbi.nlm.nih.gov/articles/PMC10609085/

12 Bayat M, Mardani H, Roghani-Mamaqani H, Hoogenboom R. Self-indicating polymers: A pathway to intelligent materials. Chem Soc Rev [Internet]. 2024 Apr 22 [cited 2025 Jun 10];53(8):4045–85. Available from: https://pubs.rsc.org/en/content/articlehtml/2024/cs/d3cs00431g

13 Chaouachi S, Elleuch S, Hamdi B, Zouari R. Experimental (FTIR, Raman, UV–visible and PL) and theoretical (DFT and TDDFT) studies on bis(8–hydroxyquinolinium) tetrachlorocobaltate(II) compound. J Mol Struct [Internet]. 2016 Dec 5 [cited 2025 Jun 10];1125:149–61. Available from: https://www.sciencedirect.com/science/article/pii/S0022286016306354?casa_token=nDkFrj4QTlsAAAAA:bZk3lv0JCeZb5nXnkQDbkfMqPJ-l-nW1toYUqQmZcHoNjtOzvXYc9KkffrHteChsldYa1cn3

14 Prasad RD, Sarvalkar PD, Prasad N, Prasad SR, Prasad RS, Prasad RB, et al. A review on spectroscopic techniques for analysis of nanomaterials and biomaterials. ES Energy & Environment [Internet]. 2024 [cited 2025 Jun 10]. Available from: https://www.espublisher.com/journals/articledetails/1264

15 Qenawy M. Ultrafast photoinduced processes in core and core–shell quantum dots for solar cell applications "Tiny Crystals for Big Applications". [Doctoral thesis] [Internet]. 2015 [cited 2025 Jun 10]. Available from: https://portal.research.lu.se/en/publications/ultrafast-photoinduced-processes-in-core-and-coreshell-quantum-do

16 Tian XD, Liu BJ, Li JF, Yang ZL, Ren B, Tian ZQ. SHINERS and plasmonic properties of Au Core SiO_2 shell nanoparticles with optimal core size and shell thickness. J Raman Spectrosc [Internet]. 2013 Jul 1 [cited 2025 Jun 10];44(7):994–8. Available from: https://doi.org/10.1002/jrs.4317

17 Chen J, Lv J, Liu X, Lin J, Chen X. A study on theoretical models for investigating time-resolved photoluminescence in halide perovskites. Phys Chem Chem Phys [Internet]. 2023 Mar 15 [cited 2025 Jun 10];25(11):7574–88. Available from: https://pubs.rsc.org/en/content/articlehtml/2023/cp/d2cp05723a

18 Zhang JZ. Optical properties and spectroscopy of nanomaterials. Google books [Internet]. World Scientific. 2009 Jul 21 [cited 2025 Jun 10].

19 Alshehawy AM, Mansour DEA, Ghali M, Lehtonen M, Darwish MMF. Photoluminescence spectroscopy measurements for effective condition

assessment of transformer insulating oil. Processes [Internet]. 2021 Apr 21 [cited 2025 Jun 6];9(5):732. Available from: https://www.mdpi.com/2227-9717/9/5/732/htm

20 Li Y, Ji S, Gao Y, Luo H, Jin P. Modification of mott phase transition characteristics in VO_2@TiO_2 core/shell nanostructures by misfit-strained heteroepitaxy. ACS Appl Mater Interfaces [Internet]. 2013 Jul 24 [cited 2025 Jun 10];5(14):6603–14. Available from: https://doi.org/10.1021/am401297g

21 Cao Y, Zhang J, Yang Y, Huang Z, Long NV, Fu C. Engineering of SERS substrates based on noble metal nanomaterials for chemical and biomedical applications. Appl Spectrosc Rev [Internet]. 2015 Jul 3 [cited 2025 Jun 10];50(6):499–525. Available from: https://scholar.google.com/scholar_url?url=https://www.tandfonline.com/doi/pdf/10.1080/05704928.2014.923901%3Fcasa_token%3DRJj7bLN0zzoAAAAA:Z78WgrGSo2Uk3ez0s7RjtFLftuDr5ulCf4lPRoN1-w02psoyhIAlcqlVsYSQBh4md6_3gl_xDTc&hl=en&sa=T&oi=ucasa&ct=ucasa&ei=OltIaOLwEOWs6rQPpf7X-Aw&scisig=AAZF9b_PrrbRqM9BPyp6wWL_Bk1C

22 Lee JH, Nam JM, Jeon KS, Lim DK, Kim H, Kwon S, et al. Tuning and maximizing the single-molecule surface-enhanced Raman scattering from DNA-tethered nanodumbbells. ACS Nano [Internet]. 2012 Nov 27 [cited 2025 Jun 10];6(11):9574–84. Available from: https://doi.org/10.1021/nn3028216

23 Hess C. New advances in using Raman spectroscopy for the characterization of catalysts and catalytic reactions. Chem Soc Rev [Internet]. 2021 Mar 15 [cited 2025 Jun 10];50(5):3519–64. Available from: https://pubs.rsc.org/en/content/articlehtml/2021/cs/d0cs01059f

24 Park S, Lee J, Khan S, Wahab A, Kim M. SERSNet: Surface-enhanced Raman spectroscopy based biomolecule detection using deep neural network. Biosensors [Internet]. 2021 Nov 30 [cited 2025 Jun 6];11(12):490. Available from: https://www.mdpi.com/2079-6374/11/12/490/htm

25 Khan SA, Khan SB, Khan LU, Farooq A, Akhtar K, Asiri AM. Fourier transform infrared spectroscopy: Fundamentals and application in functional groups and nanomaterials characterization. Handbook of Mater Charact [Internet]. 2018 Sep 18 [cited 2025 Jun 10];317–44. Available from: https://link.springer.com/chapter/10.1007/978-3-319-92955-2_9

26 Cook E, Labiento G, Chauhan BPS. Fundamental methods for the phase transfer of nanoparticles. Molecules [Internet]. 2021 Oct 13 [cited 2025 Jun 10];26(20):6170. Available from: https://www.mdpi.com/1420-3049/26/20/6170/htm

27 Zhao Y, Adiyeri Saseendran DP, Huang C, Triana CA, Marks WR, Chen H, et al. Oxygen evolution/reduction reaction catalysts: From in situ monitoring and reaction mechanisms to rational design. Chem Rev [Internet]. 2023 May 10 [cited 2025 Jun 10];123(9):6257–358. Available from: https://doi.org/10.1021/acs.chemrev.2c00515

28 Rai S, Guin M, De A, Singh NB. Functionalized nanomaterials: Basics, properties and applications. ACS Symp Ser [Internet]. 2022 Jun 13 [cited 2025 Jun 10];1418:27–66. Available from: https://pubs.acs.org/doi/abs/10.1021/bk-2022-1418.ch002

29 Bonifacio CS, Carenco S, Wu CH, House SD, Bluhm H, Yang JC. Thermal stability of core-shell nanoparticles: A combined in situ study by XPS and TEM. Chem

Mater [Internet]. 2015 Oct 27 [cited 2025 Jun 10];27(20):6960–8. Available from: https://doi.org/10.1021/acs.chemmater.5b01862

30 Jayawardena HSN, Liyanage SH, Rathnayake K, Patel U, Yan M. Analytical methods for characterization of nanomaterial surfaces. Anal Chem [Internet]. 2021 Feb 2 [cited 2025 Jun 10];93(4):1889–911. Available from: https://doi.org/10.1021/acs.analchem.0c05208

31 Gao Q, Yao B, Liu Y, Shi L, Yan Z, Xu L, et al. Strain relaxation enhances ammonia electrosynthesis from nitrate on Cu/CuAu core/shell nanocrystals with ordered intermetallic layers. Chem Catal [Internet]. 2025 [cited 2025 Jun 10];5(6):101328. Available from: https://www.cell.com/action/showFullText?pii=S2667109325000661

32 Rades S, Hodoroaba VD, Salge T, Wirth T, Lobera MP, Labrador RH, et al. High-resolution imaging with SEM/T-SEM, EDX and SAM as a combined methodical approach for morphological and elemental analyses of single engineered nanoparticles. RSC Adv [Internet]. 2014 Oct 7 [cited 2025 Jun 10];4(91):49577–87. Available from: https://pubs.rsc.org/en/content/articlehtml/2014/ra/c4ra05092d

33 Wuhrer R, Moran K, Matthews M. WDS-SD: Next generation of Wavelength Dispersive Spectrometers (WDS) with a Silicon Drift Detector (SDD) – What can it do, where are we now and where is it going? Micros Microanal [Internet]. 2023 Jul 22 [cited 2025 Jun 10];29(Supplement_1):846–8. Available from: https://dx.doi.org/10.1093/micmic/ozad067.420

34 Pomerantseva E, Bonaccorso F, Feng X, Cui Y, Gogotsi Y. Energy storage: The future enabled by nanomaterials. Science (1979) [Internet]. 2019 Nov 22 [cited 2025 Jun 10];366(6468). Available from: https://doi.org/10.1126/science.aan8285?download=true

35 Guo D, Xie G, Luo J. Mechanical properties of nanoparticles: Basics and applications. J Phys D Appl Phys [Internet]. 2013 Dec 3 [cited 2025 Jun 10];47(1):013001. Available from: https://iopscience.iop.org/article/10.1088/0022-3727/47/1/013001

36 Wu SL, Yang RX, Wey MY. Catalytic methane decomposition to hydrogen over a surface-protected core-shell Ni@SiO_2 catalyst. Chem Eng Technol [Internet]. 2018 Jul 1 [cited 2025 Jun 10];41(7):1448–56. Available from: https://doi.org/10.1002/ceat.201700315

37 Utke I, Michler J, Winkler R, Plank H. Mechanical properties of 3D nanostructures obtained by focused electron/ion beam-induced deposition: A review. Micromachines [Internet]. 2020 Apr 10 [cited 2025 Jun 10];11(4):397. Available from: https://www.mdpi.com/2072-666X/11/4/397/htm

38 Sun Y, Peng G, Hu Y, Dou G, Chen P, Zhang T. Spherical indentation model for evaluating the elastic properties of the shell of microsphere with core-shell structure. Int J Solids Struct [Internet]. 2021 Nov 1 [cited 2025 Jun 10];230–31:111159. Available from: https://www.sciencedirect.com/science/article/pii/S0020768321002493

39 Peckus D, Rong H, Stankevičius L, Juodenas M, Tamulevičius S, Tamulevičius T, et al. Hot electron emission can lead to damping of optomechanical modes in core-shell Ag@TiO_2 nanocubes. J Phys Chem C [Internet]. 2017 Nov 2 [cited 2025 Jun 10];121(43):24159–67. Available from: https://doi.org/10.1021/acs.jpcc.7b06667

40 Quinaz T, Freire TF, Olmos A, Martins M, Ferreira FBN, de Moura MFSM, et al. The influence of hydroxyapatite crystals on the viscoelastic behavior of poly(vinyl alcohol) braid systems. Biomimetics [Internet]. 2024 Feb 5 [cited 2025 Jun 10];9(2):93. Available from: https://www.mdpi.com/2313-7673/9/2/93/htm

41 Agrawal M, Antham CP, Jalal SS, Nagpal A, Rajalakshmi B, Dwivedi SP. Innovative advances and prospects in in situ materials testing: A comprehensive review. E3S Web Conf [Internet]. 2024 Mar 25 [cited 2025 Jun 10];505:01031. Available from: https://www.e3s-conferences.org/articles/e3sconf/abs/2024/35/e3sconf_icarae2023_01031/e3sconf_icarae2023_01031.html

42 Smeaton MA, Abellan P, Spurgeon SR, Unocic RR, Jungjohann KL. Tutorial on in situ and *operando* (scanning) transmission electron microscopy for analysis of nanoscale structure-property relationships. ACS Nano [Internet]. 2024 Dec 31 [cited 2025 Jun 10];18(52). Available from: https://doi.org/10.1021/acsnano.4c09256

43 Mestl G. In situ Raman spectroscopy — A valuable tool to understand operating catalysts. J Mol Catal A Chem [Internet]. 2000 Sep 8 [cited 2025 Jun 10];158(1):45–65. Available from: https://www.sciencedirect.com/science/article/pii/S138111690000042X?casa_token=GF1I8wBjsAEAAAAA:EcjUQIioQTjdsPOoVwp28_8uixGeByNkji7qrFnUftBcmbrA8Mjy_2RtKL9R91GEzsl5xQUn

44 Gopalan Sibi M, Verma D, Kim J. Magnetic core–shell nanocatalysts: Promising versatile catalysts for organic and photocatalytic reactions. Catal Rev Sci Eng [Internet]. 2020 Apr 2 [cited 2025 Jun 10];62(2):163–311.

45 Han HL, Melaet G, Alayoglu S, Somorjai GA. In situ microscopy and spectroscopy applied to surfaces at work. ChemCatChem [Internet]. 2015 Nov 16 [cited 2025 Jun 10];7(22):3625–38. Available from: https://doi.org/10.1002/cctc.201500642

46 Alanazi N, Almutairi M, Alodhayb AN. A review of quartz crystal microbalance for chemical and biological sensing applications. Sens Imaging [Internet]. 2023 Dec 1 [cited 2025 Jun 10];24(1):1–33. Available from: https://link.springer.com/article/10.1007/s11220-023-00413-w

47 Tian T, Lu H, Yang S, Qin J, Ma S, Meng J, et al. Research progress of functional atomic force microscopy at the interface of polymer nanocomposite dielectrics. Macromol Res [Internet]. 2024 Dec 1 [cited 2025 Jun 10];32(12):1159–85. Available from: https://link.springer.com/article/10.1007/s13233-024-00307-0

48 Gupta P, Rai N, Verma A, Gautam V. Microscopy based methods for characterization, drug delivery, and understanding the dynamics of nanoparticles. Med Res Rev [Internet]. 2024 Jan 1 [cited 2025 Jun 10];44(1):138–68. Available from: https://doi.org/10.1002/med.21981

49 Hao C, Wu X, Sun M, Zhang H, Yuan A, Xu L, et al. Chiral core-shell upconversion nanoparticle@MOF nanoassemblies for quantification and bioimaging of reactive oxygen species *in vivo*. J Am Chem Soc [Internet]. 2019 Dec 11 [cited 2025 Jun 10];141(49):19373–8. Available from: https://doi.org/10.1021/jacs.9b09360

50 Ramasundaram S, Jeevanandham S, Vijay N, Divya S, Jerome P, Oh TH. Unraveling the dynamic properties of new-age energy materials chemistry using advanced in situ transmission electron microscopy. Molecules [Internet]. 2024 Sep 17 [cited 2025 Jun 10];29(18):4411. Available from: https://www.mdpi.com/1420-3049/29/18/4411/htm

51 Henderson GS, de Groot FMF, Moulton BJA. X-ray absorption near-edge structure (XANES) spectroscopy. Rev Mineral Geochem. 2014 Jan 1;78(1):75–138.

52 Honecker D, Bersweiler M, Erokhin S, Berkov D, Chesnel K, Venero DA, et al. Using small-angle scattering to guide functional magnetic nanoparticle design. Nanoscale Adv [Internet]. 2022 Feb 15 [cited 2025 Jun 10];4(4):1026–59. Available from: https://pubs.rsc.org/en/content/articlehtml/2022/na/d1na00482d

53 Rezaei B, Mostufa S, Wu K. Characterization methods for magnetic nanoparticles. Magnetic Nanopart Nanomed [Internet]. 2024 Jan 1 [cited 2025 Jun 10]; 155–77. Available from: https://www.sciencedirect.com/science/article/abs/pii/B9780443216688000067

54 Chaupard M, de Frutos M, Gref R. Deciphering the structure and chemical composition of drug nanocarriers: From bulk approaches to individual nanoparticle characterization. Part Part Syst Charact [Internet]. 2021 Sep 1 [cited 2025 Jun 10];38(9):2100022. Available from: https://doi.org/10.1002/ppsc.202100022

55 Tam J, Merino D. Stochastic optical reconstruction microscopy (STORM) in comparison with stimulated emission depletion (STED) and other imaging methods. J Neurochem [Internet]. 2015 Nov 1 [cited 2025 Jun 10];135(4):643–58. Available from: https://doi.org/10.1111/jnc.13257

5

Core-shell Nanomaterials for Energy Applications

This chapter explores the role of core-shell nanomaterials in addressing critical challenges in energy conversion, storage, and environmental sustainability. Core-shell structures, with their unique combination of materials, provide significant advantages in enhancing the efficiency, stability, and functionality of various energy technologies. The chapter covers their applications in photovoltaics (PVs), where core-shell materials improve light absorption, charge carrier separation, and stability in solar cells. In energy storage systems, core-shell nanomaterials serve as advanced electrode materials, improving ion transport and cycle life in lithium-ion and sodium-ion batteries, supercapacitors, and fuel cells. The use of core-shell catalysts in hydrogen generation through water splitting and in electrocatalysis and photocatalysis for CO_2 reduction and pollutant degradation is also discussed, highlighting their improved catalytic performance. The chapter further emphasizes the potential of core-shell nanomaterials to address global challenges in renewable energy and environmental remediation. Finally, it identifies the challenges of scalability, cost, and regulation in the widespread adoption of these materials and suggests pathways for overcoming these barriers to realize their full potential in industrial applications.

5.1 Introduction

Core-shell nanomaterials have emerged as a pivotal class of advanced materials in the quest to address the global energy challenges of the 21st century. These materials, characterized by their unique architecture comprising a core material encapsulated by a shell, offer exceptional tunability in terms of structural, optical, and electronic properties. This distinctive configuration allows researchers to harness the best of both materials, making core-shell nanomaterials highly promising for applications in energy conversion and storage. Energy-related concerns such as the depletion of fossil fuels, environmental degradation, and the urgent need for sustainable alternatives have driven advancements in nanotechnology [1]. Core-shell nanomaterials have revolutionized fields like PVs, batteries, hydrogen generation, and catalysis, thanks to their superior charge transport capabilities, enhanced surface area, and

Core-Shell Nanomaterials: From Fundamentals to Applications, First Edition. Shreya Sharma et al.

stability. The ability to engineer both the core and shell components independently provides unparalleled control over material properties, enabling breakthroughs in energy efficiency and device longevity.

In energy conversion applications, such as solar cells, core-shell structures are pivotal in enhancing light absorption and charge carrier separation, thereby boosting device efficiency. Materials such as quantum dots (QDs) and perovskites with core-shell architectures have demonstrated remarkable improvements in power conversion efficiencies. The shell layer serves as a protective barrier, minimizing surface defects and improving stability under environmental stressors. Similarly, in energy storage systems, core-shell nanomaterials are employed to enhance the performance of batteries, supercapacitors, and fuel cells [2]. Their ability to offer high ion conductivity, increased active surface area, and improved cycling stability has positioned them as next-generation electrode materials. For example, core-shell carbon–metal oxide structures have shown great promise in lithium-ion and sodium-ion batteries, offering higher energy densities and longer cycle lives compared to traditional materials. Moreover, core-shell nanomaterials play a crucial role in hydrogen generation and storage systems. Their catalytic properties facilitate efficient water splitting, a clean and sustainable method for hydrogen production [3]. The tailored design of core-shell catalysts ensures high activity, selectivity, and stability, making them indispensable for hydrogen energy applications. As shown in Figure 5.1, core-shell nanomaterials find diverse applications in energy technologies, from PVs

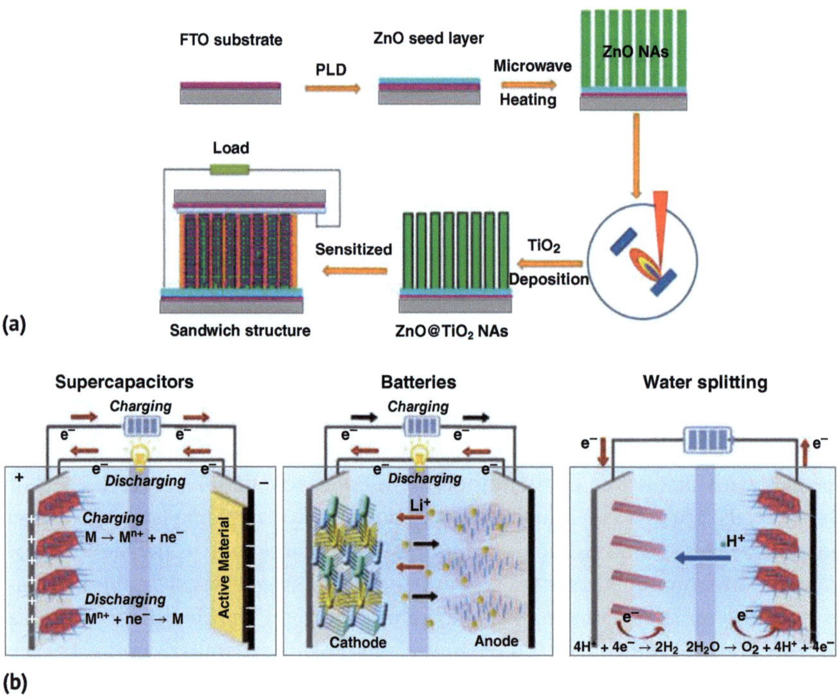

Figure 5.1 Depiction of the role of core-shell nanomaterials in energy applications, including (a) solar cells, [5]/MDPI/CC BY 4.0; (b) batteries; and catalysis, [6]/John Wiley & Sons/CC BY 4.0.

to electrocatalysis. This figure provides a conceptual illustration of their multifunctional roles in energy conversion and storage systems, highlighting their transformative impact on modern energy technologies.

In addition to their conversion and storage applications, core-shell nanomaterials are increasingly being explored for their catalytic properties. Electrocatalysis and photocatalysis are critical processes for addressing energy and environmental issues. Core-shell catalysts are highly efficient for CO_2 reduction, water splitting, and pollutant degradation, offering sustainable solutions for cleaner energy production [4]. Their structural stability and ability to promote reaction kinetics make them ideal candidates for such applications.

The chapter delves into the various energy applications of core-shell nanomaterials, starting with their role in PVs and moving on to energy storage systems, hydrogen generation, and electrocatalysis. Each section highlights the key material properties, mechanisms, and recent advancements in these areas. Furthermore, potential future directions for research and development are discussed, offering insights into how core-shell nanomaterials can continue to shape the landscape of energy technologies. As the global demand for sustainable and efficient energy solutions intensifies, core-shell nanomaterials are poised to play a transformative role. Their multifunctional properties and structural adaptability make them essential components of advanced energy systems. By optimizing their synthesis techniques and tailoring their properties, researchers can unlock new possibilities for cleaner, more efficient, and sustainable energy solutions. The versatility and functionality of core-shell nanomaterials underscore their immense potential in addressing some of the most pressing energy challenges of our time. The chapter provides a comprehensive overview of their applications, emphasizing the technological advancements and future opportunities in this dynamic field.

5.2 Photovoltaics: Enhancing Solar Cell Efficiency

The ever-increasing demand for clean and sustainable energy solutions has driven significant advancements in PV technology, which harnesses solar energy to generate electricity. However, the efficiency and stability of traditional solar cells remain major challenges. Core-shell nanomaterials have emerged as promising candidates to address these limitations, offering unique structural and functional properties that can optimize the light absorption, charge carrier dynamics, and overall performance of solar cells. By integrating these nanomaterials into various types of solar cells, significant improvements in efficiency and long-term stability have been realized, paving the way for next-generation PV technologies.

One of the primary goals in PV technology is to maximize light absorption to generate more charge carriers and increase power output. Core-shell nanomaterials are particularly suited for this task due to their tunable optical properties. By carefully selecting the core and shell materials, it is possible to design structures that capture a broader spectrum of sunlight. Core-shell nanoparticles, such as plasmonic metal nanoparticles (gold–silver combinations), exhibit strong light-scattering properties that enhance the absorption of incident light [7]. The shell can act as a dielectric layer that optimizes the interaction between light and the nanostructure. This not only increases the path length of light within the solar cell but also boosts the efficiency of photon capture. Moreover, the use of semiconductor core-shell nanowires further enhances light trapping. The core typically functions as a light-absorbing material, while the shell improves charge collection by passivating surface defects [8]. These nanowires can be vertically aligned to form dense arrays, ensuring efficient light absorption across a wide range of wavelengths.

Effective charge carrier separation and transport are crucial for minimizing recombination losses in solar cells. Core-shell nanomaterials address this challenge by providing a built-in electric field at the core-shell interface. This electric field facilitates the separation of photogenerated electron–hole pairs and directs them toward their respective electrodes. For instance, in core-shell nanostructures such as $ZnO@TiO_2$, the ZnO core functions as an electron-conducting material, while the TiO_2 shell reduces recombination by serving as an insulating layer that prevents backflow of electrons. This configuration leads to enhanced charge carrier mobility and improved photoconversion efficiency. Additionally, core-shell nanomaterials with gradient composition layers further enhance charge transport by minimizing potential barriers at interfaces. This results in smoother charge flow, thereby increasing the overall efficiency of the solar cell.

Perovskite solar cells (PSCs) have rapidly gained attention due to their remarkable power conversion efficiencies and cost-effective manufacturing processes. However, their stability under environmental conditions is a significant concern [9]. The integration of core-shell nanomaterials has proven to be a game-changer for addressing these issues. Core-shell materials, such as SiO_2-coated perovskite nanoparticles, provide a protective barrier that shields the perovskite layer from moisture and oxygen, both of which contribute to its degradation. This results in improved stability and prolonged device lifetime. Furthermore, the use of core-shell nanostructures enhances the charge carrier dynamics in PSCs. For example, PbS@CdS core-shell QDs have been employed to improve the light absorption and charge separation properties of perovskite layers. This has led to significant improvements in power conversion efficiency [10].

Quantum dot solar cells (QDSCs) represent a promising class of PV technology, leveraging the unique optical and electronic properties of QDs for efficient light harvesting. The incorporation of core-shell QDs has further advanced the performance of these solar cells. The core-shell architecture in QDs, such as CdSe@ZnS, plays a critical role in passivating surface defects and enhancing charge carrier lifetime [11]. The ZnS shell acts as a protective layer that prevents oxidation and reduces non-radiative recombination. This results in improved stability and higher photoconversion efficiency. In addition to passivation, the shell layer can be engineered to create a type II band alignment, which promotes efficient charge separation. This configuration is particularly beneficial for QDSCs, as it minimizes recombination losses and increases the generation of photocurrent. As illustrated in Figure 5.2, core-shell nanomaterials significantly improve light absorption and charge carrier separation in solar cells. The schematic shows how the core material absorbs sunlight while the shell enhances carrier dynamics and protects against degradation.

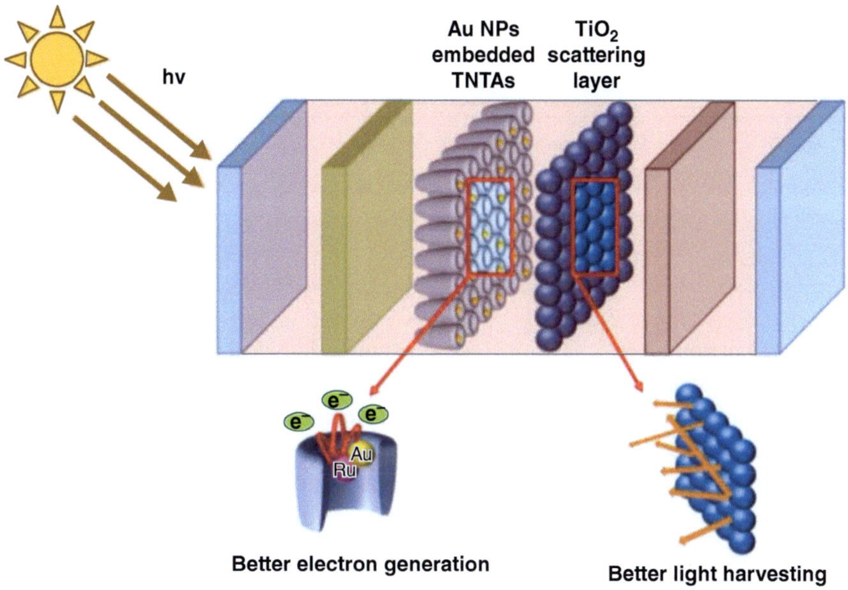

Figure 5.2 Illustration of a core-shell nanomaterial improving light absorption and charge carrier dynamics in solar cell. [15]/MDPI/CC BY 4.0.

The integration of core-shell nanomaterials in solar cells has led to substantial improvements in both efficiency and stability. PSCs incorporating core-shell nanomaterials have achieved power conversion efficiencies exceeding 25%, while QDSCs have demonstrated efficiency gains of up to 12% due to enhanced light absorption and charge carrier dynamics [12]. The ability to tune the core-shell structure for optimal optical and electronic properties has been instrumental in achieving these results. One of the most significant advantages of core-shell nanomaterials is their ability to improve the stability of solar cells. The protective shell layer shields the core from environmental degradation factors such as moisture, oxygen, and UV radiation. This has led to improved long-term performance and durability of PV devices [13]. While core-shell nanomaterials have shown immense potential in enhancing solar cell efficiency and stability, several challenges remain.

The large-scale synthesis of core-shell nanomaterials with uniform properties is a technical challenge that needs to be addressed for commercial viability. Developing cost-effective and scalable synthesis methods is crucial for their widespread adoption. The use of toxic materials, such as cadmium-based QDs, raises environmental and health concerns [14]. The development of eco-friendly alternatives is essential to ensure the sustainability of core-shell nanomaterials in PV applications. The combination of core-shell nanomaterials with advanced PV technologies, such as tandem solar cells and hybrid systems, presents exciting opportunities for further efficiency gains. The integration of these materials with transparent and flexible substrates could also enable novel applications such as building-integrated photovoltaics (BIPVs).

Table 5.1 Comparison of different core-shell materials used in solar cells.

Material composition	Solar cell type	Efficiency improvement (%)	Stability enhancement
$TiO_2@SiO_2$	Perovskite	20–30	High
CdSe@ZnS	Quantum dot	15–25	Medium
$Ag@TiO_2$	Thin film	10–20	High
ZnO@Graphene	Hybrid	12–18	Medium

Core-shell nanomaterials have emerged as a transformative technology in the field of PVs, addressing key challenges related to light absorption, charge carrier dynamics, and stability. Their unique structural and functional properties make them indispensable for the development of high-performance solar cells. As research continues to explore novel material combinations and synthesis techniques, core-shell nanomaterials are poised to play a central role in the global transition toward sustainable energy solutions. The future of PV technology will undoubtedly be shaped by the continued advancements in core-shell nanomaterial design and application.

Table 5.1 provides a comparative analysis of different core-shell materials used in solar cells, highlighting their material composition, solar cell type, efficiency improvements, and stability enhancements.

5.3 Energy Storage: Batteries, Supercapacitors, and Fuel Cells

The growing global demand for efficient and sustainable energy storage technologies has spurred extensive research into advanced materials that can significantly enhance the performance of energy storage devices. Core-shell nanomaterials have emerged as one of the most promising candidates for these applications due to their unique structural advantages, which combine the benefits of a stable shell and a functional core. These nanomaterials play a pivotal role in batteries, supercapacitors, and fuel cells by enhancing ion transport, increasing energy and power densities, and improving overall stability. Their versatility and tunability make them key enablers of next-generation energy storage systems.

Electrode materials are critical to the performance of energy storage devices, as they dictate the capacity, energy density, and cycling stability of the system. Core-shell nanomaterials provide a unique advantage by offering a well-defined interface that can enhance charge transfer and minimize degradation during cycling. The core component typically serves as an active material with high electrical conductivity or energy storage capability, while the shell acts as a protective barrier, preventing the core from degradation and improving the stability of the electrode [16]. For instance, carbon-coated metal oxides such as $Fe_3O_4@C$ and $MnO_2@C$ are widely used in energy storage applications because the carbon shell provides excellent electrical conductivity and prevents the aggregation of metal oxide particles during cycling. Additionally, core-shell nanostructures allow for the design of hierarchical architectures that increase the electrode surface area, facilitating faster ion diffusion and better contact with the electrolyte. These properties make them particularly suitable for applications in batteries, supercapacitors, and fuel cells.

Lithium-ion batteries (LIBs) have become the gold standard for portable electronics and electric vehicles due to their high energy density and long cycle life. However, challenges such as capacity fading and safety concerns limit their widespread adoption. Core-shell nanomaterials offer innovative solutions to these issues. In LIBs, core-shell structures can improve the stability and electrochemical performance of anodes and cathodes [17]. For example, silicon-based anodes, which have a high theoretical capacity, suffer from severe volume changes during lithiation and delithiation [18]. The integration of a carbon shell around silicon nanoparticles (Si@C) has proven effective in accommodating volume expansion and maintaining the structural integrity of the electrode. This results in enhanced cycling stability and higher capacity retention. Similarly, core-shell cathodes, such as $LiFePO_4$@C, demonstrate improved conductivity and faster lithium-ion transport, leading to higher energy densities. The carbon shell also acts as a buffer, preventing the dissolution of active materials and extending the battery's lifespan. Sodium-ion batteries (SIBs) have gained attention as a cost-effective alternative to LIBs, particularly for large-scale energy storage applications. Core-shell nanomaterials are equally valuable for SIBs, addressing challenges related to poor cycling stability and slow ion diffusion. For example, $Na_3V_2(PO_4)_3$@C core-shell structures exhibit enhanced ion transport and improved electrochemical performance due to the conductive carbon shell. These advancements make SIBs a viable option for grid storage and renewable energy integration.

Supercapacitors are energy storage devices that offer high power density, fast charge–discharge rates, and long cycle life. However, their relatively low energy density compared to batteries limits their application in certain fields. Core-shell nanomaterials have been instrumental in overcoming this limitation by enhancing both the energy and power densities of supercapacitors [19]. In supercapacitors, the core-shell architecture allows for the combination of materials with complementary properties. For instance, metal oxides or hydroxides can serve as the core material, providing high capacitance, while conductive polymers or carbon materials act as the shell, enhancing electrical conductivity and stability [20]. A notable example is the use of NiO@C core-shell structures, which exhibit high specific capacitance and excellent cycling stability [21]. The carbon shell not only improves conductivity but also protects the NiO core from structural degradation during repeated charge–discharge cycles. Additionally, hierarchical core-shell nanostructures, such as MnO_2@CNT (carbon nanotube), provide a large surface area for electrolyte access, enabling faster ion transport and higher energy storage capacity [22]. These advancements have led to the development of high-performance supercapacitors for applications in electric vehicles and renewable energy systems. As depicted in Figure 5.3, core-shell nanomaterials play a pivotal role in improving the performance of batteries and supercapacitors by enhancing electrode properties and facilitating efficient charge storage mechanisms.

Fuel cells are an essential component of clean energy systems, converting chemical energy into electrical energy through electrochemical reactions. Core-shell

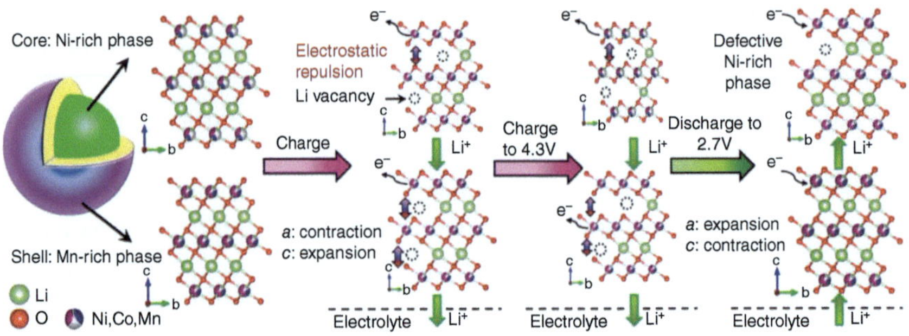

Figure 5.3 Schematic illustration of a core-shell nanomaterial enhancing energy storage performance by enabling controlled (de)lithiation in Ni-rich cathode materials and optimizing charge carrier dynamics. [28]/with permission of Elsevier.

nanomaterials have shown great promise in enhancing the efficiency and durability of fuel cells, particularly in the development of advanced catalysts for the oxygen reduction reaction (ORR) and hydrogen oxidation reaction (HOR). One of the major challenges in fuel cell technology is the high cost and poor stability of platinum-based catalysts. Core-shell nanostructures, such as Pt@Pd and Pt@Ni, have been developed to address these issues. The core material typically provides high catalytic activity, while the shell enhances stability and reduces the amount of precious metal required. This not only improves the cost-effectiveness of the fuel cell but also extends its operational lifespan. Moreover, non-precious metal core-shell catalysts, such as Fe_3O_4@C, have gained attention as environmentally friendly alternatives to platinum-based catalysts. These materials exhibit excellent ORR activity and stability, making them suitable for use in proton exchange membrane fuel cells (PEMFCs).

Core-shell nanomaterials also play a crucial role in solid oxide fuel cells (SOFCs), where they are used to enhance ionic conductivity and stability at high temperatures. For example, Y_2O_3-stabilized ZrO_2@CeO_2 core-shell structures have been shown to improve the performance of SOFCs by facilitating oxygen ion transport and minimizing interfacial resistance.

While core-shell nanomaterials have demonstrated significant potential in energy storage applications, several challenges remain.

- **Scalability:** The large-scale synthesis of core-shell nanomaterials with uniform properties is a technical challenge that needs to be addressed for commercial viability. Developing cost-effective and scalable synthesis methods is crucial for their widespread adoption [23].
- **Environmental and Health Concerns:** The use of toxic materials in certain core-shell nanostructures raises environmental and health concerns. Developing eco-friendly alternatives and implementing proper disposal methods are essential for sustainable development [24].
- **Integration with Hybrid Energy Systems:** The integration of core-shell nanomaterials with hybrid energy storage systems, such as battery–supercapacitor

Table 5.2 Performance comparison of core-shell materials in different energy storage systems.

Material type	Energy storage device	Capacity improvement (%)	Stability (cycle life)
Si@C	Lithium-ion battery (LIB)	300	500 cycles
Sn@C	Sodium-ion battery (SIB)	250	600 cycles
MnO_2@ Graphene	Supercapacitor	200	1000 cycles
Pt@Pd	Fuel cell	150	High

hybrids, presents exciting opportunities for achieving high energy and power densities simultaneously [25].

- **AI and Machine Learning Integration:** The application of AI and machine learning in the design and optimization of core-shell nanomaterials could revolutionize the field by enabling the discovery of novel material combinations and synthesis pathways [26].

Core-shell nanomaterials have revolutionized the field of energy storage by offering unique structural and functional advantages. Their ability to enhance ion transport, improve stability, and increase energy and power densities has made them indispensable for applications in batteries, supercapacitors, and fuel cells [27]. As research continues to explore novel material combinations and synthesis techniques, core-shell nanomaterials are poised to play a central role in the development of next-generation energy storage systems. Addressing scalability, environmental concerns, and integration challenges will be key to unlocking their full potential and driving the global transition toward sustainable energy solutions.

Table 5.2 provides a comparative analysis of the performance of different core-shell materials in various energy storage systems, highlighting their capacity improvement and stability benefits.

5.4 Hydrogen Generation and Storage Systems

Hydrogen, as a clean and sustainable energy carrier, has gained significant attention due to its potential to address global energy challenges and environmental concerns. However, challenges related to efficient hydrogen generation and storage remain key barriers to its widespread use. Core-shell nanomaterials have emerged as promising solutions to these challenges, particularly in water splitting for hydrogen generation and hydrogen storage applications. Their unique structural properties enable enhanced catalytic efficiency, improved stability, and increased performance, making them ideal candidates for next-generation hydrogen technologies.

Water splitting is one of the most promising methods for hydrogen production, where water (H_2O) is split into hydrogen (H_2) and oxygen (O_2) through electrochemical reactions. The reaction occurs in two steps: the oxygen evolution reaction (OER) and the hydrogen evolution reaction (HER) [29]. However, the efficiency of water splitting is heavily reliant on the performance of the catalysts used for these reactions. Traditional catalysts, such as platinum (Pt) and iridium (Ir), are expensive and scarce, limiting the scalability of hydrogen production. This has led to intense research into alternative, cost-effective materials, with core-shell nanomaterials showing exceptional promise due to their unique ability to enhance catalytic performance.

Core-shell structures offer a significant advantage by combining the strengths of two different materials. In the case of water splitting, the core material typically provides the necessary electronic and catalytic properties, while the shell material optimizes stability and conductivity. For instance, in HER, platinum-based core-shell catalysts, such as Pt@Ni, have demonstrated superior catalytic activity compared to pure platinum. The nickel shell in this structure not only enhances the overall electrochemical performance but also reduces the platinum content, making it more cost-effective [30]. Core-shell structures can also be designed with non-precious metal-based catalysts, which are more abundant and environmentally friendly. For example, iron-based core-shell catalysts like Fe@Fe_2O_3 have been shown to exhibit impressive activity for both OER and HER. The iron oxide shell provides stability, while the metallic core ensures efficient electron transfer. These core-shell nanomaterials offer a promising alternative to platinum and other precious metal catalysts, addressing both cost and scarcity issues in hydrogen production.

A significant benefit of core-shell nanomaterials in water splitting is the ability to tailor their properties by varying the thickness of the shell, the composition of the core, or the overall morphology. This tunability allows for optimization of the catalyst to achieve the best possible performance for specific reactions. By enhancing charge transfer kinetics and minimizing catalyst degradation, core-shell structures improve the efficiency of water splitting, resulting in higher hydrogen generation rates and

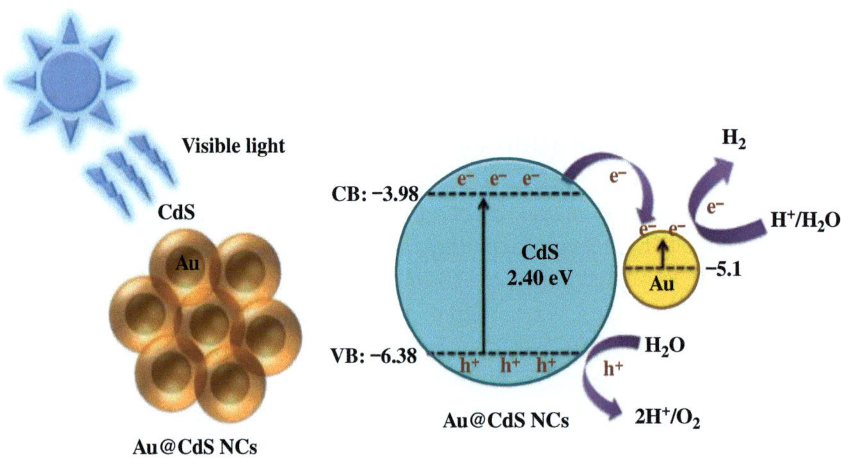

Figure 5.4 Au@CdS NCs core-shell enhancing hydrogen generation efficiency through water splitting. [33]/MDPI/CC BY 4.0.

better stability during long-term operation [31]. As depicted in Figure 5.4, core-shell catalysts can significantly enhance the hydrogen production process in water splitting by improving the efficiency of the electrochemical reactions. The schematic shows a core-shell nanomaterial where the core is an active catalytic metal, such as platinum, and the shell is a conductive or stable material, such as nickel or iron oxide, which facilitates electron transfer while protecting the core material from degradation [32]. The combined effect of the core-shell structure results in enhanced hydrogen generation efficiency, higher stability, and reduced costs, making these nanomaterials ideal candidates for large-scale hydrogen production.

The catalytic properties of core-shell nanomaterials play a critical role in enhancing the overall efficiency of hydrogen production via water splitting. The unique design of core-shell structures allows for the optimization of electronic properties, leading to better catalytic behavior. The core material in the structure typically serves as the primary active site for catalysis, while the shell material often plays a role in improving the electronic environment, stability, or conductivity [34]. One of the most notable benefits of core-shell materials in catalysis is the ability to control the electronic structure of the active sites. For instance, when platinum is used as the core material and a less expensive metal, like nickel or cobalt, is used as the shell, the electron density around the platinum core can be altered [32]. This change can increase the catalytic efficiency of the platinum core for HERs by improving the binding strength of hydrogen intermediates, thus enhancing the overall performance of the catalyst.

In addition to improving efficiency, core-shell nanomaterials also demonstrate better stability during prolonged catalytic cycles. Traditional catalysts can undergo structural degradation or leaching over time, reducing their effectiveness. Core-shell structures, however, protect the core material from such degradation. For example, the shell can prevent the leaching of precious metals and shield the core from oxidation or corrosion, resulting in a more durable catalyst.

Apart from hydrogen generation, storage is another critical component of the hydrogen economy. Hydrogen needs to be stored efficiently and safely to ensure its widespread use in energy applications, including fuel cells and portable energy systems. Core-shell nanomaterials have shown significant promise in enhancing hydrogen storage due to their high surface area, tunable porosity, and ability to accommodate hydrogen molecules at the nanoscale [35]. Core-shell structures can be used to design advanced materials that improve the absorption and desorption of hydrogen. For example, core-shell nanomaterials based on metal-organic frameworks (MOFs) or carbon-based composites are being investigated for hydrogen storage [36]. The core typically consists of a high-capacity material, such as a metal, while the shell provides structural integrity, preventing the collapse of the material during hydrogen absorption.

In addition, core-shell nanomaterials can be used to modify the storage capacity and release kinetics of hydrogen in storage systems. The shell can influence the interaction between the hydrogen molecules and the core, allowing for faster absorption and release rates. This is particularly important for applications in fuel cell-powered vehicles, where rapid hydrogen storage and release are required for optimal performance. Furthermore, the lightweight nature of core-shell nanomaterials makes them ideal for applications in portable hydrogen storage systems [37]. These materials allow for the development of compact and efficient storage devices that can be used in a variety of industries, including transportation, energy production, and electronics.

Core-shell nanomaterials offer significant advantages for both hydrogen generation and storage. By enhancing catalytic properties in water splitting, these materials can improve the efficiency and stability of hydrogen production systems, reducing costs and providing a more sustainable solution to hydrogen generation. Additionally, their use in hydrogen storage applications allows for more efficient and compact systems, which is essential for the widespread adoption of hydrogen as a clean energy carrier. As research continues to advance, core-shell nanomaterials will play an increasingly vital role in overcoming the challenges associated with hydrogen production, storage, and utilization, contributing to a cleaner and more sustainable energy future.

5.5 Electrocatalysis and Photocatalysis

Electrocatalysis and photocatalysis are two essential processes in the field of energy conversion and environmental remediation. Both processes depend on catalysts to facilitate the required reactions, such as hydrogen production, CO_2 reduction, and pollutant degradation. Core-shell nanomaterials, with their unique structural characteristics, have proven to be highly effective in enhancing the efficiency, selectivity, and stability of these catalytic processes. By combining the strengths of different materials within a single nanostructure, core-shell catalysts can be tailored to improve the overall performance of electrochemical and photocatalytic reactions, making them vital for future energy and environmental applications.

Electrocatalysis involves the acceleration of electrochemical reactions, such as the HER, OER, and CO_2 reduction reaction (CO_2RR). These reactions are central to the development of sustainable energy technologies like fuel cells, electrolyzers, and CO_2 capture and conversion systems [38]. In the context of these reactions, core-shell nanomaterials provide several advantages, primarily through enhanced catalytic efficiency and stability. One of the key benefits of core-shell structures in electrocatalysis is the ability to design catalysts that combine the optimal electronic properties of different materials. For example, platinum (Pt), a highly effective catalyst for HER, is often combined with cheaper, more abundant metals like nickel or cobalt in core-shell configurations [39]. The platinum core acts as the active site for hydrogen evolution, while the shell, made of a transition metal, can enhance the stability of the catalyst by preventing platinum degradation under harsh acidic conditions. This structure not only reduces the cost of the catalyst but also maintains high catalytic activity, even in demanding environments. Additionally, the shell can modify the electronic properties of the core, optimizing the interaction between the catalyst and reactants, leading to improved reaction rates.

Another significant area where core-shell nanomaterials have made a substantial impact is in the OER, which is crucial for water splitting to produce hydrogen as a clean energy source. Ruthenium (Ru) and iridium (Ir) are commonly used for OER catalysts, but their high cost and limited availability have led to the development of core-shell materials. For example, a RuO_2 core with an IrO_2 shell has shown exceptional catalytic activity and long-term stability. The IrO_2 shell enhances the OER performance while providing better durability than pure RuO_2 [40]. These core-shell structures prevent metal dissolution and degradation, thus extending the operational life of the catalyst. Other combinations, such as Fe@Fe_2O_3 and Ni@NiO, have also demonstrated significant improvements in OER efficiency, particularly under alkaline conditions.

In addition to water splitting, core-shell catalysts have shown promising results in CO_2 reduction, a reaction critical for mitigating climate change. The reduction of CO_2 to valuable chemicals like methane, ethylene, and formic acid can be achieved using electrocatalysts, and core-shell nanomaterials can offer high selectivity and efficiency. By carefully selecting the materials for the core and shell, researchers can tune the catalyst to selectively produce certain products while suppressing unwanted side reactions. For instance, Cu@CuO core-shell nanomaterials have been explored for CO_2RR, as the core of copper promotes the reduction of CO_2, while the CuO shell can modulate the electronic properties of the catalyst, enhancing selectivity for specific products.

Photocatalysis, driven by light energy, is another area where core-shell nanomaterials are playing a crucial role. Photocatalytic reactions, such as water splitting for hydrogen generation, pollutant degradation, and CO_2 conversion, require efficient light absorption, charge separation, and catalytic activity [41]. Core-shell nanostructures can offer substantial improvements in these processes by enhancing the interaction between the catalyst and light, as well as improving the charge carrier dynamics. One of the primary advantages of core-shell photocatalysts is their ability to enhance light absorption and charge separation. For instance, core-shell systems that combine semiconductors with different band gaps, such as CdSe@ZnS or TiO_2@Cu, have been shown to increase the efficiency of photocatalytic water splitting. In these systems, the core absorbs light and generates electron–hole pairs, while the shell prevents the recombination of these charge carriers, allowing for more efficient charge transfer to the surface where the catalytic reactions occur. In the case of CdSe@ZnS, the core-shell structure helps to stabilize the core material, preventing photodegradation and enhancing its long-term performance under visible light.

Photocatalytic pollutant degradation is another application where core-shell catalysts have shown great promise. Pollutants in wastewater, such as organic dyes, heavy metals, and pesticides, can be effectively removed using photocatalytic processes. Core-shell nanomaterials, such as Ag@TiO_2, have been explored for the degradation of these pollutants [42]. The silver core enhances the photocatalytic performance through surface plasmon resonance (SPR), which boosts the generation of electron–hole pairs upon light exposure. The TiO_2 shell is a well-known photocatalyst that can degrade organic contaminants, and it also protects the silver from leaching or oxidation, ensuring the catalyst's stability [43]. This combination of properties results in enhanced catalytic efficiency and durability, making it suitable for large-scale environmental remediation applications. In photocatalytic CO_2 reduction, core-shell systems like Cu@Cu_2O and Ni@$NiFe_2O_4$ have shown potential to reduce CO_2 to valuable chemicals such as methane and methanol. The core material, such as copper, is active for CO_2 reduction, while the shell material, such as copper oxide or nickel ferrite, can modify the electronic structure of the catalyst, enhancing selectivity for desired products and preventing the catalyst from deactivating over time [44]. These core-shell photocatalysts have demonstrated improved efficiency compared to single-component materials and show promise in addressing the challenges of CO_2 mitigation. Figure 5.5 illustrates the mechanism of a core-shell photocatalyst for pollutant degradation.

The versatility of core-shell nanomaterials in electrocatalysis and photocatalysis extends to a wide range of applications. In CO_2 reduction, core-shell catalysts can be designed to selectively produce valuable chemicals, thereby contributing to carbon recycling and mitigating climate change. In water splitting, core-shell nanomaterials are playing a crucial role in improving the efficiency and stability of hydrogen production, which is essential for the development of hydrogen-based clean energy systems [45]. Additionally, in the degradation of pollutants, core-shell photocatalysts are offering a promising solution for the removal of organic contaminants and hazardous substances from water and air. These applications are all crucial for advancing renewable energy technologies, environmental sustainability, and addressing global energy challenges. Core-shell nanomaterials provide a unique opportunity to enhance catalytic performance, reduce costs, and extend the

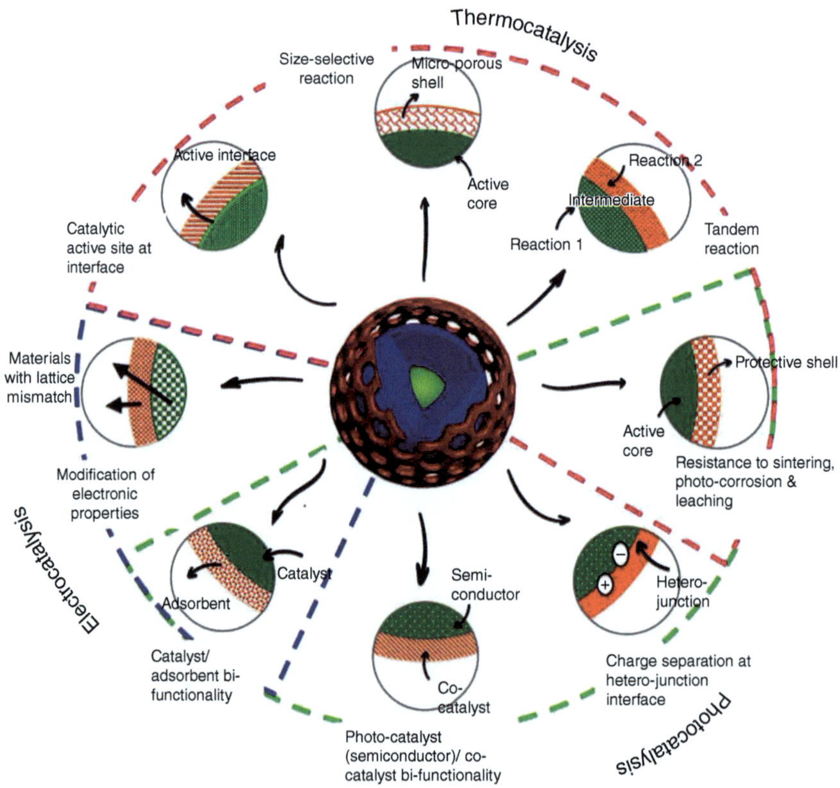

Figure 5.5 Core-shell nanomaterials as efficient catalysts for thermocatalysis, electrocatalysis, and photocatalysis processes. [44]/with permission of Royal Society of Chemistry.

longevity of catalysts, making them a key component of future energy systems and environmental remediation strategies [16, 46, 47].

In conclusion, the integration of core-shell nanomaterials into electrocatalysis and photocatalysis is revolutionizing the efficiency, selectivity, and durability of catalysts for energy and environmental applications. Through strategic design and material selection, core-shell structures offer significant advantages in reactions such as CO_2 reduction, water splitting, and pollutant degradation. These materials are poised to play a pivotal role in addressing global energy and environmental challenges, providing solutions that are both sustainable and efficient. Table 5.3 shows the comparison of core-shell catalysts for various electrocatalytic and photocatalytic reactions.

Table 5.3 Comparison of core-shell catalysts for various electrocatalytic and photocatalytic reactions.

Material composition	Reaction type	Efficiency	Stability
Pt@Ni (Platinum core, Nickel shell)	Hydrogen Evolution Reaction (HER)	High catalytic activity due to enhanced hydrogen binding strength	High stability, reduced platinum degradation in acidic conditions
RuO_2@IrO_2 (Ruthenium oxide core, Iridium oxide shell)	Oxygen Evolution Reaction (OER)	Excellent catalytic efficiency, particularly under high voltages	High stability, minimal degradation over long-term operation
Cu@Cu_2O (Copper core, Copper oxide shell)	Oxygen Evolution Reaction (OER)	Moderate efficiency, lower voltage requirements for OER compared to conventional catalysts	Good stability under alkaline conditions, susceptible to oxidation under acidic conditions
Fe@Fe_2O_3 (Iron core, Iron oxide shell)	Oxygen Evolution Reaction (OER)	High catalytic activity and lower cost compared to platinum-based catalysts	Excellent stability in alkaline media, but less stable under acidic conditions
Ni@NiO (Nickel core, Nickel oxide shell)	Hydrogen Evolution Reaction (HER)	Improved hydrogen production efficiency at lower overpotentials	Good stability under basic conditions, moderate performance in acidic environments
Pt@Co (Platinum core, Cobalt shell)	Hydrogen Evolution Reaction (HER)	High efficiency, enhanced hydrogen adsorption due to the core-shell structure	Excellent stability, reduced cost by lowering platinum content
TiO_2@Cu (Titanium oxide core, Copper shell)	Photocatalytic Water Splitting	Enhanced light absorption and charge separation efficiency	Moderate stability; copper shell prone to oxidation over time

(Continued)

Table 5.3 (Continued)

Material composition	Reaction type	Efficiency	Stability
$CdSe@ZnS$ (Cadmium selenide core, Zinc sulfide shell)	Photocatalytic Water Splitting	High quantum efficiency for hydrogen production under visible light	Stable under UV light, reduced photodegradation due to ZnS shell
$Ag@TiO_2$ (Silver core, Titanium dioxide shell)	Photocatalytic Organic Pollutant Degradation	Enhanced photocatalytic efficiency due to surface plasmon resonance (SPR) of silver	Excellent long-term stability, minimal leaching of silver
$Co@Co_3O_4$ (Cobalt core, Cobalt oxide shell)	Oxygen Reduction Reaction (ORR)	High efficiency, lower overpotentials compared to conventional platinum catalysts	Stable in acidic media, offers good corrosion resistance
$Cu@CuO$ (Copper core, Copper oxide shell)	Electrocatalytic CO_2 Reduction	Moderate efficiency, but reduced overpotentials for CO_2 reduction	Stable under reaction conditions, susceptible to oxidation over time
$Ni@NiFe_2O_4$ (Nickel core, Nickel ferrite shell)	Water Splitting (OER and HER)	High efficiency for both HER and OER, improved stability over single-component catalysts	High stability under both acidic and basic conditions, minimal catalyst deactivation

5.6 Conclusion

Core-shell nanomaterials have proven to be transformative in the fields of energy conversion and storage, offering novel solutions to some of the most pressing challenges in energy sustainability, environmental remediation, and the development of efficient technologies. Throughout the chapter, we have explored the diverse applications of core-shell structures in PVs, energy storage systems, hydrogen generation, and electrocatalysis and photocatalysis. In PVs, core-shell materials have demonstrated their ability to enhance light absorption, improve charge carrier separation, and increase the stability of solar cells. These advancements contribute to the production of more efficient and durable solar energy technologies, which are crucial for the global transition to renewable energy. In the domain of energy storage, core-shell nanomaterials have emerged as advanced electrode materials, significantly enhancing the performance of lithium-ion and sodium-ion batteries, supercapacitors, and fuel cells. Their ability to facilitate efficient ion transport and maintain stability under repeated cycling makes them vital for next-generation energy storage solutions.

Hydrogen generation, through processes such as water splitting, is another area where core-shell catalysts have shown considerable promise. The catalytic efficiency and durability of these materials significantly improve the overall performance of hydrogen production systems, providing a cleaner and more sustainable method of energy generation. Furthermore, core-shell materials have proven to be invaluable in electrocatalytic and photocatalytic reactions, offering enhanced efficiency in processes like CO_2 reduction and pollutant degradation, which play critical roles in environmental sustainability.

While the potential of core-shell nanomaterials is immense, challenges such as scalability, cost, and regulatory concerns remain. Addressing these issues will require continued innovation in synthesis methods, as well as the development of standardized protocols for their use in industrial applications. Moreover, the integration of these materials into real-world systems will require close collaboration between material scientists, engineers, and policymakers to ensure their safe and effective deployment.

In summary, core-shell nanomaterials are at the forefront of advancing technologies in energy conversion, storage, and environmental protection. Their unique structural and functional properties enable significant improvements in efficiency, stability, and selectivity, making them indispensable in tackling the world's energy and environmental challenges. As research in this area progresses, we can expect to see even more groundbreaking applications that will further accelerate the shift toward a sustainable and energy-efficient future.

References

1 Khurana LK, Acharya AB, editors. Energy and infrastructure management in post covid-19 era - Google books [Internet]. Allied Publishers Private Limited. 2022 [cited 2025 Mar 7]. Available from: https://books.google.co.in/books?hl=en&lr=&id=1KGEEAAAQBAJ&oi=fnd&pg=PA3&dq=Energy-related+concerns+such+as+the+depletion+of+fossil+fuels,+environmental+degradation,+and+the+urgent+need+for+sustainable+alternatives+have+driven+advancements+in+nanotechnology&ots=c6OSezlky4&sig=iKxHCoSk3y7Ia9BHKX8JDkzOBq0&redir_esc=y#v=onepage&q&f=false

2 Feng HP, Tang L, Zeng GM, Tang J, Deng YC, Yan M, et al. Carbon-based core–shell nanostructured materials for electrochemical energy storage. J Mater Chem A Mater [Internet]. 2018 May 1 [cited 2025 Mar 7];6(17):7310–37. Available from: https://pubs.rsc.org/en/content/articlehtml/2018/ta/c8ta01257a

3 Shafiee A, Rabiee N, Ahmadi S, Baneshi M, Khatami M, Iravani S, et al. Core–shell nanophotocatalysts: Review of materials and applications. ACS Publications [Internet]. 2022 Jan 28 [cited 2025 Mar 7];2022(1):86. Available from: https://pubs.acs.org/doi/abs/10.1021/acsanm.1c03714?casa_token=lhGXa6w9vJoAAAAA:8xJMxkxb5G0TjKXanzVpSX9omnvwFcxvrieFdEoUm_n4Ey-dvkab1HSzmfthaQElC6j0Eu18TGcapI9yUw

4 Darwish MSA, El Naggar AMA, Morshedy AS, Haneklaus N. Increased production of hydrogen with in situ CO_2 capture through the process of water splitting using magnetic core/shell structures as novel photocatalysts. Environ Sci Pollut Res [Internet]. 2021 Jan 1 [cited 2025 Mar 7];28(3):3566–78. Available from: https://link.springer.com/article/10.1007/s11356-020-10752-w

5 Liu L, Wang H, Wang D, Li Y, He X, Zhang H, et al. ZnO@TiO_2 Core/shell nanowire arrays with different thickness of TiO_2 shell for dye-sensitized solar cells. Crystals [Internet]. 2020 Apr 21 [cited 2025 Mar 7];10(4):325. Available from: https://www.mdpi.com/2073-4352/10/4/325/htm
6 Li X, Wang J. One-dimensional and two-dimensional synergized nanostructures for high-performing energy storage and conversion. InfoMat [Internet]. 2020 Jan 1 [cited 2024 Sep 22];2(1):3–32. Available from: https://onlinelibrary.wiley.com/doi/full/10.1002/inf2.12040
7 Pustovalov VK, Astafyeva LG, Fritzsche W. Optical properties of core-shell nanoparticles and their application for effective absorption, scattering, extinction and filtering solar and optical radiation. Photonics Nanostruct. 2024 Dec 1;62:101312.
8 Yeo RJ, Yeo JCC, Yu Tan S, Sng A, Tomczak N, Muiruri JK, et al. Core-shell micro- and nano-structures for the modification of light-surface interactions. Adv Opt Mater [Internet]. 2024 Feb 1 [cited 2025 Mar 7];12(4):2301955. Available from: https://onlinelibrary.wiley.com/doi/full/10.1002/adom.202301955
9 Zhou Q, Liu X, Liu Z. The balance between efficiency, stability and environmental impacts in perovskite solar cells: A review. J Phys Energy [Internet]. 2020 Feb 10 [cited 2025 Mar 7];2(2):022001. Available from: https://iopscience.iop.org/article/10.1088/2515-7655/ab5eee
10 Li Z. High performance PbS and PbS/CdS quantum dots for luminescence solar concentrators [PhD thesis]. [cited 2025 Mar 7]. Available from: https://doi.org/10.26190/unsworks/28797
11 Singh Selopal G, Zhao H, Wang ZM, Rosei F, Selopal GS, Wang ZM, et al. Core/shell quantum dots solar cells. Adv Funct Mater [Internet]. 2020 Mar 1 [cited 2025 Mar 7];30(13):1908762. Available from: https://onlinelibrary.wiley.com/doi/full/10.1002/adfm.201908762
12 Xu L, Yuan S, Ma L, Zhang B, Fang T, Li X, et al. All-inorganic perovskite quantum dots as light-harvesting, interfacial, and light-converting layers toward solar cells. J Mater Chem A Mater [Internet]. 2021 Sep 14 [cited 2025 Mar 7];9(35):18947–73. Available from: https://pubs.rsc.org/en/content/articlehtml/2021/ta/d1ta02786g
13 Zhang S, Liu Z, Zhang W, Jiang Z, Chen W, Chen R, et al. Barrier designs in perovskite solar cells for long-term stability. Adv Energy Mater [Internet]. 2020 Sep 1 [cited 2025 Mar 7];10(35):2001610. Available from: https://onlinelibrary.wiley.com/doi/full/10.1002/aenm.202001610
14 Sharma SN, Semalti P, Rajpal B, Rao AS. Pioneering advancements in quantum dot solar cells: Innovations in synthesis and cutting-edge applications. Curr Opin Colloid Interface Sci. 2025 Jun 1;77:101905.
15 Lee KH, Han SH, Chuquer A, Yang HY, Kim J, Pham XH, et al. Effect of Au nanoparticles and scattering layer in dye-sensitized solar cells based on freestanding TiO_2 nanotube arrays. Nanomaterials [Internet]. 2021 Jan 27 [cited 2025 Mar 7];11(2):328. Available from: https://www.mdpi.com/2079-4991/11/2/328/htm
16 Li W, Elzatahry A, Aldhayan D, Zhao D. Core–shell structured titanium dioxide nanomaterials for solar energy utilization. Chem Soc Rev [Internet]. 2018 Nov 12

[cited 2025 Mar 7];47(22):8203–37. Available from: https://pubs.rsc.org/en/content/articlehtml/2018/cs/c8cs00443a

17 Su L, Jing Y, Zhou Z. Li ion battery materials with core–shell nanostructures. Nanoscale [Internet]. 2011 Oct 5 [cited 2025 Mar 7];3(10):3967–83. Available from: https://pubs.rsc.org/en/content/articlehtml/2011/nr/c1nr10550g

18 Feng K, Li M, Liu W, Kashkooli AG, Xiao X, Cai M, et al. Silicon-based anodes for lithium-ion batteries: From fundamentals to practical applications. Small [Internet]. 2018 Feb 1 [cited 2025 Mar 7];14(8):1702737. Available from: https://onlinelibrary.wiley.com/doi/full/10.1002/smll.201702737

19 Malavekar D, Pujari S, Jang S, Bachankar S, Kim JH. Recent development on transition metal oxides-based core–shell structures for boosted energy density supercapacitors. Small [Internet]. 2024 Aug 1 [cited 2025 Mar 7];20(31):2312179. Available from: https://onlinelibrary.wiley.com/doi/full/10.1002/smll.202312179

20 Lokhande VC, Lokhande AC, Lokhande CD, Kim JH, Ji T. Supercapacitive composite metal oxide electrodes formed with carbon, metal oxides and conducting polymers. J Alloys Compd. 2016 Oct 15;682:381–403.

21 Wang X, Wang X, Yi L, Liu L, Dai Y, Wu H. Preparation and capacitive properties of the core–shell structure carbon aerogel microbeads- nanowhisker-like NiO composites. J Power Sources. 2013 Feb 15;224:317–23.

22 Han A, Liu G. Recent advances in metal-based electrocatalysts: From fundamentals and structural regulations to applications in anion-exchange membrane fuel cells. Mater Chem Front [Internet]. 2024 Feb 12 [cited 2025 Mar 7];8(4):903–29. Available from: https://pubs.rsc.org/en/content/articlehtml/2024/qm/d3qm00947e

23 Wang Z, Tang M. The cytotoxicity of core-shell or non-shell structure quantum dots and reflection on environmental friendly: A review. Environ Res. 2021 Mar 1;194:110593.

24 Zou Y, Sun Z, Wang Q, Ju Y, Sun N, Yue Q, et al. Core-shell magnetic particles: Tailored synthesis and applications. Chem Rev [Internet]. 2025 Jan 22 [cited 2025 Mar 7];125(2):972–1048. Available from: https://pubs.acs.org/doi/full/10.1021/acs.chemrev.4c00710

25 Gohar O, Zubair Khan M, Bibi I, Bashir N, Tariq U, Bakhtiar M, et al. Nanomaterials for advanced energy applications: Recent advancements and future trends. Mater Des. 2024 May 1;241:112930.

26 Zhu X, Li Y, Gu N. Application of artificial intelligence in the exploration and optimization of biomedical nanomaterials. Nano Biomed Eng. 2023 [cited 2025 Mar 7];15(3):342–353. Available from: http://doi.org/10.26599/NBE.2023.9290035

27 Ma X, Li Y, Hussain I, Shen R, Yang G, Zhang K, et al. Core–shell structured nanoenergetic materials: Preparation and fundamental properties. Adv Mater [Internet]. 2020 Jul 1 [cited 2025 Mar 7];32(30):2001291. Available from: https://onlinelibrary.wiley.com/doi/full/10.1002/adma.202001291

28 Hua W, Schwarz B, Azmi R, Müller M, Dewi Darma MS, Knapp M, et al. Lithium-ion (de)intercalation mechanism in core-shell layered Li(Ni,Co,Mn)O_2 cathode materials. Nano Energy. 2020 Dec 1;78:105231.

29 Thakur J, Phogat P, Sharma S. Catalyst design for efficient water splitting: A comprehensive review of challenges and opportunities. Fuel [Internet]. 2025

Jul [cited 2025 Mar 7];392:134954. Available from: https://www.researchgate.net/publication/389599720_Catalyst_design_for_efficient_water_splitting_A_comprehensive_review_of_challenges_and_opportunities

30 Zhao X, Sasaki K. Advanced Pt-based core-shell electrocatalysts for fuel cell cathodes. Acc Chem Res. 2022 May 3;55(9):1226–36.

31 Yin X, Yang L, Gao Q. Core–shell nanostructured electrocatalysts for water splitting. Nanoscale [Internet]. 2020 Aug 6 [cited 2025 Mar 7];12(30):15944–69. Available from: https://pubs.rsc.org/en/content/articlehtml/2020/nr/d0nr03719b

32 Gawande MB, Goswami A, Asefa T, Guo H, Biradar AV, Peng DL, et al. Core–shell nanoparticles: Synthesis and applications in catalysis and electrocatalysis. Chem Soc Rev [Internet]. 2015 Oct 19 [cited 2025 Mar 7];44(21):7540–90. Available from: https://pubs.rsc.org/en/content/articlehtml/2015/cs/c5cs00343a

33 Lin YR, Chang YC, Chiao YC, Ko FH. Au@CdS nanocomposites as a visible-light photocatalyst for hydrogen generation from tap water. Catalysts [Internet]. 2022 Dec 24 [cited 2025 Mar 6];13(1):33. Available from: https://www.mdpi.com/2073-4344/13/1/33/htm

34 Zhang Q, Lee I, Joo JB, Zaera F, Yin Y. Core-shell nanostructured catalysts. Acc Chem Res [Internet]. 2013 Aug 20 [cited 2025 Mar 7];46(8):1816–24. Available from: https://pubs.acs.org/doi/full/10.1021/ar300230s

35 Xu Y, Li Y, Gao L, Liu Y, Ding Z. Advances and prospects of nanomaterials for solid-state hydrogen storage. Nanomaterials [Internet]. 2024 Jun 16 [cited 2025 Mar 7];14(12):1036. Available from: https://www.mdpi.com/2079-4991/14/12/1036/htm

36 Hu W, Zheng M, Xu B, Wei Y, Zhu W, Li Q, et al. Design of hollow carbon-based materials derived from metal–organic frameworks for electrocatalysis and electrochemical energy storage. J Mater Chem A Mater [Internet]. 2021 Feb 23 [cited 2025 Mar 7];9(7):3880–917. Available from: https://pubs.rsc.org/en/content/articlehtml/2021/ta/d0ta10666f

37 Ghosh Chaudhuri R, Paria S. Core/shell nanoparticles: Classes, properties, synthesis mechanisms, characterization, and applications. Chem Rev [Internet]. 2012 Apr 11 [cited 2025 Mar 7];112(4):2373–433. Available from: https://pubs.acs.org/doi/full/10.1021/cr100449n

38 Yuan Y, Li J, Zhu Y, Qiao Y, Kang Z, Wang Z, et al. Water in electrocatalysis. Angew Chem Int Ed [Internet]. 2025 Feb 21 [cited 2025 Mar 7];64(18):e202425590. Available from: https://onlinelibrary.wiley.com/doi/full/10.1002/anie.202425590

39 Gong S, Zhang YX, Niu Z. Recent advances in earth-abundant core/noble-metal shell nanoparticles for electrocatalysis. ACS Catal [Internet]. 2020 Oct 2 [cited 2025 Mar 7];10(19):10886–904. Available from: https://pubs.acs.org/doi/full/10.1021/acscatal.0c02587

40 Liu SH, Wu CC. Three-dimensional nitrogen-doped graphene quantum dot/Ag leaf/Cu_2O-Cu trunk/nickel foam nanocomposites as ultra-high-efficiency cathode electrocatalysts for hydrogen evolution reaction applications. J Energy Storage. 2025 Jan 15;106:114727.

41 Acar C, Dincer I, Naterer GF. Review of photocatalytic water-splitting methods for sustainable hydrogen production. Int J Energy Res [Internet]. 2016 Sep 1 [cited 2025 Mar 7];40(11):1449–73. Available from: https://onlinelibrary.wiley.com/doi/full/10.1002/er.3549

42 Altaf NUH, Naz MY, Shukrullah S, Bhatti HN. Growth of Ag, TiO_2 and $AgTiO_2$ nanoparticles through liquid-plasma interaction for bacterial and wastewater treatment. Pak J Agric Sci [Internet]. 2024 May 1 [cited 2025 Mar 7];61(3):889. Available from: https://openurl.ebsco.com/contentitem/doi:10.21162%2FPAKJAS%2F24.262?sid=ebsco:plink:crawler&id=ebsco:doi:10.21162%2FPAKJAS%2F24.262

43 Ramesh N, Lai CW, Johan MRB, Mousavi SM, Badruddin IA, Kumar A, et al. Progress in photocatalytic degradation of industrial organic dye by utilising the silver doped titanium dioxide nanocomposite. Heliyon [Internet]. 2024 Dec 30 [cited 2025 Mar 7];10(24):e40998. Available from: https://www.cell.com/action/showFullText?pii=S2405844024170296

44 Das S, Pérez-Ramírez J, Gong J, Dewangan N, Hidajat K, Gates BC, et al. Core–shell structured catalysts for thermocatalytic, photocatalytic, and electrocatalytic conversion of CO_2. Chem Soc Rev [Internet]. 2020 May 26 [cited 2025 Mar 7];49(10):2937–3004. Available from: https://pubs.rsc.org/en/content/articlehtml/2020/cs/c9cs00713j

45 Elsapagh RM, Sultan NS, Mohamed FA, Fahmy HM. The role of nanocatalysts in green hydrogen production and water splitting. Int J Hydrogen Energy. 2024 May 20;67:62–82.

46 Wu ZP, Shan S, Zang SQ, Zhong CJ. Dynamic core-shell and alloy structures of multimetallic nanomaterials and their catalytic synergies. Acc Chem Res. 2020 Dec 15;53(12):2913–24.

47 Hu P, Morabito JV, Tsung CK. Core-shell catalysts of metal nanoparticle core and metal-organic framework shell. ACS Catal. 2014 Dec 5;4(12):4409–19.

6

Core-shell Nanomaterials in Environmental Applications

Core-shell nanomaterials have garnered significant attention for their potential to address a wide range of environmental challenges, offering innovative solutions for pollution control, resource recovery, and sustainability. These materials, composed of a distinct core and functional shell, exhibit unique properties that enable efficient pollutant removal, energy recovery, and environmental remediation. This chapter explores the diverse applications of core-shell nanomaterials in environmental contexts, focusing on water purification, air quality improvement, carbon capture, waste management, and resource recycling. The chapter highlights the mechanisms by which core-shell materials operate, their advantages over conventional materials, and their potential to revolutionize waste treatment and pollution control processes. However, the widespread adoption of these materials is hindered by scalability issues, high production costs, and concerns regarding their environmental impact and toxicity. Looking to the future, the development of biodegradable, multifunctional, and smart core-shell nanomaterials, coupled with advancements in artificial intelligence (AI)-driven design and sustainable manufacturing processes, holds the promise of overcoming these challenges. The chapter concludes by emphasizing the need for continued research to optimize the environmental benefits of core-shell nanomaterials while minimizing their ecological footprint.

6.1 Introduction

The 21st century has witnessed unparalleled technological advancement and industrial growth, improving the quality of life for millions. However, this progress has come at a considerable environmental cost, threatening both human health and the stability of ecosystems. One of the most pressing concerns is pollution in its various forms—air, water, and soil contamination—which continues to degrade environmental quality across the globe. The World Health Organization (WHO) estimates that air pollution alone accounts for millions of premature deaths annually. Water sources are increasingly burdened with industrial effluents, agricultural runoffs, and hazardous waste, leading to severe health risks for communities.

Core-Shell Nanomaterials: From Fundamentals to Applications, First Edition. Shreya Sharma et al.

Additionally, the inefficient management of solid waste and the over-reliance on fossil fuels have exacerbated carbon emissions, contributing to climate change and environmental degradation. These multifaceted challenges require innovative solutions that go beyond conventional methods to safeguard the environment for future generations.

Conventional environmental remediation techniques, while valuable, often face limitations. Filtration systems, chemical treatments, and catalytic converters, for example, can be resource-intensive, costly, and sometimes generate secondary pollutants [1]. Furthermore, the increasing complexity of pollutants necessitates advanced materials with enhanced functional properties that can operate efficiently across diverse environmental conditions. To address these challenges, the integration of nanotechnology into environmental remediation has emerged as a game-changing approach. Nanotechnology enables the manipulation of materials at the atomic and molecular levels, leading to superior physical, chemical, and electronic properties. This capability makes nanomaterials particularly effective for capturing, degrading, and sensing pollutants, positioning them as vital components in modern environmental protection strategies.

Nanomaterials, defined by their nanoscale dimensions (typically between 1 and 100 nanometers), exhibit unique properties such as a high surface area-to-volume ratio, enhanced reactivity, and tunable electronic characteristics. These features make them highly versatile for applications in adsorption, filtration, catalysis, and sensing. For instance, metal oxide nanoparticles have shown great promise in removing heavy metals from water, while carbon-based nanomaterials have been used for capturing volatile organic compounds (VOCs) in the air [2]. Nanostructured catalysts, on the other hand, can break down complex organic pollutants into benign by-products. Figure 6.1 provides a conceptual illustration of how nanomaterials, including core-shell structures, address key environmental issues such as air pollution, water contamination, and waste management. Figure 6.1 shows a conceptual diagram showing environmental issues (pollution, water contamination, air quality, etc.) being tackled by core-shell nanomaterials. Among the various classes of nanomaterials, core-shell nanostructures have garnered significant attention due to their exceptional multifunctionality and structural flexibility. A core-shell nanomaterial typically consists of a central core surrounded by one or more shell layers, each serving specific roles. This design offers several advantages, including improved stability, enhanced surface interactions, and the ability to integrate complementary properties within a single structure. By carefully selecting the core and shell materials, researchers can tailor the overall performance of the nanostructure to meet the requirements of specific environmental applications.

The role of core-shell nanomaterials in environmental remediation is expansive and transformative. In water purification, core-shell materials have demonstrated remarkable capabilities in adsorption, filtration, and photocatalytic degradation. Metal-core@oxide-shell nanomaterials, for instance, can effectively capture heavy metals such as arsenic and lead while providing stability and reusability. The shell layer, typically made of silica or alumina, protects the reactive core from environmental degradation, thereby prolonging the lifespan of the material. Photocatalytic core-shell nanostructures, such as TiO_2-based composites, harness solar energy to break down organic pollutants and disinfect water [3]. This approach offers a

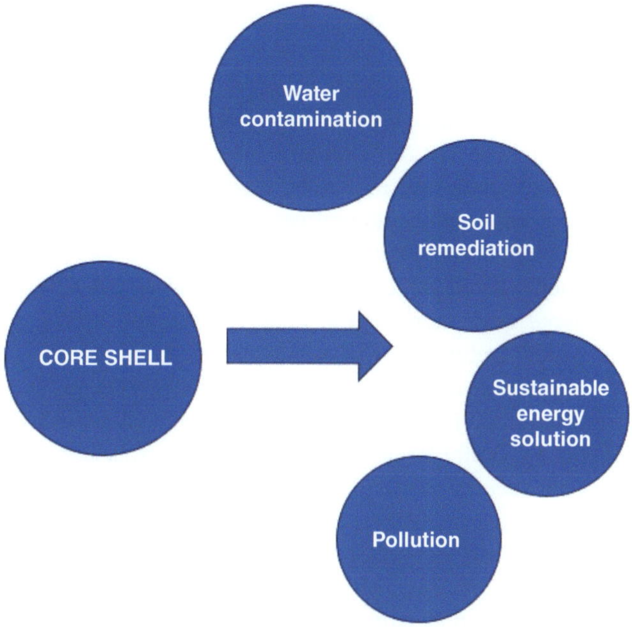

Figure 6.1 A conceptual diagram showing environmental issues (pollution, water contamination, air quality, etc.) being tackled by core-shell nanomaterials.

sustainable alternative to traditional chemical treatments that often involve toxic reagents. Air quality improvement is another critical area where core-shell nanomaterials are making significant contributions. Urban environments are plagued by harmful gases such as nitrogen oxides (NOx), carbon monoxide (CO), and sulfur dioxide (SO_2), along with particulate matter (PM) [4]. Core-shell nanomaterials are being employed in catalytic converters and air filters to capture and neutralize these pollutants. Their high surface area and tunable chemical properties make them highly effective in adsorbing and degrading airborne contaminants. Additionally, gas sensors based on core-shell nanostructures have demonstrated enhanced sensitivity and selectivity, enabling real-time monitoring of air quality. For example, ZnO@carbon core-shell materials have shown excellent performance in detecting low concentrations of toxic gases.

Beyond water and air remediation, core-shell nanomaterials are playing a pivotal role in carbon capture and sequestration (CCS) technologies. The rising levels of atmospheric CO_2 are a major driver of global warming, necessitating the development of efficient carbon capture systems. Core-shell nanomaterials, particularly those based on porous carbon cores and metal-organic framework (MOF) shells, have shown great potential for selectively adsorbing and storing CO_2 [5]. These materials not only offer high adsorption capacities but also exhibit stability under harsh conditions, making them suitable for large-scale industrial applications. Furthermore, some core-shell structures can catalyze the conversion of captured CO_2 into useful chemicals, contributing to a circular carbon economy.

Waste management and recycling are additional domains where core-shell nanomaterials are driving innovation. Traditional waste treatment methods often fall short in recovering valuable resources or neutralizing hazardous components.

Core-shell catalysts have emerged as powerful tools for transforming waste into valuable products such as biofuels and specialty chemicals. For instance, metal@carbon core-shell materials have been used to catalyze the conversion of biomass into biofuels, offering a sustainable solution to the global energy crisis. In the context of e-waste recycling, core-shell nanomaterials can facilitate the selective recovery of precious metals such as gold and platinum. Moreover, biodegradable core-shell materials are being developed as eco-friendly alternatives to conventional plastics, addressing the growing problem of plastic pollution.

The importance of core-shell nanomaterials for environmental sustainability extends beyond their functional advantages. These materials are also enabling greener and more sustainable manufacturing practices. Green synthesis approaches, which involve the use of natural precursors and benign reaction conditions, are gaining popularity as alternatives to traditional chemical methods that often involve toxic solvents and high energy consumption. For example, plant extracts and biodegradable templates have been successfully used to fabricate core-shell nanomaterials with excellent environmental remediation properties [6]. Another key factor contributing to the sustainability of core-shell nanomaterials is their scalability and adaptability for large-scale applications. Advances in manufacturing techniques, such as spray drying, self-assembly, and electrospinning, have made it possible to produce these materials in cost-effective and scalable ways. This scalability is crucial for deploying core-shell nanomaterials in real-world environmental remediation scenarios, where large quantities of material are often required.

Despite their immense potential, core-shell nanomaterials face certain challenges that need to be addressed to unlock their full environmental benefits. One major challenge is the complexity of synthesis and the need for precise control over the core-shell structure. Achieving uniformity and reproducibility at the nanoscale remains a significant hurdle [7]. Additionally, the environmental impact of some core-shell nanomaterials, particularly those containing heavy metals, raises concerns about their long-term sustainability. Therefore, efforts must be made to develop safe and biodegradable alternatives that do not pose risks to human health or the environment. Looking ahead, the integration of core-shell nanomaterials with emerging technologies such as AI and machine learning (ML) holds great promise. AI-driven approaches can accelerate the discovery and optimization of core-shell nanomaterials by predicting their performance and identifying optimal synthesis conditions. Moreover, interdisciplinary collaborations involving materials science, environmental engineering, and computational modeling are likely to drive the development of next-generation core-shell nanomaterials with unprecedented capabilities.

6.2 Core-shell Nanomaterials for Water Purification

6.2.1 Adsorption Techniques

The increasing presence of hazardous pollutants in water and air has necessitated the development of effective remediation strategies. Among these, adsorption is one of the most promising methods due to its simplicity, cost-effectiveness, and ability to handle various contaminants. Adsorption involves the accumulation of substances (adsorbates) on the surface of a solid or liquid (adsorbent). Core-shell nanomaterials have gained significant attention as advanced adsorbents owing to their unique structure, high surface area, and tunable surface chemistry [8]. These materials provide enhanced adsorption efficiency, stability, and reusability, making them ideal for tackling environmental pollution.

The mechanism of adsorption using core-shell nanomaterials primarily relies on the interplay between the core and the shell components. The core typically provides structural stability and may contribute to the overall adsorption process by offering additional active sites. The shell, on the other hand, is engineered to enhance the surface area, functional groups, and selectivity of the material.

The adsorption process generally involves physical or chemical interactions between the adsorbent and the target contaminant molecules.

a. **Physical Adsorption (Physisorption):** This occurs due to weak van der Waals forces between the adsorbate and the surface of the core-shell material [9]. Physisorption is generally reversible and less specific but allows for high adsorption rates due to large surface areas provided by nanomaterials.
b. **Chemical Adsorption (Chemisorption):** Chemisorption involves the formation of chemical bonds between the adsorbate and functional groups present on the shell surface [10]. This process is more specific and often irreversible, providing stronger adsorption interactions for target pollutants.

In many cases, hybrid mechanisms involving both physical and chemical adsorption are observed, which further enhance the performance of core-shell materials. Figure 6.2 illustrates the adsorption process using a core-shell nanomaterial, highlighting the interactions between the material surface and pollutant molecules.

Numerous core-shell nanomaterials have demonstrated remarkable efficiency in removing heavy metals, dyes, and other pollutants from water and air. Below are some notable examples:

- **Fe_3O_4@SiO_2 for heavy metal removal:** Iron oxide (Fe_3O_4) core-shell nanostructures coated with silica (SiO_2) have proven highly effective in adsorbing toxic heavy metals such as arsenic, lead, and mercury from water. The magnetic Fe_3O_4 core facilitates easy separation of the adsorbent from the treated water using an external magnetic field [12], while the SiO_2 shell provides a chemically stable surface with abundant hydroxyl groups for binding heavy metals.

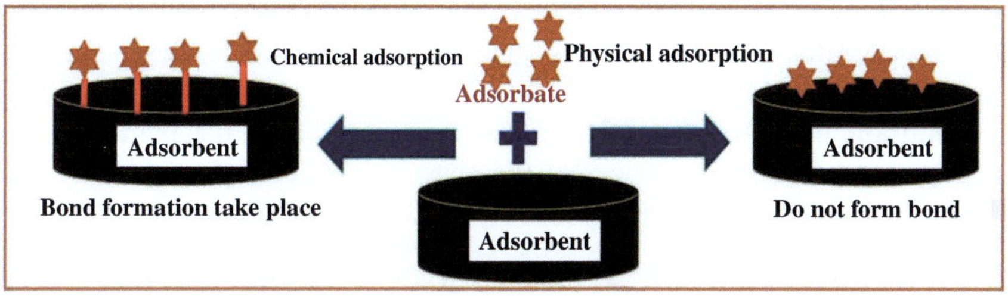

Figure 6.2 Schematic representation of the adsorption process using core-shell nanomaterials. [11]/De Gruyter/CC BY 4.0.

- **ZnO@Carbon for organic dye adsorption:** Core-shell structures comprising a zinc oxide (ZnO) core and a carbon shell have shown excellent performance in adsorbing organic dyes from wastewater [13]. The carbon shell enhances the adsorption capacity due to its porous nature and strong interactions with dye molecules.
- **TiO_2@Polymer for multifunctional adsorption:** Titanium dioxide (TiO_2) coated with polymer shells has demonstrated effective removal of both inorganic and organic pollutants [14]. The polymer shell can be functionalized with specific groups to target particular contaminants, while TiO_2 contributes to the photocatalytic degradation of adsorbed pollutants.
- **Magnetic core-shell nanoparticles for arsenic removal:** Core-shell materials with magnetic cores and functionalized polymer or metal oxide shells have been extensively used for arsenic removal [15]. These materials exhibit high adsorption efficiency and can be rapidly recovered using external magnets, making them suitable for real-world applications.

Several factors influence the adsorption efficiency of core-shell nanomaterials:

- **Surface area and pore structure:** High surface area and a well-defined pore structure facilitate greater contact between the adsorbent and the target pollutants, enhancing adsorption capacity. Core-shell materials with mesoporous or hierarchical pore structures are particularly effective in this regard [16].
- **Surface functional groups:** The presence of specific functional groups on the shell surface can significantly influence adsorption efficiency. For instance, hydroxyl, carboxyl, and amine groups provide active binding sites for heavy metals and organic pollutants [17].
- **Core-shell composition:** The choice of core and shell materials plays a crucial role in determining the adsorption performance. Magnetic cores enable easy separation, while functional shells can be tailored for selectivity and stability.
- **pH of the solution:** The pH of the solution affects the ionization state of both the adsorbent and the adsorbate, influencing the adsorption efficiency. For example, metal oxides exhibit higher adsorption capacity for heavy metals at specific pH ranges due to changes in surface charge.
- **Temperature:** Temperature can influence the kinetics and thermodynamics of the adsorption process. Higher temperatures may increase adsorption rates but can also lead to desorption in some cases.
- **Contact time:** Adequate contact time between the adsorbent and the pollutant is essential for achieving optimal adsorption efficiency. Core-shell nanomaterials often demonstrate rapid adsorption kinetics due to their nanoscale dimensions and high surface reactivity.
- **Pollutant concentration:** The initial concentration of the pollutant can impact the adsorption capacity of the material. Higher pollutant concentrations may saturate the available binding sites, reducing efficiency.

Core-shell nanomaterials have revolutionized adsorption techniques for environmental remediation by offering enhanced surface properties, functional tunability, and structural stability. Their ability to selectively target and capture various pollutants makes them indispensable for sustainable water and air treatment technologies. By carefully optimizing factors, such as core-shell composition, surface functionalization, and operating conditions, these materials can play a pivotal role in mitigating the adverse impacts of environmental pollution.

6.2.2 Filtration Applications

Filtration is one of the most widely used techniques for removing contaminants from water, air, and industrial effluents. In environmental and industrial applications, filtration processes must be highly efficient, durable, and capable of handling a wide variety of pollutants. Membrane technologies, a cornerstone of modern filtration systems, have benefited immensely from advancements in nanotechnology, particularly through the use of core-shell nanomaterials. These materials enhance the performance of filtration membranes, improving their selectivity, permeability, and longevity. Core-shell nanomaterials, with their unique structure consisting of a core material surrounded by a protective or functional shell, provide enhanced properties that make them ideal candidates for improving filtration technologies [18]. The core material generally imparts structural integrity, while the shell can be engineered for specific properties such as hydrophilicity, antifouling, or chemical resistance. By combining these features, core-shell nanomaterials help to tackle the challenges associated with filtration, such as pore clogging, reduced flux, and limited selectivity for particular contaminants.

Filtration membranes are used in various applications, including water purification, wastewater treatment, and air filtration. The key to improving these systems lies in enhancing the performance of the membranes themselves. Core-shell nanomaterials have emerged as a transformative solution due to their ability to fine-tune the surface properties of membranes. One of the main areas where core-shell materials contribute is in improving the permeability and selectivity of the filtration process. For instance, in reverse osmosis (RO) membranes, core-shell nanoparticles can be incorporated into the membrane matrix to increase the porosity and surface area, allowing for higher water flux without sacrificing contaminant rejection efficiency [19]. The core material, such as silica or carbon, provides stability and strength, while the shell can be modified with functional groups to improve the interaction between the membrane and the contaminants, leading to better separation efficiency. Moreover, core-shell nanomaterials can be used to create membrane surfaces that are selectively permeable to certain ions or molecules. This selectivity is crucial for processes such as desalination, where the membrane must reject salt and other solutes while allowing water molecules to pass through. By adjusting the core-shell architecture, the membrane's interaction with different pollutants, whether organic or inorganic, can be precisely controlled, enhancing the efficiency of filtration systems. Figure 6.3 shows an illustration of a filtration membrane.

One of the persistent challenges in membrane filtration is membrane fouling, a process where contaminants accumulate on the membrane surface, reducing its performance over time. Fouling leads to clogging of pores, reduced flow rates, and higher energy costs for filtration systems. Traditional antifouling strategies have limited success and often involve the use of toxic chemicals or frequent membrane

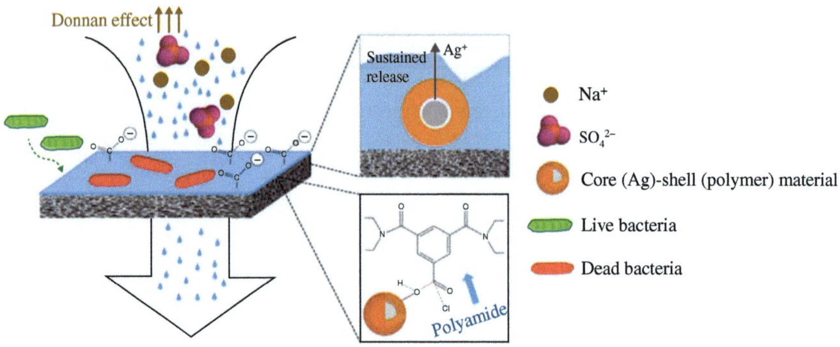

Figure 6.3 Cross-sectional illustration of a filtration membrane incorporating core-shell nanomaterials. Reproduced from [20] copyright © 2022 Elsevier B.V. All rights reserved.

replacements. However, core-shell nanomaterials offer a promising solution to this problem. The incorporation of core-shell nanoparticles into filtration membranes can significantly enhance their antifouling properties. The shell material, often made from hydrophilic polymers or functionalized materials like polyethylene glycol (PEG), can effectively resist the adsorption of foulants such as proteins, bacteria, and organic matter [21]. The core, typically made from materials like silica or carbon, provides structural stability and helps in maintaining the integrity of the membrane during prolonged use.

For example, in water treatment applications, core-shell nanocomposites can create a membrane surface that resists biofouling by preventing bacterial adhesion. The hydrophilic shell material forms a barrier that reduces the attraction between the membrane surface and microorganisms [22]. Additionally, the core-shell structure can facilitate self-cleaning properties through electrostatic repulsion or by enhancing the shear force of the fluid flow, which helps in preventing the buildup of fouling layers. Another advantage of using core-shell nanomaterials is their ability to increase the overall filtration capacity of membranes. Core-shell composites can improve the mechanical properties of the membranes, making them more resistant to physical stresses, which extends their lifespan and enhances filtration efficiency. Furthermore, the incorporation of core-shell nanomaterials into the membrane matrix can reduce the energy consumption of filtration systems by lowering the required pressure for the filtration process, making them more sustainable and cost-effective.

Core-shell nanomaterials have found significant applications in both industrial and domestic filtration systems. In the industrial sector, the use of core-shell nanomaterials in filtration membranes has been crucial for handling large-scale water purification and wastewater treatment. For instance, in the oil and gas industry, water treatment plants face challenges related to the removal of emulsified oil, heavy metals, and other hazardous substances. Core-shell nanomaterials have been used to develop membranes that not only remove these contaminants efficiently but also resist fouling by oil particles, extending the operational life of the filtration systems [23]. In one case study, a core-shell material consisting of a silica core and a hydrophilic polymer shell was incorporated into a microfiltration membrane. This membrane exhibited enhanced oil–water separation, with the shell material preventing oil droplets from adhering to the surface, thus improving the membrane's performance over time. The incorporation of the core-shell material led to a significant reduction in fouling and increased the lifespan of the filtration unit, demonstrating the effectiveness of these materials in industrial applications. In domestic applications, core-shell nanomaterials have been used in household water filtration systems to improve the removal of heavy metals, bacteria, and organic pollutants. Core-shell membranes in portable water filters have been shown to effectively reduce lead and other harmful substances from drinking water. The antifouling properties of these membranes also ensure that they maintain high efficiency over extended use, making them more reliable for long-term domestic applications [24]. For example, a core-shell membrane incorporating a silver nanoparticle core and a polymer shell has been developed for use in home water filters. The silver nanoparticles provide antimicrobial properties, preventing bacterial growth on the membrane surface, while the polymer shell enhances the membrane's chemical resistance and fouling resistance. This combination of properties allows the filter to remain effective over longer periods without frequent maintenance or replacement.

6.2.3 Photocatalysis for Water Treatment

Water pollution is one of the most pressing environmental issues worldwide, threatening both public health and the ecosystem. Among the various techniques developed for water treatment, photocatalysis has emerged as a promising solution due to its ability to degrade a wide range of organic pollutants under the influence of light. Photocatalysis involves the use of semiconductor materials that, when illuminated with light (usually UV or visible), activate electrons and holes on their surface. These activated species can then interact with water molecules and pollutants, leading to the degradation of harmful substances [25]. Core-shell nanomaterials, in particular, have shown great potential in enhancing photocatalytic activity, owing to their unique structural properties that improve the efficiency of pollutant degradation and increase the stability of the photocatalysts.

The photocatalytic degradation process is primarily driven by the generation of electron–hole pairs in the semiconductor material upon exposure to light. When photons with energy greater than or equal to the bandgap of the photocatalyst are absorbed, they excite electrons from the valence band to the conduction band, leaving behind holes in the valence band. These photogenerated electrons and holes are highly reactive and can participate in several reactions to break down pollutants. The excited electrons in the conduction band can reduce oxygen molecules (O_2) to form superoxide anions (O_2•−), while the holes in the valence band can oxidize water molecules (H_2O) to form hydroxyl radicals (•OH) [26]. These reactive oxygen species (ROS) are highly potent oxidizing agents that can degrade organic pollutants, bacteria, and other contaminants in water. The ROS generated by the photocatalyst attack the organic contaminants, breaking them down into smaller, less toxic compounds. For instance, aromatic compounds, dyes, pesticides, and pharmaceuticals can undergo oxidative degradation, leading to their mineralization into non-toxic by-products like CO_2, water, and other inorganic molecules [26]. One of the challenges in photocatalysis is the rapid recombination of the electron–hole pairs before they can participate in pollutant degradation, which reduces the overall efficiency of the process. To improve photocatalytic efficiency, it is essential to develop strategies that minimize this recombination and extend the lifetime of the reactive species. Core-shell nanomaterials have shown great promise in this regard, as the core material can absorb the light and generate electrons, while the shell material can stabilize the excited states and prevent recombination.

Core-shell nanomaterials have shown significant improvements in photocatalytic efficiency compared to traditional single-phase photocatalysts. By combining two semiconductors with complementary properties, core-shell structures can optimize the charge separation and enhance the photocatalytic performance. For example, ZnO@ZnS core-shell photocatalysts have gained considerable attention due to their ability to combine the advantages of both ZnO and zinc sulfide (ZnS) for photocatalytic applications [27].

ZnO is a widely studied photocatalyst known for its high electron mobility, wide bandgap (3.37 eV), and ability to absorb UV light efficiently. However, one of the limitations of ZnO is its tendency to undergo rapid recombination of the electron–hole pairs, which reduces its efficiency in photocatalytic processes. Additionally, its absorption spectrum is limited to UV light, which only accounts for a small portion of the solar spectrum. ZnS, on the other hand, has a lower bandgap (3.68 eV) and exhibits excellent photocatalytic activity in the visible light region [28]. By coating the ZnO core with a ZnS shell, the core-shell structure can overcome some of the limitations of ZnO. The ZnS shell helps to extend the absorption range into the visible spectrum, making the photocatalyst more effective under natural sunlight. Additionally, the ZnS shell acts as a passivation layer, reducing the recombination of electron–hole pairs and enhancing the stability of the photocatalyst. The core-shell architecture of ZnO@ZnS offers several advantages for photocatalysis, including improved charge separation, enhanced photocatalytic activity, and better light absorption. The combination of ZnO and ZnS allows the photocatalyst to operate efficiently under both UV and visible light, making it an ideal candidate for water treatment applications that require the degradation of a wide range of pollutants. Figure 6.4 (a) shows FESEM and (b) HR-TEM image of ZnO@ZnS core-shell, whereas (c) shows the photocatalytic activity.

The effectiveness of ZnO@ZnS core-shell photocatalysts in real-world water treatment scenarios has been demonstrated in several studies. The performance evaluation typically involves assessing the photocatalytic degradation of model pollutants, such as organic dyes, pesticides, and industrial effluents, under controlled conditions. These evaluations provide insights into the efficiency, stability, and scalability of the photocatalyst in practical applications.

a. **Degradation of Organic Pollutants:** One of the primary applications of ZnO@ZnS photocatalysts is in the degradation of organic pollutants in wastewater. Studies have shown that these core-shell materials exhibit high photocatalytic degradation rates for a variety of organic contaminants, including

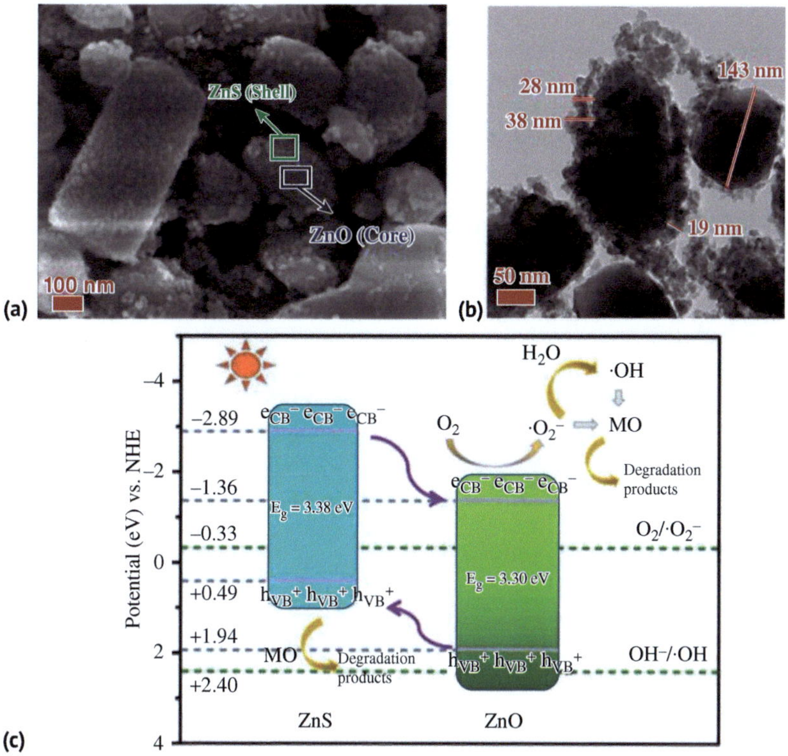

Figure 6.4 (a) FESEM image of ZnO@ZnS core-shell, (b) HR-TEM image of ZnO@ZnS core-shell, reproduced with permission from [29]/John Wiley & Sons, and (c) diagram showing photocatalytic pollutant degradation using ZnO@ZnS core-shell materials. [30]/MDPI/CC BY 4.0.

dyes (e.g., methylene blue, rhodamine B), phenolic compounds, and pesticides. Under UV and visible light irradiation, the ZnO@ZnS photocatalyst efficiently breaks down these pollutants into smaller, non-toxic compounds. The photocatalyst's ability to work under visible light makes it a more sustainable and energy-efficient option compared to UV-only photocatalysts [31].

b. **Antibacterial Activity:** In addition to degrading organic pollutants, ZnO @ZnS photocatalysts also exhibit antibacterial properties. The ROS generated during photocatalysis can effectively kill bacteria and other microorganisms in contaminated water. This makes ZnO@ZnS photocatalysts an attractive option for disinfecting drinking water and treating wastewater in regions where microbial contamination is a major concern.
c. **Real-world Applications and Challenges:** While ZnO@ZnS core-shell materials show great potential in laboratory settings, their performance in real-world water treatment scenarios can be influenced by factors such as pollutant concentration, pH, and the presence of competing ions or organic matter [32]. In addition, the long-term stability and recyclability of the photocatalysts are crucial for their commercial viability. Ongoing research is

focused on optimizing the synthesis of these core-shell materials, improving their performance under realistic environmental conditions, and scaling up their production for large-scale applications.

Photocatalysis using core-shell materials, particularly ZnO@ZnS, holds significant promise for sustainable water treatment. The enhanced photocatalytic activity, stability, and visible light response make these materials ideal candidates for addressing global water pollution. By optimizing the core-shell architecture and further exploring their real-world applications, these photocatalysts can play a critical role in improving water quality and contributing to environmental sustainability.

Core-shell nanomaterials have demonstrated significant potential in environmental applications, particularly for pollutant removal through techniques such as adsorption, filtration, and photocatalysis. To provide a clearer comparison of their efficiency and applications, Table 6.1 summarizes various core-shell nanomaterials used for these processes, highlighting the core-shell composition, target pollutants, and reported efficiency rates.

Table 6.1 Comparative table of different core-shell nanomaterials used for adsorption, filtration, and photocatalysis, listing core-shell composition, target pollutant, and efficiency rates.

Core-shell composition	Target pollutant	Application type	Efficiency rate
Fe_3O_4@SiO_2	Heavy metals (e.g., Pb^{2+}, Cd^{2+})	Adsorption	>90% removal efficiency
ZnO@ZnS	Organic dyes, phenolic compounds	Photocatalysis	~85% degradation rate under UV/Visible light
Ag@TiO_2	Bacteria, organic contaminants	Photocatalysis	95% bactericidal efficiency
CuO@SiO_2	Organic dyes (e.g., methyl orange)	Photocatalysis	~80% degradation in four hours
Ag@C	Bacteria, waterborne pathogens	Filtration (antibacterial)	90–95% bacterial removal efficiency
ZnO@Carbon	Pesticides, aromatic compounds	Adsorption	~85% adsorption rate in 30 minutes
$Cu_3(PO_4)_2$@TiO_2	Heavy metals, organics	Adsorption & Photocatalysis	>85% removal efficiency under sunlight
SiO_2@Ag/Polymer	Oil spills, hydrophobic pollutants	Filtration & Adsorption	>90% removal efficiency
TiO_2@Al_2O_3	Inorganic contaminants (e.g., nitrates)	Filtration & Photocatalysis	75–80% efficiency in one hour under UV light

6.3 Core-shell Nanomaterials for Air Quality Improvement

6.3.1 Pollutant Removal

Air pollution remains a significant threat to human health and environmental sustainability, with PM and VOCs being two of the most concerning pollutants. PM, including fine particles (PM2.5 and PM10), originates from industrial processes, vehicle emissions, and natural events like wildfires. These particles can penetrate deep into the respiratory system, causing severe health issues. VOCs, on the other hand, are organic chemicals that vaporize at room temperature, often originating from industrial solvents, paints, and household products [33, 34]. They contribute to ground-level ozone formation and pose toxic threats when inhaled over time. To combat these pollutants, core-shell nanostructures have emerged as effective materials for catalytic degradation and pollutant trapping.

The removal of PM and VOCs involves different strategies based on physical, chemical, and catalytic processes. For PM, filtration and adsorption techniques are commonly employed. Advanced filtration systems use nanomaterials to capture and immobilize fine particles due to their high surface area and nanoscale pore structures. The interaction between the particles and the nanomaterial surface enhances particle capture efficiency. For VOCs, the degradation mechanism typically involves catalytic oxidation, where VOC molecules are broken down into harmless by-products such as CO_2 and water (H_2O). Catalytic processes are often facilitated by semiconductor or metal-based nanomaterials that provide active sites for chemical reactions [35]. Upon exposure to light or heat, these catalysts generate reactive species such as hydroxyl radicals (•OH) and superoxide ions (O2•−) that attack and break down the VOC molecules.

Core-shell nanostructures have demonstrated superior performance in pollutant removal due to their tailored architectures. By combining the properties of both the core and shell materials, these nanostructures provide enhanced catalytic activity, stability, and pollutant trapping capabilities.

a. **Catalytic Degradation:** Core-shell nanostructures such as TiO_2@SiO_2 and ZnO@CeO_2 have been widely used for photocatalytic degradation of VOCs. The core material (e.g., TiO_2) absorbs light and generates reactive species, while the shell (e.g., SiO_2) enhances stability and prevents recombination of electron–hole pairs. This synergy results in efficient breakdown of VOCs under light irradiation [36]. Similarly, Pt@TiO_2 core-shell structures offer exceptional catalytic efficiency for the oxidation of VOCs at lower temperatures, making them suitable for industrial air purification systems.
b. **Pollutant Trapping:** Core-shell materials such as Fe_3O_4@Carbon and SiO_2@Ag have shown great potential in trapping PM due to their high surface area and functionalized surfaces. The magnetic properties of Fe_3O_4 cores allow for easy recovery and regeneration of the adsorbent material, while the carbon shell provides excellent adsorption capacity for PM and toxic gases. Additionally,

hydrophobic shells on certain core-shell materials can prevent moisture interference, further improving pollutant trapping efficiency.

c. **Hybrid Functions:** Some core-shell nanostructures combine catalytic degradation and adsorption functionalities, making them versatile in air purification systems. For example, $CuO@TiO_2$ not only traps PM but also degrades VOCs through photocatalytic reactions. These hybrid materials are ideal for use in air filtration units for both industrial and domestic environments. Core-shell nanostructures offer a multifunctional approach to pollutant removal by enabling both efficient catalytic degradation of VOCs and effective trapping of PM [37]. Their tunable properties and superior performance make them promising candidates for next-generation air purification technologies, contributing to cleaner and healthier air quality.

6.3.2 Gas Sensors

Gas sensors play a critical role in monitoring and controlling air quality, detecting toxic gases, and ensuring industrial safety. They are essential for detecting hazardous gases such as NOx, CO, and hydrogen (H_2), which are harmful to both health and the environment. Recent advancements in nanotechnology have introduced core-shell nanomaterials as promising candidates for gas-sensing applications due to their enhanced sensitivity, selectivity, and stability. The unique architecture of core-shell structures provides tunable surface properties, improved gas adsorption capabilities, and faster response times, making them ideal for advanced gas sensor systems.

Gas sensing primarily involves the interaction between gas molecules and the surface of a sensing material, which leads to measurable changes in electrical, optical, or thermal properties. In semiconducting gas sensors, for example, the adsorption of gas molecules alters the resistance of the material, which can be monitored to detect the presence and concentration of the target gas. Core-shell nanomaterials enhance gas-sensing performance by offering a combination of properties from the core and shell components. The core often serves as the primary sensing material, while the shell provides protection, stability, and additional functionalization [38]. For instance, a metal oxide core (such as SnO_2) can offer high gas adsorption and conductivity changes, while a shell made of porous silica or noble metals (like Au or Pt) can improve selectivity and catalytic activity. One key advantage of core-shell structures is the prevention of unwanted chemical reactions or environmental degradation. The shell can act as a passivation layer, enhancing the long-term stability of the gas sensor. Moreover, the core-shell interface creates additional active sites for gas adsorption, improving sensitivity.

Examples of core-shell nanomaterials for NOx, CO, and H_2 detection:

a. **NOx Detection:** Core-shell nanomaterials like $SnO_2@ZnO$ have demonstrated excellent performance in detecting NOx. The SnO_2 core provides high conductivity and gas adsorption, while the ZnO shell improves stability and response to NO_2 gas. The synergy between the core and shell materials results in faster response and recovery times [39].

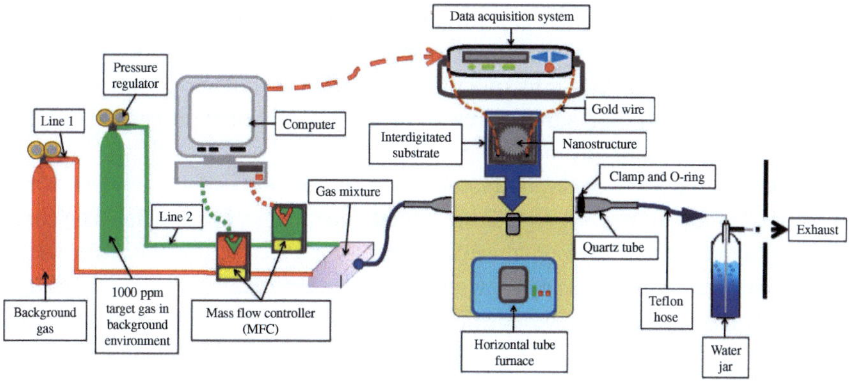

Figure 6.5 A gas sensor setup incorporating core-shell nanomaterials. Reproduced from [41] copyright © 2016 Elsevier B.V. All rights reserved.

b. **CO Detection:** For CO detection, Fe_2O_3@SiO_2 and TiO_2@Pt nanostructures are highly effective. The Fe_2O_3 core offers a large surface area for gas adsorption, while the SiO_2 shell enhances selectivity by preventing interference from other gases. Additionally, the catalytic properties of Pt in the TiO_2@Pt structure improve the oxidation of CO, leading to more accurate and sensitive detection [40].
c. **H_2 Detection:** H_2 gas detection is critical for fuel cell applications and safety monitoring. Core-shell materials like Pd@TiO_2 have shown remarkable sensitivity to H_2 due to the high affinity of palladium (Pd) for H_2 adsorption. The TiO_2 shell provides stability and enhances the sensor's response time by facilitating electron transfer during gas interactions. Figure 6.5 illustrates core-shell material and usage in gas sensor.

Core-shell nanomaterials significantly improve the sensitivity and selectivity of gas sensors. Sensitivity refers to the sensor's ability to detect low concentrations of gas, while selectivity indicates its capacity to distinguish between different gases. The high surface area, abundant active sites, and tailored shell functionalities of core-shell structures contribute to enhanced gas adsorption and reaction rates, leading to better sensitivity. The shell material plays a crucial role in improving selectivity by acting as a molecular sieve or catalytic filter. For example, noble metal shells (such as Au or Pd) can selectively catalyze the adsorption of specific gases, reducing interference from other gas species. Additionally, functionalized shells with dopants or coatings can further fine-tune the sensor's selectivity [42]. Moreover, core-shell nanomaterials exhibit faster response and recovery times due to the efficient charge transfer at the core-shell interface. This is particularly important for real-time monitoring applications where quick detection of gas leaks or hazardous emissions is crucial.

6.4 Core-shell Nanomaterials for CCS

The increasing concentration of atmospheric CO_2 due to industrial activities and fossil fuel combustion has raised global concerns about climate change and environmental sustainability. CCS technologies are essential for mitigating CO_2 emissions and reducing their impact on global warming. Among emerging solutions, core-shell nanomaterials have gained significant attention for their ability to efficiently capture, store, and convert CO_2 into valuable products. The tailored architecture of these materials, with a high surface area and tunable functional properties, makes them ideal candidates for advanced CCS applications.

Core-shell nanomaterials play a pivotal role in the adsorption and catalytic conversion of CO_2 due to their synergistic structural and chemical properties. The core component typically provides a high surface area for gas adsorption, while the shell enhances chemical stability, selectivity, and reactivity. One example is Fe_3O_4@SiO_2, where the Fe_3O_4 core offers magnetic properties for easy material recovery, and the SiO_2 shell improves the material's thermal and chemical stability. Such structures can selectively adsorb CO_2 while remaining resistant to moisture and other impurities [43]. Another promising system is ZnO@Carbon, where the ZnO core interacts with CO_2 molecules, and the porous carbon shell boosts adsorption capacity and prevents the agglomeration of active sites. Additionally, core-shell materials such as TiO_2@MOF have demonstrated dual functionalities, enabling not only CO_2 adsorption but also its photochemical conversion into fuels like methane and methanol. The catalytic activity of TiO_2 combined with the high adsorption capacity of the MOF shell creates a highly efficient system for carbon capture and utilization (CCU) technologies (as shown in Figure 6.6).

To further enhance CO_2 capture efficiency, hybrid core-shell materials integrating porous cores and functionalized shells have been developed. Porous core materials, such as silica and zeolites, offer abundant adsorption sites, while functional shells can be engineered to improve selectivity and adsorption kinetics. MOF@Graphene Oxide hybrid structures exemplify this approach, where the MOF core provides high CO_2 adsorption capacity, and the graphene oxide shell enhances electron conductivity and gas diffusion [44]. Another example is Carbon@Amine-Functionalized Polymers,

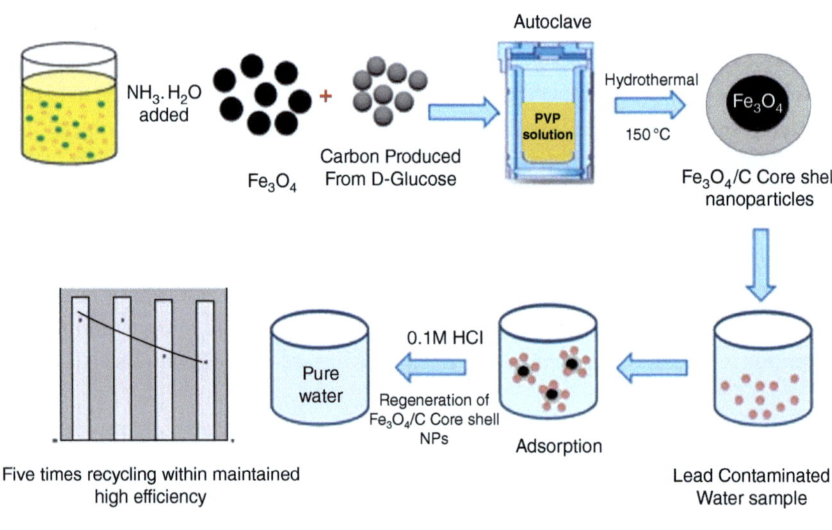

Figure 6.6 Schematic showing the CO_2 adsorption and conversion process using core-shell materials. Reproduced from [48] copyright © 2019 Wiley-VCH Verlag GmbH & Co. KGaA, Weinhei.

where the carbon core ensures structural stability, and the amine-functionalized shell selectively binds to CO_2 molecules through chemisorption [45]. These hybrid materials not only exhibit high adsorption capacities but also demonstrate excellent recyclability and stability over multiple adsorption–desorption cycles, making them suitable for practical deployment in CCS systems. By combining porous cores with functional shells, researchers have achieved significant advancements in both adsorption efficiency and material longevity.

Direct air capture (DAC) technology involves capturing CO_2 directly from ambient air, which poses unique challenges due to the low concentration of CO_2 (approximately 400 ppm) [46]. Core-shell nanomaterials have emerged as promising candidates for DAC applications due to their high surface areas, tunable porosities, and functionalized surfaces that enhance CO_2 binding. Recent advancements include the development of MgO@Silica Aerogel materials, which combine the high reactivity of MgO for CO_2 adsorption with the thermal stability and lightweight nature of silica aerogels. These materials have demonstrated excellent performance in capturing CO_2 from dilute gas streams. Similarly, Metal Nitride@Carbon nanostructures have been explored for DAC applications, where the metal nitride core interacts with CO_2 molecules, and the carbon shell prevents material degradation under harsh environmental conditions [47]. The incorporation of catalytic cores, such as Ni@MOF, has further advanced DAC systems by enabling in situ conversion of captured CO_2 into fuels or valuable chemicals, thus closing the carbon loop. Such integrated approaches are paving the way for more sustainable and economically viable CCS technologies. Figure 6.6 shows the CO_2 adsorption and conversion process using core-shell materials. Table 6.2 compares different core-shell materials for carbon capture, showing CO_2 adsorption capacities and structural properties.

Core-shell nanomaterials offer a transformative approach to CCS, addressing the urgent need for innovative solutions to combat climate change. Their unique structural and functional properties enable efficient CO_2 adsorption, storage, and conversion, making them indispensable for future CCS and DAC technologies. By integrating porous cores with functional shells, researchers continue to push the boundaries of material design, unlocking new possibilities for sustainable carbon management solutions.

6.5 Core-shell Nanomaterials in Waste Management and Recycling

6.5.1 Waste Treatment and Resource Recovery

Effective waste management and resource recovery are critical for addressing environmental challenges and promoting sustainable development. Core-shell nanomaterials, with their versatile catalytic and functional properties, are emerging as transformative solutions in waste treatment and resource recovery processes [49]. Their ability to catalyze complex reactions and selectively bind to target components makes them valuable for waste-to-energy (WTE) conversion and e-waste recycling.

Table 6.2 Comparative table of core-shell nanomaterials for carbon capture.

Core-shell composition	Core material	Shell material	Target application	CO_2 adsorption capacity (mmol/g)	Key structural properties
Fe_3O_4@SiO_2	Fe_3O_4	SiO_2	CO_2 adsorption	4.8	High stability and moisture resistance
ZnO@Carbon	ZnO	Porous Carbon	CO_2 capture and storage	5.5	Enhanced surface area, prevents agglomeration
TiO_2@MOF	TiO_2	MOF	Photocatalytic CO_2 conversion	7.2	High adsorption capacity, catalytic sites for conversion
MOF@ Graphene Oxide	MOF	Graphene Oxide	DAC	6.4	High gas diffusion and stability
Carbon@ Amine-Functionalized Polymer	Carbon	Amine-Polymer	Selective CO_2 capture	8.1	Chemisorption with strong CO_2 binding
MgO@Silica Aerogel	MgO	Silica Aerogel	DAC	5.9	Lightweight, excellent thermal stability
Ni@MOF	Ni	MOF	CO_2 conversion	6.8	Integrated catalytic function

WTE technologies aim to convert organic and industrial waste into valuable forms of energy, such as biogas, syngas, and biofuels. Core-shell nanomaterials, particularly metal-oxide-based systems, play a pivotal role in enhancing the efficiency and sustainability of these processes. For example, Ni@SiO_2 and CeO_2@ZrO_2 core-shell catalysts are widely used in gasification and pyrolysis processes due to their high thermal stability and catalytic activity. The core component often provides the primary catalytic function, while the shell protects against sintering and deactivation under harsh operating conditions. In the case of Ni@SiO_2, the nickel core catalyzes the breakdown of organic molecules, while the silica shell prevents agglomeration and prolongs the catalyst's lifespan. These nanomaterials not only improve energy yield but also reduce harmful emissions during the conversion process. Moreover, core-shell structures can facilitate photocatalytic degradation of organic pollutants in industrial wastewater, simultaneously producing H_2 gas as a clean energy by-product. This dual-functionality underscores their potential for integrating waste treatment and energy recovery in a single system.

Electronic waste (e-waste) contains valuable metals, such as gold, silver, and platinum, along with hazardous substances that require careful disposal. Core-shell nanomaterials offer advanced solutions for the selective recovery of precious metals from complex e-waste matrices [50]. For instance, Fe_3O_4@Au core-shell nanoparticles are highly effective for magnetically assisted gold recovery due to their dual functionalities: the Fe_3O_4 core enables easy separation, while the gold shell selectively binds to target metal ions. Another promising application is the use of TiO_2@Carbon nanocomposites for photocatalytic degradation of toxic components in e-waste, rendering them safer for disposal. These materials not only enhance recovery efficiency but also contribute to reducing the environmental footprint of e-waste processing.

6.5.2 Biodegradable and Sustainable Core-shell Materials

The rising concerns about environmental pollution and the depletion of non-renewable resources have driven the search for biodegradable and sustainable materials. In this context, core-shell nanomaterials are emerging as an innovative solution, offering tunable properties while being environmentally friendly. The development of biodegradable and sustainable core-shell structures holds immense potential for applications in compostable packaging and green technologies, contributing to a circular economy.

Eco-friendly core-shell nanomaterials are typically composed of biodegradable polymers, natural biopolymers, or hybrid materials. Common biodegradable core-shell systems include polylactic acid (PLA)@chitosan and starch@cellulose nanostructures. The core often imparts structural integrity, while the shell provides protective barriers, functional coatings, or specific surface characteristics [51]. The development of these materials focuses on creating high-performance systems that degrade naturally without releasing harmful by-products into the environment. Researchers are also exploring bio-based materials, such as polyhydroxyalkanoates (PHA) and alginate, to build sustainable core-shell architectures. These systems are engineered to possess desirable properties such as mechanical strength, moisture resistance, and antimicrobial capabilities, all while being compostable. In addition to natural polymers, hybrid systems integrating inorganic biodegradable materials (e.g., magnesium oxide) with biopolymers have shown promising applications in environmental sensing and biodegradable packaging. The integration of functional shells further enhances their stability, performance, and recyclability.

Compostable packaging is one of the most promising applications for biodegradable core-shell materials. Traditional packaging materials, often derived from fossil fuels, contribute significantly to environmental pollution and waste accumulation. Core-shell structures based on biopolymers are now being used to develop films and coatings for food packaging. These materials not only protect food from spoilage but also degrade naturally in composting environments, offering an eco-friendly alternative to conventional plastics. For example, PLA@chitosan films have demonstrated excellent antimicrobial properties and moisture resistance, making them ideal for extending the shelf life of perishable products [52]. Similarly, starch-based core-shell films are being explored for biodegradable pouches and wrappers. Beyond packaging, biodegradable core-shell materials have applications in agriculture, where they serve as carriers for slow-release fertilizers and pesticides. These materials degrade over time, releasing nutrients in a controlled manner without leaving toxic residues in the soil. In sustainable technologies, energy storage devices and biodegradable electronics are increasingly adopting core-shell architectures made from eco-friendly materials. For instance, core-shell supercapacitors using cellulose-based shells have shown remarkable performance while remaining fully degradable after their lifecycle.

6.6 Challenges and Future Perspectives

6.6.1 Scalability and Economic Considerations

- **High production costs:** The synthesis of core-shell nanomaterials often involves multi-step processes, such as precise deposition techniques, controlled chemical reactions, and high-temperature treatments, which contribute to elevated manufacturing costs. Scaling up these processes while maintaining structural precision is a key challenge.
- **Batch-to-batch inconsistencies:** Ensuring uniformity in core-shell structures at large scales remains problematic, as variations in size, composition, and shell thickness can affect material properties and performance.
- **Need for scalable synthesis methods:** Techniques such as hydrothermal and coprecipitation must be optimized for industrial production to ensure reproducibility and cost efficiency. Continuous-flow synthesis and automated production systems offer potential solutions.
- **Economic feasibility:** Despite their high performance, the cost–benefit ratio of core-shell nanomaterials needs to be justified for widespread adoption in commercial applications, particularly in resource-constrained sectors like water purification and waste management.

6.6.2 Environmental Impact and Toxicity Concerns

- **Toxicity of nanomaterials:** Many core-shell nanomaterials, such as metal and metal oxide-based systems, may release toxic ions or degrade into harmful by-products under environmental conditions. This poses risks to aquatic life and soil ecosystems.
- **Accumulation in the environment:** The persistence of non-biodegradable nanomaterials in ecosystems can lead to bioaccumulation and long-term environmental consequences. Strategies to develop biodegradable and eco-friendly alternatives are essential.
- **Safe disposal and recycling:** The lack of standardized protocols for the safe disposal and recycling of nanomaterials remains a critical concern. Developing recovery and reuse technologies for core-shell structures could mitigate environmental risks.
- **Life cycle assessment (LCA):** Comprehensive LCA studies are necessary to evaluate the environmental footprint of core-shell nanomaterials, from synthesis to disposal, and guide the development of greener synthesis methods.

6.6.3 Future Trends in Core-shell Environmental Nanotechnology

- **Integration with AI and ML:** AI-driven design and ML algorithms can accelerate the discovery of novel core-shell materials with optimized properties for environmental applications. Predictive modeling can help identify the best material compositions and synthesis conditions.
- **Biodegradable core-shell systems:** The future will see an increasing shift toward biodegradable and naturally sourced materials for core-shell architectures. These materials will offer eco-friendly solutions for environmental remediation and waste management.
- **Hybrid multifunctional materials:** The development of hybrid core-shell systems that combine multiple functionalities, such as simultaneous pollutant adsorption, photocatalysis, and energy recovery, will be a major research focus.
- **DAC technologies:** Core-shell nanomaterials will play a critical role in DAC applications by enabling efficient and scalable CO_2 capture and conversion systems.
- **Industrial adoption and policy support:** As regulatory frameworks evolve, there will be a greater emphasis on the commercialization of safe and sustainable core-shell nanomaterials for environmental applications. Policy support and public–private partnerships will drive innovation and market adoption.
- **Smart and responsive materials:** Core-shell nanomaterials with stimuli-responsive properties, such as pH, temperature, or light responsiveness, will enable advanced environmental monitoring and adaptive remediation technologies.

6.7 Conclusion

Core-shell nanomaterials have emerged as a transformative force in environmental applications, offering innovative solutions to some of the most pressing challenges of our time, including pollution control, resource recovery, and sustainable development. Their unique structure, combining the distinct properties of the core and shell, allows them to perform multiple functions simultaneously, such as adsorption, filtration, and photocatalysis, making them highly efficient for tackling environmental contaminants. The ability to tailor these nanomaterials for specific applications, ranging from water purification and air quality improvement to carbon capture and waste treatment, holds great promise for advancing environmental nanotechnology.

However, despite their impressive capabilities, challenges remain in scaling up the production of core-shell materials while keeping costs manageable. The potential environmental and toxicity risks of these materials, especially in large quantities, also require careful consideration and comprehensive studies on their long-term impact on ecosystems. Addressing these challenges will be crucial to ensure the safe and sustainable integration of core-shell nanomaterials into real-world applications.

Looking ahead, the future of core-shell environmental nanotechnology appears promising, with trends pointing toward the development of biodegradable, multi-functional, and smart nanomaterials. The integration of AI and ML in the design process will accelerate the discovery of new, efficient materials, while advances in manufacturing techniques will help reduce costs and improve scalability. Moreover, as policies evolve to support the commercialization of green technologies, core-shell nanomaterials are expected to play a pivotal role in shaping a more sustainable future.

In conclusion, while core-shell nanomaterials are already proving to be valuable tools in the fight for environmental sustainability, continued innovation and research are essential to fully unlock their potential in addressing global environmental challenges. Their integration into various sectors, coupled with advancements in eco-friendly production methods and regulatory frameworks, will pave the way for a cleaner, more sustainable world.

References

1 Mulvaney D, Richards RM, Bazilian MD, Hensley E, Clough G, Sridhar S. Progress towards a circular economy in materials to decarbonize electricity and mobility. Renew Sustain Energy Rev. 2021 Mar 1;137:110604.

2 Nawaz F, Ali M, Ahmad S, Yong Y, Rahman S, Naseem M, et al. Carbon based nanocomposites, surface functionalization as a promising material for VOCs (volatile organic compounds) treatment. Chemosphere. 2024 Sep 1;364:143014.

3 Arif M, Hannan Zahid A, Saghir S, Mahsud A. From sunlight to sustainability: The properties, evolution, and prospects of photocatalytic building materials. Catal Rev [Internet]. 2025 Jan 8 [cited 2025 Jan 31];1–72. Available from: https://www.tandfonline.com/doi/abs/10.1080/01614940.2024.2445044

4 Dutta J, Singh PP. Air pollutants and acid precipitation: Impact on ecology and human health. In: Multidimensional Approaches to Impacts of Changing Environment on Human Health [Internet]. CRC Press. 2021 Sep 1 [cited 2025 Jan 31];3–25. Available from: https://www.taylorfrancis.com/chapters/edit/10.1201/9781003095422-1/air-pollutants-acid-precipitation-joystu-dutta-prem-prakash-singh

5 Hu P, Morabito JV, Tsung CK. Core-shell catalysts of metal nanoparticle core and metal-organic framework shell. ACS Catal [Internet]. 2014 Dec 5 [cited 2025 Jan 31];4(12):4409–19. Available from: https://pubs.acs.org/doi/full/10.1021/cs5012662

6 Ndoke PN, Jimoh OD. Impact of traffic emission on air quality in a developing city. 2004. AU J Technol.

7 Chaouiki A, Chafiq M, Ko YG. The art of controlled nanoscale lattices: A review on the self-assembly of colloidal metal–organic framework particles and their multifaceted architectures. Mater Sci Eng R: Rep. 2024 Jun 1;159:100785.

8 Mane PV, Rego RM, Yap PL, Losic D, Kurkuri MD. Unveiling cutting-edge advances in high surface area porous materials for the efficient removal of toxic metal ions from water. Prog Mater Sci. 2024 Dec 1;146:101314.

9 Hosseini SR, Omidkhah M, Mehri Lighvan Z, Norouzbahari S, Ghadimi A. Synthesis, characterization, and gas adsorption performance of an efficient hierarchical ZIF-11@ZIF-8 core–shell metal–organic framework (MOF). Sep Purif Technol. 2023 Feb 15;307:122679.

10 Norsko JK. Chemisorption on metal surfaces. Rep Prog Phys [Internet]. 1990 [cited 2025 Jan 31];53(10):1253. Available from: https://iopscience.iop.org/article/10.1088/0034-4885/53/10/001/meta

11 Khan M, Ali SW, Shahadat M, Sagadevan S. Applications of polyaniline-impregnated silica gel-based nanocomposites in wastewater treatment as an efficient adsorbent of some important organic dyes. Green Process Synth [Internet]. 2022 Jan 1 [cited 2025 Jan 31];11(1):617–30. Available from: https://www.degruyter.com/document/doi/10.1515/gps-2022-0063/html

12 Esmaeili Lashkarian E, Ahmadi S, Beigmohammadi F. Recent application of nanomaterials-based magnetic solid phase micro-extraction for heavy metals food toxicity. Int J New Chem [Internet]. 2024 Mar 1 [cited 2025 Jan 31];11(4):362–92. Available from: https://www.ijnc.ir/article_711568.html

13 Tzvetkov G, Kaneva N, Spassov T. Room-temperature fabrication of core-shell nano-ZnO/pollen grain biocomposite for adsorptive removal of organic dye from water. Appl Surf Sci. 2017 Apr 1;400:481–91.

14 Gopinath KP, Madhav NV, Krishnan A, Malolan R, Rangarajan G. Present applications of titanium dioxide for the photocatalytic removal of pollutants from water: A review. J Environ Manage. 2020 Sep 15;270:110906.

15 Nnadozie EC, Ajibade PA, Bontempi E, Quina M, Bogush A. Multifunctional magnetic oxide nanoparticle (MNP) core-shell: Review of synthesis, structural studies and application for wastewater treatment. Molecules [Internet]. 2020 Sep 9 [cited 2025 Jan 31];25(18):4110. Available from: https://www.mdpi.com/1420-3049/25/18/4110/htm

16 Zou H, Luo Z, Yang X, Xie Q, Zhou Y. Toward emerging applications using core–shell nanostructured materials: A review. J Mater Sci [Internet]. 2022 Jun 6 [cited 2025 Jan 31];57(24):10912–42. Available from: https://link.springer.com/article/10.1007/s10853-022-07328-z

17 Yang X, Wan Y, Zheng Y, He F, Yu Z, Huang J, et al. Surface functional groups of carbon-based adsorbents and their roles in the removal of heavy metals from aqueous solutions: A critical review. Chem Eng J. 2019 Jun 15;366:608–21.

18 Mélinon P, Begin-Colin S, Duvail JL, Gauffre F, Boime NH, Ledoux G, et al. Engineered inorganic core/shell nanoparticles. Phys Rep. 2014 Oct 20;543(3):163–97.

19 Ahmed MA, Mahmoud SA, Mohamed AA. Nanomaterials-modified reverse osmosis membranes: A comprehensive review. RSC Adv [Internet]. 2024 Jun 12 [cited 2025 Jan 31];14(27):18879–906. Available from: https://pubs.rsc.org/en/content/articlehtml/2024/ra/d4ra01796j

20 Tong Y, Wang Y, Bian S, Ge H, Xiao F, Li L, et al. Incorporating Ag@RF core-shell nanomaterials into the thin film nanocomposite membrane to improve permeability and long-term antibacterial properties for nanofiltration. Sci Total Environ. 2022 Sep 15;839:156231.

21 Soltannia B, Islam MA, Cho JY, Mohammadtabar F, Wang R, Piunova VA, et al. Thermally stable core-shell star-shaped block copolymers for antifouling enhancement of water purification membranes. J Memb Sci. 2020 Mar 15;598:117686.

22 Selim MS, El-Safty SA, Shenashen MA, Higazy SA, Elmarakbi A. Progress in biomimetic leverages for marine antifouling using nanocomposite coatings. J Mater Chem B [Internet]. 2020 May 6 [cited 2025 Jan 31];8(17):3701–32. Available from: https://pubs.rsc.org/en/content/articlehtml/2020/tb/c9tb02119a

23 Putatunda S, Bhattacharya S, Sen D, Bhattacharjee C. A review on the application of different treatment processes for emulsified oily wastewater. Int J Environ Sci Technol [Internet]. 2019 May 1 [cited 2025 Jan 31];16(5):2525–36. Available from: https://link.springer.com/article/10.1007/s13762-018-2055-6

24 Thatai S, Khurana P, Boken J, Prasad S, Kumar D. Nanoparticles and core–shell nanocomposite based new generation water remediation materials and analytical techniques: A review. Microchem J. 2014 Sep 1;116:62–76.

25 Zhang J, Tian B, Wang L, Xing M, Lei J. Photocatalysis fundamentals, materials and applications. Springer. 2018 [cited 2025 Jan 31];100. Available from: http://link.springer.com/10.1007/978-981-13-2113-9

26 Jia H, He W, Wamer WG, Han X, Zhang B, Zhang S, et al. Generation of reactive oxygen species, electrons/holes, and photocatalytic degradation of rhodamine B by photoexcited CdS and Ag_2S micro-nano structures. J Phys Chem C [Internet]. 2014 Sep 18 [cited 2025 Jan 31];118(37):21447–56. Available from: https://pubs.acs.org/doi/full/10.1021/jp505783y

27 Ojha A, Tiwary D, Oraon R, Singh P. Degradations of endocrine-disrupting chemicals and pharmaceutical compounds in wastewater with carbon-based nanomaterials: A critical review. Environ Sci Pollut Res [Internet]. 2021 Jun 1 [cited 2025 Jan 31];28(24):30573–94. Available from: https://link.springer.com/article/10.1007/s11356-021-13939-x

28 Rana P, Soni V, Kumar R, Sonu, Chawla A, Parwaz Khan AA, et al. Advances in photocatalytic hydrogen production with ZnO/ZnS-Based nanostructured materials. Fuel. 2025 Apr 15;386:134286.

29 Phogat P, Shreya, Jha R, Singh S. Photocatalytic performance of ZnO@ZnS core–shell heterostructures for malachite green and rhodamine B dye degradation. Phys Status Solidi (a) [Internet]. 2024 Oct 1 [cited 2025 Jan 31];221(20):2400404. Available from: https://onlinelibrary.wiley.com/doi/full/10.1002/pssa.202400404

30 Jin X, Chen J, Chen F, Duan H, Wang Z, Li J. Solid-state synthesis of ZnO/ZnS photocatalyst with efficient organic pollutant degradation performance. Catalysts [Internet]. 2022 Sep 1 [cited 2025 Jan 31];12(9):981. Available from: https://www.mdpi.com/2073-4344/12/9/981/htm

31 Alwany AB, Youssef GM, Saleh EE, Samir OM, Algradee MA, Alnehia A. Structural, optical and radiation shielding properties of ZnS nanoparticles QDs. Optik (Stuttg). 2022 Jun 1;260:169124.

32 Khan JA, Ahamad S, Ansari MAH, Tauqeer M, Park CH, Park JP, et al. State-of-the-art in ZnS-based nanoarchitects for visible-light photocatalytic degradation of antibiotics and organic dyes. J Water Process Eng. 2024 Nov 1;67:106151.

33 Phogat P, Jha R, Scripta SSP. Harnessing ZnO morphologies in energy application and sustainable development. Phys Scr [Internet]. 2024

[cited 2025 Jan 31];99(10):102004. Available from: https://iopscience.iop.org/article/10.1088/1402-4896/ad7990/meta?casa_token=HTl6TO05UksAAAAA:inJa6dRLau8rC7H18Ub37VBKcnqqta-HXPLGLSlhMDZ4To6gStInlXzNRmow4u9hUOhMcKSAwq5fhGxnygMDzbPToDe3

34 Halios CH, Landeg-Cox C, Lowther SD, Middleton A, Marczylo T, Dimitroulopoulou S. Chemicals in European residences – Part I: A review of emissions, concentrations and health effects of volatile organic compounds (VOCs). Sci Total Environ. 2022 Sep 15;839:156201.

35 Vandenbroucke A. Abatement of volatile organic compounds by combined use of non-thermal plasma and heterogeneous catalysis. Ghent University. 2015 [cited 2025 Jan 31]. Available from: https://biblio.ugent.be/publication/7225448/file/7225449

36 Cuong ND, Quang DT. Progress through synergistic effects of heterojunction in nanocatalysts - Review. Vietnam J Chem [Internet]. 2020 Aug 1 [cited 2025 Jan 31];58(4):434–63. Available from: https://onlinelibrary.wiley.com/doi/full/10.1002/vjch.202000072

37 Bai X, Qi X, Liu Y, Sun J, Shen T, Pan L. Photothermal catalytic degradation of VOCs: Mode, system and application. Chem Asian J [Internet]. 2025 Jan 2 [cited 2025 Jan 31];20(1):e202400993. Available from: https://aces.onlinelibrary.wiley.com/doi/abs/10.1002/asia.202400993?casa_token=mmkKJJocjosAAAAA:mrauQiK_3qh2TXOsb7gvBZOdGWOCKKJa13c2rv-VCv1hWmpnklOTrLlluzaTumb6yUBAQjExEtBaEXxD

38 Kalambate PK, Dhanjai, Huang Z, Li Y, Shen Y, Xie M, et al. Core@shell nanomaterials based sensing devices: A review. TrAC Trends Anal Chem. 2019 Jun 1;115:147–61.

39 Liu N, Yu H, Yang Y, Dong XT. Fabrication of 3D multi-edges cube porous SnO2/ZnO composites as high-performance NOx gas sensor. IEEE Sens J. 2020 Mar 15;20(6):2852–9.

40 Ma H, Guo Y, Ye L, Cheng Y, Wang X. Janus micro/nanomotors for enhanced disease treatment through their deep penetration capability. Acta Biomater. 2025;196:50–77. https://doi.org/10.1016/j.actbio.2025.02.055.

41 Arafat MM, Haseeb ASMA, Akbar SA, Quadir MZ. In-situ fabricated gas sensors based on one dimensional core-shell TiO_2-Al_2O_3 nanostructures. Sens Actuators B Chem. 2017 Jan 1;238:972–84.

42 Zhu M, Zhang H, Zhang S, Yao H, Shi X, Xu S. Chemoresistive gas sensors based on noble-metal-decorated metal oxide semiconductors for H2 detection. Materials [Internet]. 2025 Jan 19 [cited 2025 Jan 31];18(2):451. Available from: https://www.mdpi.com/1996-1944/18/2/451/htm

43 Ibeniaich M, Elansary M, Minaoui K, Mouhib Y, El Haj YA, Belaiche Y, et al. Exploring the effect of Hf (IV) doping in spinel ferrite $CoHf_xFe_2$-xO_4 on magnetic properties, electrochemical impedance, and photocatalytic activity: In-depth structural study. J Mol Struct. 2024 Dec 15;1318:139395.

44 Ehrnst Y, Ahmed H, Komljenovic R, Massahud E, Shepelin NA, Sherrell PC, et al. Acoustotemplating: Rapid synthesis of freestanding quasi-2D MOF/graphene oxide heterostructures for supercapacitor applications. J Mater Chem A Mater

[Internet]. 2022 Mar 30 [cited 2025 Jan 31];10(13):7058–72. Available from: https://pubs.rsc.org/en/content/articlehtml/2022/ta/d1ta10493d

45 Demirel Y, Matzen M, Winters C, Gao X. Capturing and using CO_2 as feedstock with chemical looping and hydrothermal technologies. Int J Energy Res [Internet]. 2015 Jun 25 [cited 2025 Jan 31];39(8):1011–47. Available from: https://onlinelibrary.wiley.com/doi/full/10.1002/er.3277

46 Wang J, Fu R, Wen S, Ning P, Helal MH, Salem MA, et al. Progress and current challenges for CO_2 capture materials from ambient air. Adv Compos Hybrid Mater [Internet]. 2022 Oct 20 [cited 2025 Jan 31];5(4):2721–59. Available from: https://link.springer.com/article/10.1007/s42114-022-00567-3

47 Mao X, Guo R, Chen Q, Zhu H, Li H, Yan Z, et al. Recent advances in graphitic carbon nitride based electro-catalysts for CO_2 reduction reactions. Molecules [Internet]. 2023 Apr 7 [cited 2025 Jan 31];28(8):3292. Available from: https://www.mdpi.com/1420-3049/28/8/3292/htm

48 Sayed M, Abbas M, Abdel Moniem SM, Ali MEM, Naga SM. Facile and room temperature synthesis of superparamagnetic Fe_3O_4/C core/shell nanoparticles for efficient removal of Pb(II) from aqueous solution. ChemistrySelect [Internet]. 2019 Feb 14 [cited 2025 Jan 31];4(6):1857–65. Available from: https://onlinelibrary.wiley.com/doi/full/10.1002/slct.201803939

49 Zhou Y, Li Y, Hou Y, Wang C, Yang Y, Shang J, et al. Core-shell catalysts for the elimination of organic contaminants in aqueous solution: A review. Chem Eng J. 2023 Jan 1;455:140604.

50 Xia J, Ghahreman A. Sustainable technologies for the recycling and upcycling of precious metals from e-waste. Sci Total Environ. 2024 Mar 15;916:170154.

51 Mehandole A, Walke N, Mahajan S, Aalhate M, Maji I, Gupta U, et al. Core–Shell type lipidic and polymeric nanocapsules: The transformative multifaceted delivery systems. AAPS PharmSciTech [Internet]. 2023 Jan 26 [cited 2025 Jan 31];24(1):1–37. Available from: https://link.springer.com/article/10.1208/s12249-023-02504-z

52 Harahap H, Julianti E, Safitri A, Jaafar M, Nasution H, Harahap H, et al. Smart packaging based on polylactic acid: The effects of antibacterial and antioxidant agents from natural extracts on physical–mechanical properties, colony reduction, perishable food shelf life, and future prospective. Polymers [Internet]. 2023 Oct 16 [cited 2025 Jan 31];15(20):4103. Available from: https://www.mdpi.com/2073-4360/15/20/4103/htm

7

Biomedical Applications of Core-shell Nanostructures

Core-shell nanostructures have emerged as versatile and innovative materials with significant potential in biomedical applications, offering a wide range of functionalities for drug delivery, diagnostics, and theranostics. These nanomaterials, consisting of a core material enveloped by a shell, combine the unique properties of both components, enabling enhanced therapeutic efficacy, targeted treatment, and real-time monitoring of disease progression. This chapter explores the various biomedical applications of core-shell nanostructures, focusing on their role in targeted drug delivery systems, bioimaging, diagnostic techniques, and theranostics. Key advantages of core-shell systems, including their multifunctionality, tunability, and biocompatibility, are discussed, along with examples of their use in cancer therapy, disease detection, and personalized medicine. Additionally, the chapter addresses the importance of biocompatibility and safety assessments, including regulatory aspects, toxicity studies, and long-term stability, ensuring the clinical translation of these promising nanomaterials. By integrating therapeutic and diagnostic functions, core-shell nanostructures hold the potential to revolutionize personalized medicine, offering new avenues for the treatment and monitoring of various diseases. As research continues to progress, core-shell nanostructures are poised to play a critical role in advancing the future of biomedical technologies.

7.1 Introduction to Biomedical Applications

With its immense potential to alter medicine, nanotechnology has emerged as a transformational field. Fundamentally, nanotechnology is the design, manufacture, and use of materials and technologies at the nanoscale, which is usually between 1 and 100 nm, where the properties of materials differ significantly from their bulk counterparts due to the unique physical, chemical, and biological phenomena that occur at this scale. This innovation has opened new frontiers in various domains, particularly in biomedical sciences, where it offers promising solutions for diagnosing and treating diseases, enhancing the precision and effectiveness of medical interventions, and improving overall healthcare outcomes.

Core-Shell Nanomaterials: From Fundamentals to Applications, First Edition. Shreya Sharma et al.

One of the primary reasons for the growing interest in nanotechnology in medicine is its ability to manipulate and control matter at a molecular and atomic level, allowing for the creation of highly targeted and efficient therapies. Traditional approaches in medicine often struggle to deliver drugs in a controlled and precise manner, leading to side effects and inefficiencies [1]. Nanotechnology offers a solution by enabling the creation of nanoparticles and nanostructures that can be engineered to interact with biological systems in a highly specific way. These materials can be designed to encapsulate therapeutic agents, protect them from degradation, and release them at precise locations in response to stimuli such as pH, temperature, or light. As a result, nanotechnology allows for more targeted and personalized treatments, with the potential to reduce adverse side effects, improve therapeutic efficacy, and enable earlier diagnosis of diseases. Within this rapidly advancing field, core-shell nanostructures stand out for their ability to combine multiple functionalities in a single material. A core-shell nanostructure typically consists of a core material that is surrounded by one or more shell layers. The core material often serves as the structural foundation, while the shell material can impart additional properties or functionalities, such as enhanced stability, targeting ability, or controlled drug release [2]. The versatility of core-shell nanostructures makes them ideal candidates for a wide range of biomedical applications, from drug delivery to bioimaging and theranostics.

As shown in Figure 7.1, core-shell nanostructures play a multifunctional role in drug delivery, diagnostics, and therapy. These structures can be designed to encapsulate medications within their cores and shield them from early deterioration in drug administration. To enable the nanoparticles to target specific cells or tissues, the shell layer can be functionalized with particular ligands, such as peptides, antibodies, or small molecules.

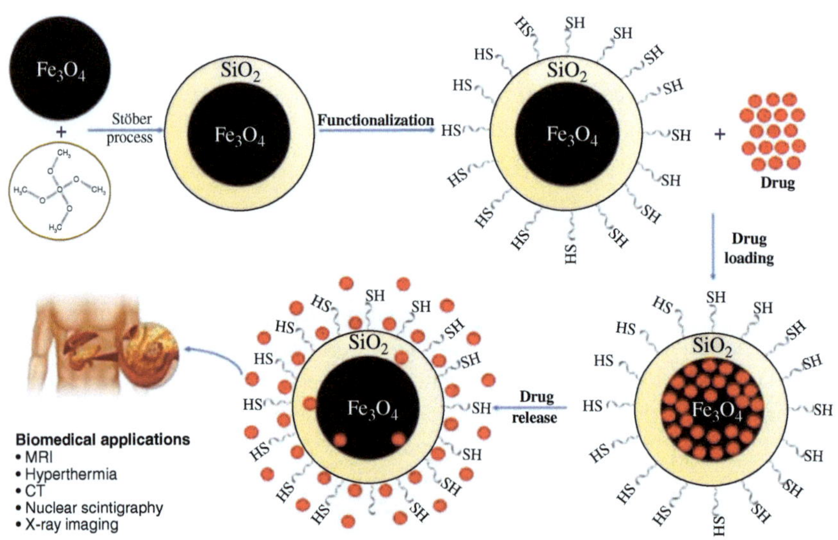

Figure 7.1 Conceptual diagram showing core-shell nanostructures in biomedical applications, highlighting their multifunctional roles in drug delivery, diagnostics, and therapy. [10]/MDPI/CC BY 4.0.

This targeted approach enables localized delivery of therapeutic agents, which can improve the effectiveness of treatments, especially in diseases like cancer where traditional drug delivery methods often result in systemic toxicity. Furthermore, the shell material can be designed to respond to environmental stimuli (e.g., changes in pH or temperature) for controlled drug release, which is crucial for maintaining therapeutic concentrations of drugs over extended periods [3]. In addition to their role in drug delivery, core-shell nanostructures are widely utilized in bioimaging and diagnostics. Nanomaterials' distinct optical, magnetic, and electrical characteristics at the nanoscale make them perfect for use as contrast agents in a variety of imaging modalities, including computed tomography (CT), magnetic resonance imaging (MRI), and fluorescence imaging. Core-shell nanostructures can be engineered to enhance the contrast in imaging, allowing for better visualization of biological structures and diseases at early stages [4]. For example, iron oxide (Fe_3O_4)-based core-shell nanoparticles are often used as contrast agents in MRI due to their superparamagnetic properties, which improve signal strength and provide clearer images of internal structures. Additionally, quantum dots and gold-based core-shell nanoparticles are commonly employed in fluorescence imaging due to their unique emission characteristics, enabling high-resolution imaging of cellular processes in real time.

Another exciting application of core-shell nanostructures is theranostics, which refers to the integration of both therapeutic and diagnostic functions within a single nanostructure. The ability to deliver drugs while simultaneously monitoring their effects through imaging is a significant advancement in personalized medicine. Theranostic core-shell nanostructures can enable real-time monitoring of therapeutic outcomes, such as tumor response to treatment or drug release kinetics, while also delivering the therapy [5]. For example, a core-shell nanoparticle might be loaded with chemotherapy drugs in the core, while the shell contains imaging agents to track the location and distribution of the nanoparticles in the body. This dual functionality allows for a more precise and effective treatment approach, as doctors can monitor the progress of the therapy in real time and adjust the treatment as needed.

The versatility of core-shell nanostructures is due in large part to their tunable properties, which can be modified at both the core and shell levels. The core material can be chosen based on its inherent properties, such as its ability to carry a therapeutic payload, interact with specific cells, or enhance imaging signals [6]. The shell, on the other hand, can be engineered to provide additional functionality, such as protection from environmental factors, selective targeting, or controlled release mechanisms. By tailoring both the core and shell materials, researchers can create nanostructures with a wide range of properties that are optimized for specific biomedical applications. This tunability allows for the development of personalized medical treatments that can be tailored to the needs of individual patients. In addition to tunability, core-shell nanostructures offer several key advantages in biomedical applications, including multifunctionality, biocompatibility, and stability. The multifunctionality of core-shell materials refers to their ability to perform multiple tasks simultaneously, such as drug delivery, bioimaging, and therapy. This multifunctionality is particularly valuable in complex diseases such as cancer, where a single treatment may not be sufficient. By incorporating both therapeutic and diagnostic functions into a single nanostructure, core-shell materials enable more effective and targeted treatments [7]. Moreover, the ability to combine multiple functionalities in one material can reduce the need for multiple therapeutic interventions, simplifying treatment regimens and improving patient outcomes.

Biocompatibility is another crucial advantage of core-shell materials. For any nanomaterial used in medical applications, it is essential that it does not induce toxic reactions or immune responses when introduced into the body. Core-shell nanostructures can be designed with biocompatible materials that minimize potential adverse effects [8]. The shell material, in particular, can be selected to enhance biocompatibility by preventing interactions between the core material and the surrounding biological environment. For example, biocompatible polymers or silica can be used as shell materials to prevent the release of toxic substances from the core and to ensure that the nanostructure does not cause harm to healthy tissues. The stability of core-shell nanostructures is also a critical factor in their success as biomedical materials. In drug delivery applications, for example, it is important that the nanostructures remain stable during circulation in the body and only release their payload at the target site. The shell material plays a vital role in maintaining the stability of the core and ensuring that the drug is not released prematurely. Furthermore, the stability of the nanostructures is essential for ensuring their long-term effectiveness, especially in applications such as bioimaging, where the nanomaterials must remain intact for extended periods of time to provide reliable imaging results [9].

Core-shell nanostructures represent a promising and versatile class of materials with significant potential to advance the fields of drug delivery, diagnostics, and therapy. Their multifunctionality, tunability, and biocompatibility make them ideal candidates for a wide range of biomedical applications, and their ability to integrate both therapeutic and diagnostic functions in a single nanostructure sets them apart from other materials. As research in nanomedicine continues to progress, core-shell nanostructures are expected to play an increasingly important role in the development of personalized, targeted therapies, ultimately improving patient outcomes and transforming the future of healthcare.

7.2 Targeted Drug Delivery Systems

7.2.1 Mechanism of Targeted Drug Delivery

Targeted drug delivery is a cutting-edge technique in contemporary medicine that enables the exact and regulated release of therapeutic agents to particular bodily regions or cells. By limiting the amount of time that the drug is exposed to healthy cells, this technique seeks to increase the effectiveness of medications while minimizing negative effects. One of the most promising strategies for achieving targeted drug delivery involves the use of core-shell nanostructures [11]. These nanostructures consist of a core material that encapsulates the therapeutic drug, surrounded by a shell material that can be engineered to provide additional functionalities, such as targeting specific cells or tissues, controlling the release of the drug, and improving the stability of the delivery system.

Core-shell materials are ideal for targeted drug delivery because they offer several advantages over traditional drug delivery systems. A range of therapeutic agents, such as tiny molecules, proteins, or nucleic acids, can be placed into the nanoparticle's core, while the shell acts as a barrier to keep the medicine from releasing too soon and to increase its stability while it is in the body [12]. This encapsulation ensures that the drug remains intact until it reaches the target site, where it can be released in a controlled manner, maximizing its therapeutic effect. One important characteristic of core-shell nanostructures that allows them to target particular cells or tissues is surface functionalization. Using ligands like peptides, antibodies, or other macromolecules that have a strong affinity for particular receptors on the surface of target cells, the shell's surface can be altered. These ligands can recognize and bind to particular cellular markers or antigens, allowing the nanoparticle to specifically interact with the desired tissue or cell type. For example, in cancer therapy, core-shell nanoparticles can be functionalized with antibodies that target tumor-specific antigens, ensuring that the drug is delivered primarily to cancer cells while sparing healthy cells from exposure to toxic drugs.

In addition to antibodies, peptides and small molecules are also commonly used for surface functionalization. Peptides can be designed to bind to specific receptors on cancer cells or other diseased tissues, providing a more personalized approach to drug delivery [13]. For instance, peptides that recognize the overexpressed receptors on tumor cells can be conjugated to the surface of the nanoparticle, allowing for targeted therapy that reduces the risk of side effects and enhances the efficacy of the treatment. Similarly, small molecules such as folic acid or transferrin can be used to target cells with specific surface markers, providing an additional level of precision in drug delivery.

One of the key advantages of core-shell nanostructures in drug delivery is their ability to provide controlled release of the therapeutic agent. The shell material can be engineered to respond to specific environmental stimuli, such as changes in pH, temperature, or light, to trigger the release of the drug at the target site [14]. To guarantee that the medication is only delivered where it is required, pH-sensitive nanoparticles, for instance, can be engineered to release their payload in the acidic environment of a tumor. Similarly, temperature-sensitive nanoparticles can be activated by the localized heat generated by tumor tissues, allowing for controlled drug release in response to specific treatment conditions.

The core-shell design also allows for the encapsulation of a wide variety of drugs, including poorly water-soluble compounds, which can be challenging to deliver using conventional drug delivery methods. The core material can be chosen based on its ability to solvate or encapsulate the drug, while the shell can be designed to improve the stability, circulation time, and biocompatibility of the nanoparticle [15]. This flexibility in design makes core-shell nanostructures a versatile and effective platform for targeted drug delivery.

Core-shell nanostructures offer a promising and highly effective approach to targeted drug delivery. Their ability to encapsulate drugs and deliver them directly to specific cells or tissues, combined with surface functionalization that enables precise targeting, makes them ideal candidates for applications in personalized medicine. By engineering the core and shell materials to respond to specific environmental

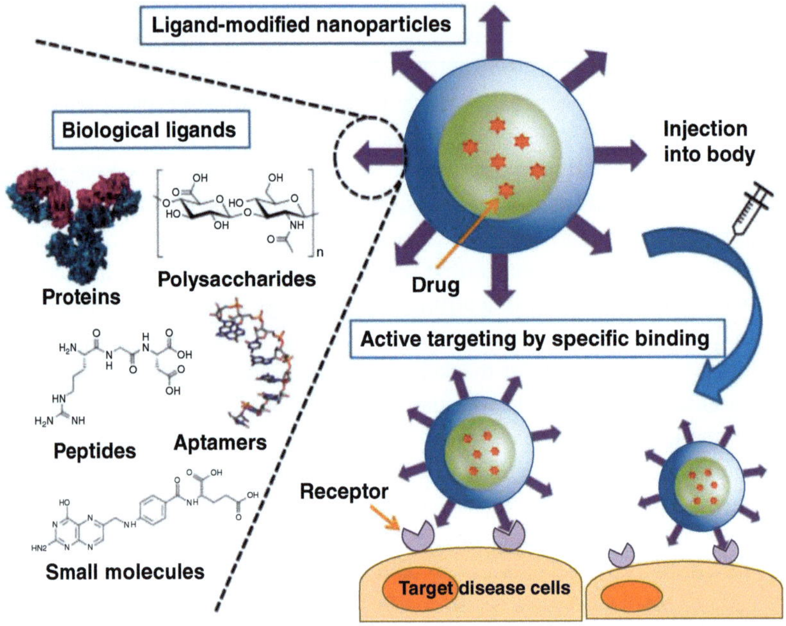

Figure 7.2 A diagram depicting a core-shell nanoparticle with a therapeutic drug payload, targeting specific cells for localized treatment. [16]/MDPI/CC BY 4.0.

cues, researchers can develop drug delivery systems that maximize therapeutic efficacy while minimizing side effects. As the field of nanomedicine continues to advance, core-shell nanostructures are expected to play an increasingly important role in the development of more effective and targeted treatments for a variety of diseases, particularly cancer and other complex conditions. As shown in Figure 7.2, these nanoparticles can be designed to not only carry and release drugs but also to serve as diagnostic agents or therapeutic agents in a single platform, thus offering a combination of therapeutic efficacy and diagnostic monitoring in a single treatment regimen. This ability to design multifunctional nanoparticles makes core-shell materials a promising tool in the evolving field of nanomedicine.

7.2.2 Types of Core-shell Nanostructures for Drug Delivery

Core-shell nanostructures offer a variety of configurations for drug delivery, with each type designed to improve the therapeutic efficacy, stability, and controlled release of drugs. Among the most commonly used types of core-shell nanostructures for drug delivery are liposomes, polymeric nanoparticles, and inorganic core-shell systems, each of which presents unique advantages for specific applications.

Liposomes are lipid-based core-shell nanostructures with a hydrophilic core and a lipid bilayer shell, mimicking cellular membranes. These nanostructures are highly biocompatible and can encapsulate both hydrophilic and hydrophobic drugs within their core and shell, respectively. Liposomes are particularly useful for the delivery of anticancer drugs, antibiotics, and gene therapies [17]. Their composition allows them to fuse with cell membranes, releasing the drug payload into the cytoplasm. Liposomes' capacity to target particular cells or tissues is further improved by functionalizing them with targeting ligands (such as peptides or antibodies), which increases the accuracy of drug delivery and reduces side effects.

Polymeric nanoparticles are another popular class of core-shell structures used in drug delivery. These nanostructures consist of a biodegradable polymeric core surrounded by a shell made from either the same or different polymers. Polymeric nanoparticles offer several advantages, including the ability to encapsulate a wide range of drugs, prolonged drug release, and the ability to control the release rate over time [18]. They are highly versatile and can be tailored to specific needs by adjusting the polymer type, molecular weight, and the functionality of the surface shell. Examples include polylactic-co-glycolic acid (PLGA) nanoparticles, which are commonly used for drug delivery due to their biodegradability and biocompatibility [19]. The polymeric shell can also be modified with targeting agents, allowing for the selective delivery of drugs to specific cells or tissues, such as cancerous tumors.

Inorganic core-shell systems typically consist of inorganic nanoparticles like gold, silica, or Fe_3O_4 as the core, with a shell made of organic or inorganic materials. These systems are especially beneficial for applications where the core material imparts specific properties, such as magnetic properties (in the case of Fe_3O_4) or optical properties (in the case of gold). For example, gold nanoparticles (AuNPs) can be used for both drug delivery and bioimaging due to their surface plasmon resonance (SPR) properties [20], which enable real-time tracking of the nanoparticles in the body. The shell can be functionalized with various ligands to achieve specific targeting, while the core provides stability and protection to the drug payload.

The ability to control the release of drugs from core-shell nanostructures is a key advantage in therapeutic applications. Controlled release mechanisms can be engineered to respond to external stimuli, such as pH, temperature, or light, ensuring that drugs are released only in specific environments or upon external triggering. pH-sensitive core-shell systems are commonly used in cancer therapy, as the tumor microenvironment is often more acidic than normal tissues. These nanostructures release their drug payload in response to the lower pH of the tumor site, minimizing the drug's exposure to healthy tissues. Temperature-sensitive systems can release drugs when exposed to localized heat, which can be applied using methods like hyperthermia, making them useful for cancer treatment [21]. Finally, light-responsive core-shell materials can release their payload when exposed to light of specific wavelengths, offering the potential for highly localized drug release and precise control over treatment timing. These various types of core-shell nanostructures, with their customizable properties and controlled release mechanisms, represent a promising approach for improving the delivery and efficacy of therapeutic agents in the body.

7.2.3 Examples of Effective Core-shell Systems

Table 7.1 presents a comparison of two effective core-shell nanostructures commonly used in targeted drug delivery: gold nanoparticle-based systems and mesoporous silica core-shell nanomaterials.

Table 7.1 Comparison of effective core-shell nanostructures for targeted drug delivery.

Core-shell nanostructure	Core material	Shell material	Target application	Mechanism	Advantages	Example drugs
Gold nanoparticle-based systems	Gold (Au)	Polymers, silica, or biomolecules (e.g., antibodies)	Targeted cancer therapy	The gold core enhances cellular uptake and enables imaging; surface functionalization with ligands targets cancer cells	High biocompatibility, ease of functionalization, and capability for dual therapy (drug delivery and imaging)	Doxorubicin, paclitaxel, and other chemotherapeutic drugs
Mesoporous silica core-shell nanomaterials	Mesoporous silica (SiO_2)	Organic polymers or biocompatible materials	Controlled drug release systems	The mesoporous core allows for the encapsulation of a large drug payload; the shell provides controlled release based on environmental stimuli (e.g., pH, temperature)	High drug-loading capacity, controlled and sustained release, tunable pore sizes	Anticancer drugs, antibiotics, and peptides
Hybrid gold-silica core-shell systems	Gold (Au)	Silica (SiO_2)	Drug delivery and imaging (cancer therapy)	Core provides optical properties for imaging, while the silica shell allows for controlled drug release	Dual functionality (drug delivery + imaging), high stability, and tunability	Doxorubicin, 5-fluorouracil (5-FU), and others

Gold nanoparticle-based systems use gold as the core material, which is highly biocompatible and capable of enhancing the uptake of therapeutic agents by tumor cells. The gold core's optical properties also enable real-time imaging and tracking of nanoparticles inside the body, facilitating theranostics. The shell is typically made of polymers, silica, or biomolecules such as antibodies, which are functionalized to target cancer cells specifically, enabling precise drug delivery.

Mesoporous silica core-shell nanomaterials use a mesoporous silica core, which provides a large surface area for loading drugs. These systems are perfect for targeted drug administration in the body since they are made to release medications in a regulated manner based on environmental variables like pH or temperature. The organic shell surrounding the silica core can be tailored to control the rate of drug release, allowing for sustained therapeutic effects. Both of these core-shell systems offer significant advantages in the realm of drug delivery [22], including the ability to improve the efficacy and reduce the side effects of drugs, as well as provide enhanced imaging capabilities for monitoring treatment progress. These nanostructures hold great potential for future clinical applications, particularly in the fields of cancer therapy and personalized medicine.

7.3 Bioimaging and Diagnostics

7.3.1 Role of Core-shell Materials in Bioimaging

Core-shell materials play a significant role in bioimaging by enhancing the sensitivity, contrast, and resolution of several imaging methods, such as fluorescence imaging, CT, and MRI. The unique properties of core-shell nanostructures make them highly suitable for these applications, offering improvements in imaging capabilities compared to conventional agents. One of the primary uses of core-shell nanostructures in bioimaging is as contrast agents. In MRI, the core-shell nanoparticles typically use magnetic materials such as Fe_3O_4 in the core, which enhance the contrast of the images by altering the magnetic properties of the surrounding tissues [23]. The shell material, often composed of biocompatible substances like silica, helps to stabilize the magnetic core and can also be functionalized with targeting agents, allowing for specific imaging of tissues or cells. This dual function of providing both imaging contrast and targeting capability makes these materials invaluable for early disease diagnosis, particularly in oncology, where they can help in visualizing tumors and metastasis.

Similarly, in CT imaging, core-shell materials can significantly improve the contrast by incorporating high atomic number materials in the core, such as gold (Au) or barium (Ba), which strongly attenuate X-rays. These materials allow for better differentiation between tissues of different densities, leading to higher resolution images [24]. The shell, often made of biocompatible polymers or silica, ensures that the core material is protected from degradation in the body and can be functionalized with molecules like antibodies to target specific cells or tissues. This targeted approach is particularly useful in diagnostic imaging of cancers, infections, and cardiovascular diseases (CVDs), where precise localization is essential.

In fluorescence imaging, core-shell nanostructures exhibit unique optical properties that are highly advantageous for bioimaging. Core-shell nanoparticles, such as those made from semiconductor materials like cadmium selenide (CdSe) or quantum dots, can be engineered to emit light at specific wavelengths [25, 26]. The core material is responsible for the fluorescence emission, while the shell material, often made of a wide band-gap semiconductor like zinc sulfide (ZnS), acts to stabilize the core and prevent quenching of the fluorescence. This results in brighter, more stable fluorescence signals, making it easier to track nanoparticles in biological systems over extended periods. Furthermore, the emission wavelength of quantum dots can be tuned by adjusting the core size, enabling multiplexed imaging, where multiple targets can be labeled and tracked simultaneously with distinct colors.

The core-shell architecture also allows for enhanced surface functionality, which is key in improving imaging sensitivity. The shell can be modified with functional groups such as biomolecules that target particular receptors or antigens on the surface of cells or tissues including peptides, antibodies, and others. This targeting capability ensures that the core-shell nanoparticles accumulate in the desired location, providing more accurate and detailed imaging results. For example, in cancer imaging, core-shell nanoparticles can be functionalized with ligands that bind specifically to cancer cell receptors, allowing for high-resolution imaging of tumors and facilitating early detection [27]. In addition to improving imaging sensitivity, core-shell materials offer the advantage of minimizing toxicity. The shell acts as a protective barrier, preventing the core material from leaching into the body, which is especially important for metals or other potentially toxic materials. This biocompatible shell ensures that the nanoparticles remain stable in the bloodstream and are less likely to cause harmful side effects, making them safer for clinical applications. As shown in Figure 7.3, core-shell nanostructures

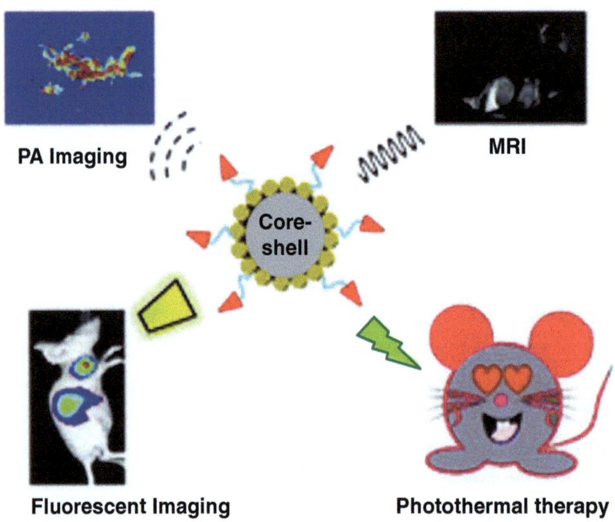

Figure 7.3 A schematic of core-shell nanostructures used in bioimaging, highlighting their use as contrast agents in MRI, CT, and fluorescence imaging for enhanced disease detection and monitoring. Reproduced with permission from [28]/Royal Society of Chemistry.

are increasingly used as contrast agents in various imaging modalities, including MRI, CT, and fluorescence imaging. Their ability to combine the unique properties of both the core and shell materials allows for improved performance and greater versatility in biomedical imaging.

7.3.2 Core-shell Materials in Diagnostics

Core-shell materials have become invaluable in diagnostics due to their unique ability to enhance sensitivity, selectivity, and versatility, particularly in the detection of biomarkers. These materials, often composed of a functionalized core surrounded by a biocompatible shell, provide a platform for the integration of diagnostic markers, enabling highly specific and efficient detection of various disease markers, including those associated with cancers, infections, and cardiovascular conditions.

One of the most prominent applications of core-shell materials in diagnostics is in the detection of biomarkers using nanoparticles, such as gold (Au)- or silica (SiO_2)-based nanoparticles. AuNPs are widely used due to their excellent optical properties, including surface plasmon resonance (SPR) [29], which can be exploited for highly sensitive detection of biological targets. The core of these nanoparticles is typically composed of gold, which is known for its stability, ease of functionalization, and biocompatibility. The shell, which can be made from materials like silica, polyethylene glycol (PEG), or other polymers, serves to protect the gold core from environmental factors and facilitates the attachment of targeting molecules, such as antibodies, peptides, or oligonucleotides, that specifically bind to disease-related biomarkers.

For example, in cancer diagnostics, gold core-shell nanoparticles functionalized with cancer-specific antibodies can bind to specific tumor antigens, enabling the detection of early-stage tumors. The SPR phenomenon associated with AuNPs allows for the detection of these biomarker-antibody interactions through simple optical techniques like colorimetric assays or localized surface plasmon resonance (LSPR) sensors [30]. The core-shell structure enhances the sensitivity of these detection methods by increasing the surface area available for biomolecule conjugation, thereby improving the overall detection limits of the assay.

Similarly, silica-based core-shell nanoparticles are increasingly used in diagnostic applications due to their high surface area, tunability, and versatility. Silica nanoparticles are non-toxic and can be easily modified to carry a wide range of functional groups on their surface. The core is typically composed of silica, which provides stability and durability, while the shell can be functionalized with various bioactive molecules, such as antibodies or DNA molecules that specifically target biomarkers [31]. A variety of biomarkers, including those linked to illnesses like cancer, can be found using silica nanoparticles, Alzheimer's, and infections. The high surface-to-volume ratio of silica nanoparticles allows for the attachment of multiple biomolecules, which can further enhance detection sensitivity. For example, in the detection of viral infections, silica-based nanoparticles can be functionalized with viral antibodies, allowing for rapid and accurate detection of the pathogen [32]. These nanoparticles can also be designed to carry fluorescent dyes or quantum dots, which provide a strong optical signal that can be easily detected using fluorescence microscopy or flow cytometry. In addition to their use in detecting single biomarkers, core-shell materials are particularly useful for multiplexed detection, where multiple biomarkers are detected simultaneously. This capability is achieved by integrating diagnostic markers into the core-shell system, allowing for the detection of multiple targets in a single assay. For instance, core-shell nanoparticles can be functionalized with a combination of antibodies or other ligands that target different biomarkers. When exposed to a sample containing multiple disease-related biomarkers, the nanoparticles will bind to the respective targets, enabling the simultaneous detection of multiple biomarkers [33].

Multiplexed detection is especially valuable in early-stage diagnostics, where the presence of multiple biomarkers can provide a more accurate and comprehensive picture of the disease. For example, in cancer diagnostics, detecting multiple tumor markers at once can improve the sensitivity and specificity of the test, allowing for earlier and more accurate diagnosis. Similarly, in infectious disease diagnostics, multiplexed core-shell nanoparticle assays can detect multiple pathogens or strains of a pathogen in a single test, reducing the time and cost associated with traditional diagnostic methods. The ability to perform multiplexed diagnostics with core-shell materials is also beneficial in point-of-care (POC) testing, where quick and reliable results are essential [34]. Core-shell nanoparticles can be integrated into portable diagnostic devices that provide rapid results at the patient's bedside or in remote areas with limited access to advanced medical facilities. This makes core-shell materials particularly useful in resource-limited settings, where timely and accurate diagnostics can have a significant impact on patient outcomes.

As illustrated in Figure 7.4, core-shell nanostructures are being increasingly used in diagnostic applications, where their special qualities make it possible to integrate several diagnostic markers within a single system and improve biomarker detection. These substances not only enhance diagnostic tests' sensitivity and specificity but also provide a versatile platform for the development of advanced diagnostic assays.

7.3.3 Applications in Early-stage Disease Detection

Core-shell nanomaterials are increasingly being utilized for early-stage disease detection, offering significant improvements in the diagnosis of a number of illnesses, including neurological problems, cardiovascular ailments, and cancer. Early diagnosis is critical for effective treatment and improved patient outcomes, and core-shell nanoparticles provide a unique platform for detecting disease biomarkers at much earlier stages than traditional diagnostic methods. In cancer detection, core-shell nanostructures have shown remarkable promise in identifying tumor biomarkers even before clinical symptoms appear. These nanoparticles, often functionalized with antibodies, peptides, or other targeting ligands, can specifically bind to cancer cell markers or antigens present on the surface of tumor cells or circulating tumor cells (CTCs) [36]. For instance, gold- or silica-based

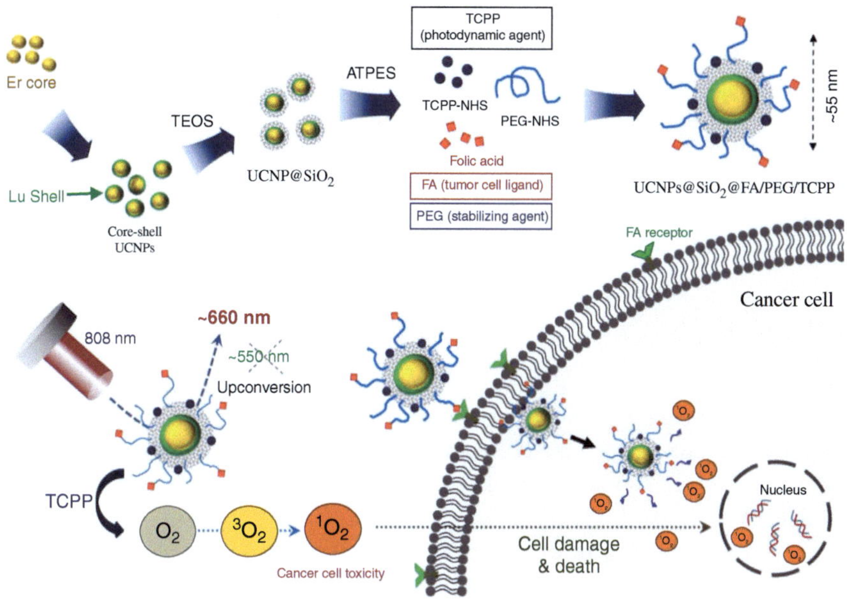

Figure 7.4 Schematic showing core-shell nanostructures in diagnostics, highlighting their role in the detection of biomarkers and the integration of multiple diagnostic markers for multiplexed detection, leading to more accurate and comprehensive disease diagnosis. [35]/MDPI/CC BY 4.0.

core-shell nanoparticles, modified with cancer-specific antibodies, can selectively bind to overexpressed tumor markers such as HER2, PSA, or CEA. These interactions can be detected using various techniques like SPR, fluorescence imaging, or MRI, providing highly sensitive detection. This ability to detect cancer at its earliest stages—before the onset of noticeable symptoms—greatly increases the chances of successful treatment, particularly in cancers like breast, prostate, and ovarian.

For CVDs, core-shell nanoparticles are also proving beneficial in detecting early biomarkers related to heart attacks, strokes, and other vascular conditions. Nanoparticles, such as Fe_3O_4 or gold core-shell nanostructures, are used to detect specific biomarkers like troponin I or C-reactive protein (CRP), which are associated with inflammation or myocardial injury [37]. These biomarkers can be detected in blood samples with high sensitivity, allowing for the detection of heart conditions at their inception. Early detection of CVDs enables timely intervention, such as medication or lifestyle changes, to prevent more severe outcomes like heart failure or stroke. Similarly, core-shell nanomaterials are making significant strides in the early detection of neurological disorders, including Alzheimer's and Parkinson's diseases. Biomarkers for these conditions, such as amyloid plaques in Alzheimer's or α-synuclein aggregates in Parkinson's, can be identified through the use of core-shell nanoparticles. For example, silica-based nanoparticles functionalized with specific ligands that bind to amyloid plaques can be used in MRI or fluorescence imaging to visualize these abnormal protein deposits in the brain long before clinical symptoms manifest [38]. Early diagnosis of neurological conditions allows for more effective treatment strategies, helping to slow disease progression and manage symptoms. Overall, core-shell nanomaterials enable highly sensitive and specific early-stage disease detection for cancer, CVDs, and neurological disorders. Their ability to target biomarkers with precision and provide real-time diagnostic information makes them powerful tools in improving patient outcomes through timely intervention.

7.3.4 Examples of Core-shell Materials in Bioimaging and Diagnostics

Core-shell nanomaterials have shown great promise in the fields of bioimaging and diagnostics, particularly through the use of materials like Fe_3O_4 core-shell nanoparticles and quantum dot–based core-shell systems. These materials enable highly sensitive and specific imaging, making them invaluable tools for early disease detection.

Fe_3O_4 core-shell nanoparticles are among the most widely used nanostructures for MRI. The core of these nanoparticles is composed of Fe_3O_4, which possesses strong magnetic properties, allowing for high-quality imaging with high contrast [39]. The shell, typically made of biocompatible materials like silica or polymers, protects the core from degradation and enhances stability in biological environments. These nanoparticles are functionalized with ligands, such as antibodies or peptides, to target specific biomarkers associated with diseases like cancer or cardiovascular conditions. When injected into the body, these functionalized Fe_3O_4 nanoparticles can accumulate at the target site, where their magnetic properties provide enhanced contrast during MRI scans, making it easier to detect abnormalities at very early stages. This capability makes Fe_3O_4 core-shell nanoparticles a crucial tool in non-invasive imaging, offering detailed insights into disease pathogenesis, tumor localization, and vascular abnormalities [40].

In addition to Fe_3O_4-based systems, quantum dot–based core-shell nanostructures are gaining traction in fluorescence imaging. Quantum dots are semiconductor nanocrystals with unique optical properties, including size-tunable fluorescence emission, which allows for high sensitivity and multiplexed detection. The core of a quantum dot is typically made of materials like CdSe, while the shell is composed of a protective layer such as ZnS, which enhances the stability and optical properties of the quantum dot [41]. These core-shell quantum dots are used as fluorescent probes in various bioimaging techniques, offering high resolution and the ability to track cellular events in real time. Functionalization with specific targeting agents enables quantum dots to selectively bind to disease markers, such as cancer antigens or biomarkers of neurological disorders. When illuminated with UV or visible light, these quantum dots emit fluorescence, allowing for the precise visualization of cellular and molecular processes. Their versatility, coupled with their ability to detect multiple biomarkers simultaneously, makes them a powerful tool for early-stage diagnostics, particularly in oncology and neurology [42]. As shown in Figure 7.5, core-shell nanomaterials like Fe_3O_4 and quantum dots have revolutionized the field of bioimaging, providing clinicians with highly detailed and accurate diagnostic information. These materials offer significant advantages, such as non-invasive, real-time imaging, which is essential for early detection and treatment planning.

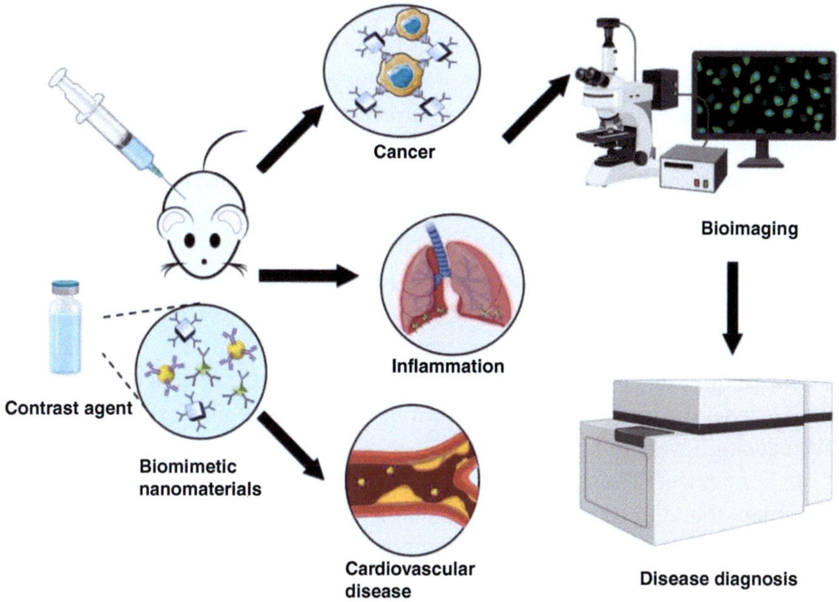

Figure 7.5 Schematic showing core-shell nanomaterials as contrast agents in bioimaging for early disease detection. [43]/Frontiers Media S.A/CC BY 4.0.

7.4 Theranostics: Combining Therapy and Diagnostics

7.4.1 Concept of Theranostics

Theranostics is a cutting-edge field that merges therapeutic and diagnostic functions into a single integrated platform. This innovative approach aims to enhance personalized medicine by enabling real-time monitoring of disease progression and the effectiveness of treatment while simultaneously delivering therapeutic agents. The integration of both diagnostic and therapeutic capabilities in one system can provide significant advantages over traditional approaches, where diagnosis and treatment are often separate processes. Theranostic systems allow for the precise targeting of specific tissues or cells, ensuring that the treatment is not only more effective but also minimizes side effects by limiting exposure to non-target areas [44].

Core-shell nanomaterials play a pivotal role in theranostics due to their unique structural properties, which allow for the encapsulation of therapeutic agents in the core and diagnostic functionalities on the shell. This configuration makes core-shell nanostructures highly versatile and well-suited for theranostic applications. The core typically houses the therapeutic payload, such as drugs, genes, or other bioactive molecules, while the shell is engineered with functional materials that enable imaging or sensing [45]. The shell may also include ligands or antibodies that allow the nanomaterial to specifically target certain receptors on diseased cells, such as cancer cells, thereby improving the precision of both drug delivery and diagnostic imaging.

The ability of core-shell nanoparticles to provide therapy and provide feedback regarding the efficacy of the treatment is one of the most important benefits of using them for theranostics. This feedback is often achieved through imaging modalities, such as MRI, fluorescence imaging, or positron emission tomography (PET), which can monitor the nanomaterials in real time. By tracking the distribution of core-shell nanoparticles in the body, clinicians can assess whether the therapeutic agents are reaching their target and whether the therapy is producing the desired effect [46]. This real-time monitoring reduces the need for invasive biopsies and can help avoid unnecessary treatments by identifying non-responsive areas early on. Moreover, the use of core-shell systems allows for controlled release mechanisms, where the therapeutic agent is released in response to specific stimuli, such as changes in pH, temperature, or light. By ensuring that the therapeutic drug is only released when it reaches the intended site, this controlled release maximizes therapy efficacy while reducing adverse effects. For example, in cancer therapy, core-shell nanoparticles can be engineered to release their drug payload in response to the acidic microenvironment found in tumors, ensuring that the drug is released precisely where it is needed, sparing healthy tissues from unnecessary exposure [47].

The concept of theranostics is particularly beneficial in cancer treatment, where it is crucial to deliver the right dose of therapy to the right location while monitoring its impact. For instance, gold–silver core-shell nanomaterials can be designed to act as both imaging agents and therapeutic agents for cancer cells. The gold core allows for imaging through techniques like MRI or CT, while the silver shell can be loaded with a chemotherapeutic agent that is released upon reaching the target site. This dual functionality makes theranostic systems highly efficient, reducing the number of treatments and improving the overall outcome. Theranostic systems are not limited to cancer treatment but have broad applications across various diseases, including cardiovascular disorders, neurological diseases, and infectious diseases. For instance, theranostic nanomaterials could be used in neurodegenerative diseases, where nanoparticles are designed to cross the blood–brain barrier and deliver therapeutic agents directly to the brain while simultaneously imaging brain activity or disease biomarkers [48]. Similarly, in CVDs, core-shell nanoparticles could be used to target atherosclerotic plaques, delivering therapeutic agents to reduce plaque size while providing real-time feedback through imaging techniques. Theranostics represents a paradigm shift in the treatment of diseases, combining the diagnostic and therapeutic functions into one platform. Core-shell nanomaterials are particularly well-suited for theranostics due to their ability to encapsulate therapeutic agents and integrate diagnostic functionalities [49]. This combination allows for highly targeted, efficient treatments that are monitored in real time, ultimately improving the precision and effectiveness of therapies while minimizing side effects. The future of theranostics lies in the continued development of advanced core-shell systems that can address a wide range of medical conditions with greater accuracy and efficiency.

7.4.2 Mechanisms in Theranostics

In theranostics, the concept of dual-functional nanostructures represents an advanced approach to combining therapeutic and diagnostic functionalities into a single system. This integration enables simultaneous treatment and monitoring, a significant step toward precision medicine. The core-shell nanostructures utilized in theranostics are designed to perform both roles—delivering therapeutic agents while allowing real-time monitoring of their efficacy using various imaging techniques. The core of a core-shell nanostructure typically carries the therapeutic payload, while the shell is functionalized to provide targeting capabilities and enable imaging or monitoring. For example, in cancer therapy, a dual-functional nanoparticle can encapsulate a chemotherapeutic drug in its core, while the shell is designed with imaging agents such as magnetic nanoparticles, fluorescent dyes, or radioactive isotopes [50]. This setup allows the delivery of the drug directly to the tumor site and simultaneously enables the use of techniques like MRI, fluorescence, or PET to track the distribution and response to the therapy. As shown in Figure 7.6, the core-shell nanostructure's ability to integrate both the therapeutic and diagnostic properties provides a seamless mechanism for treatment and monitoring, enhancing the precision and efficacy of therapy.

The regulated release of the therapeutic drug is one of theranostics' primary processes. It is possible to engineer the core-shell structure to react to particular triggers at the target site, including pH, temperature, or light. This allows the drug to be released in a controlled manner when the nanoparticle reaches the desired location, minimizing side effects and enhancing the therapeutic effect. For

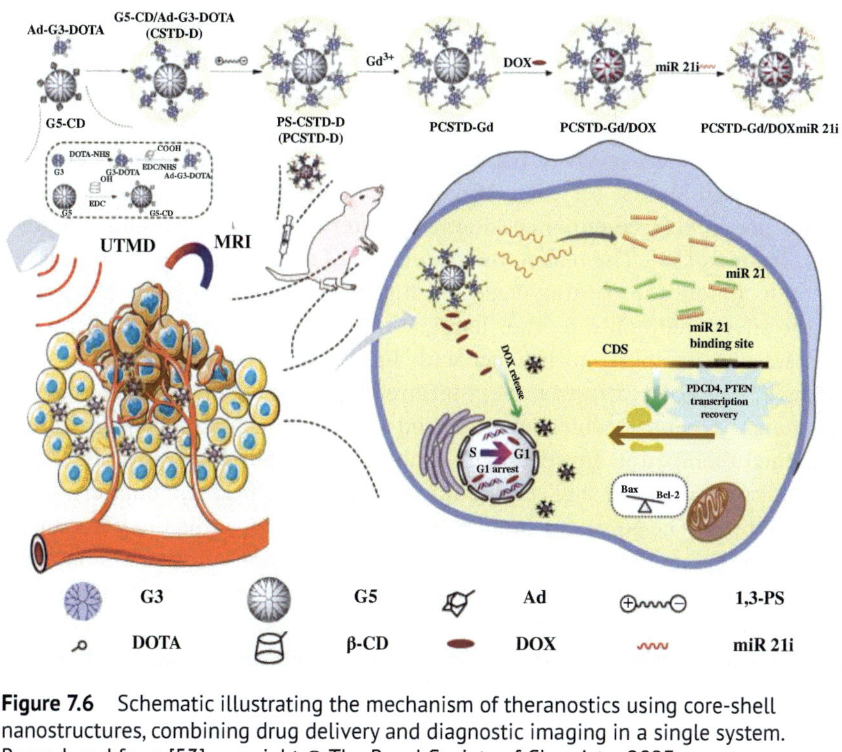

Figure 7.6 Schematic illustrating the mechanism of theranostics using core-shell nanostructures, combining drug delivery and diagnostic imaging in a single system. Reproduced from [53] copyright © The Royal Society of Chemistry 2023.

instance, in the case of chemotherapy, the nanostructure can release the drug only in the acidic environment of a tumor, avoiding unnecessary systemic exposure. Simultaneously, the imaging agents in the shell allow clinicians to monitor the localization of the nanoparticles and evaluate the response to treatment in real time [51]. For example, MRI can be used to track the movement of nanoparticles inside the body, providing insights into the efficacy of the therapy [52]. Similarly, fluorescence or PET imaging can provide high-resolution imaging of the target site, enabling the precise monitoring of drug accumulation and tumor shrinkage over time. This dual functionality of drug delivery and diagnostic imaging allows for real-time feedback, providing crucial information about the treatment progress. It can also help in adjusting therapeutic strategies promptly, ensuring that patients receive the most effective treatment based on real-time data. As the field advances, the integration of artificial intelligence and machine learning with theranostic systems will further enhance their predictive capabilities, improving personalized treatment plans.

7.4.3 Examples of Nanomaterials Used and Applications in Cancer Theranostics

Hybrid nanomaterials, such as gold–silver core-shell nanoparticles, are becoming increasingly popular in the field of theranostics due to their unique properties that combine both imaging and therapeutic capabilities. These core-shell structures are designed to take advantage of the distinct properties of the individual materials in the core and shell. In the case of gold–silver core-shell nanoparticles, gold is often used for the core because of its excellent stability, biocompatibility, and ability to enhance the imaging signal, while silver, which is highly reactive, is often employed as the shell to facilitate therapeutic interactions [54].

The gold core offers several advantages, particularly in imaging applications. AuNPs are well-known for their strong plasmonic properties, making them ideal for use in surface-enhanced Raman scattering (SERS), CT, and photoacoustic imaging [55]. The plasmonic resonance of gold enhances the signal in imaging techniques, providing clear and high-resolution images of the tumor site. The silver shell, on the other hand, is often utilized for its antimicrobial properties and its ability to interact with light, particularly in photothermal therapy (PTT). When exposed to near-infrared light, the silver shell can convert light energy into heat, which can be used to destroy cancer cells, providing a combined therapeutic approach alongside imaging. The integration of these two materials in a core-shell configuration enables hybrid nanomaterials to serve dual roles. The gold core supports effective imaging, while the silver shell provides therapeutic functionality [56]. This combination allows clinicians to monitor the distribution and localization of the nanoparticles in real time using advanced imaging modalities like CT and photoacoustic imaging, while simultaneously using the heat generated by the silver shell for cancer cell destruction through PTT. This multifaceted approach to treatment and monitoring is a hallmark of theranostic systems, allowing for more personalized and effective cancer therapies. One of the most promising applications of these hybrid nanomaterials is in cancer theranostics. Cancer is a complex disease that requires personalized treatment strategies, and theranostic platforms, such as gold–silver core-shell nanoparticles, offer a way to address this challenge by providing both real-time diagnostic and therapeutic capabilities in one platform [57]. By utilizing these nanoparticles, clinicians can monitor the tumor's response to treatment in real time, enabling them to adjust treatment plans as needed to improve patient outcomes.

In cancer theranostics, real-time monitoring of tumor response is critical. Nanoparticles like gold–silver core-shells can be tracked throughout the body using imaging techniques like CT or photoacoustic imaging. These imaging modalities allow doctors to observe how the nanoparticles accumulate in the tumor site, ensuring that the therapeutic agents are delivered exactly where they are needed. Moreover, the ability to monitor the effectiveness of the therapy in real time helps guide treatment decisions, such as whether to continue, modify, or halt the treatment based on the tumor's response. In addition to imaging, the therapeutic function of the gold–silver core-shell nanoparticles is essential for treating cancer. The silver shell's photothermal properties allow for the selective destruction of cancer cells by converting absorbed light into heat. By applying near-infrared light to the tumor site, the temperature can be raised to a level that causes localized damage to the tumor cells, resulting in tumor shrinkage or cell death [58]. This focused strategy reduces harm to nearby healthy tissues, which lessens the adverse effects of conventional cancer therapies like radiation therapy or chemotherapy. Furthermore, the real-time monitoring of drug release is another important aspect of cancer theranostics. Because the gold–silver core-shell nanoparticles are designed to release therapeutic compounds in response to particular triggers, like pH changes or exposure to light, physicians can use imaging tools to observe how the drug is released and distributed at the tumor site. This controlled release ensures that the drug is delivered to the target area, reducing systemic exposure and improving the overall efficacy of the treatment [59].

7.5 Biocompatibility and Safety Assessment

Table 7.2 summarizes the various aspects of biocompatibility and safety assessment necessary for the development and clinical translation of core-shell materials in biomedicine. It outlines the importance of non-toxicity, the risks involved, and the testing required, as well as the regulatory framework that governs the approval process for nanomaterials used in medical applications.

Table 7.3 provides a comparative analysis of various core-shell nanomaterials, outlining their key characteristics, therapeutic uses, imaging techniques, biocompatibility, and safety profiles. The materials listed are all utilized in biomedical applications, but their effectiveness, safety, and overall suitability for clinical use depend on several factors, including their composition and functionalization.

Table 7.2 Importance of biocompatibility and safety assessment.

Aspect	Details
Importance of biocompatibility	**Core-shell materials must be non-toxic, non-immunogenic, and biocompatible for medical applications.** Core-shell nanomaterials need to be carefully designed to ensure they do not trigger immune responses or cause toxicity in the body. The biocompatibility of these materials is critical to their successful application in biomedicine, particularly for long-term use. The materials must integrate well with biological systems without causing adverse reactions. **Evaluation of long-term stability, degradation, and metabolic clearance of core-shell nanomaterials in biological systems.** To assess the long-term safety and potential effectiveness of core-shell nanomaterials, their degradation in biological systems, as well as their clearance and metabolism, must be studied thoroughly. Materials that degrade into non-toxic by-products are preferred, ensuring they do not accumulate in tissues or organs over time, which could lead to long-term health issues.
Safety concerns and potential risks	**Leaching of toxic materials, accumulation in organs, and unintended biological interactions.** Core-shell nanomaterials could potentially leach toxic substances, which may lead to organ damage or dysfunction. It is essential to evaluate how materials interact with biological systems to avoid harmful accumulation in organs such as the liver or kidneys. Furthermore, the unintended interaction of nanoparticles with biological molecules may lead to cellular toxicity or other undesirable effects. **Impact on cellular systems and tissues.** The nanomaterials' impact on cellular systems needs to be assessed to understand the biological consequences of exposure. This includes studying the materials' ability to cross cellular membranes, interact with cell surface receptors, and cause any form of cellular damage, such as oxidative stress or inflammation.
In vitro and In vivo toxicity studies	**Overview of current testing methodologies (cell viability assays, animal models).** In vitro (cell culture) and in vivo (animal) models are extensively used to study the toxicity of core-shell nanomaterials. In vitro testing includes assays like MTT, LDH, or Annexin V assays to determine cell viability and apoptosis. In vivo studies involve evaluating the biodistribution, organ accumulation, and systemic toxicity of the nanoparticles in animal models. These models help assess both short- and long-term risks. **Long-term safety evaluation for clinical translation.** For clinical application, long-term safety must be rigorously evaluated. This entails conducting in-depth research in animal models and, eventually, human trials to evaluate the long-term impacts of core-shell nanomaterials, including their metabolism, biodegradability, and potential for organ accumulation over time.
Regulatory aspects of core-shell materials in biomedicine	**FDA and EMA regulations for the authorization of medications and devices based on nanomaterials**. Regulatory agencies such as the European Medicines Agency (EMA) and the Food and Drug Administration (FDA) provide specific guidelines for the approval of nanomaterials used in biomedicine. These include ensuring that core-shell nanomaterials meet stringent safety standards, such as biocompatibility and lack of toxicity, before being used in drug delivery systems or medical devices. Both organizations require thorough preclinical and clinical evaluations to assess the safety and efficacy of nanomaterials. **Compliance with regulatory standards.** In addition to biocompatibility and safety, regulatory bodies evaluate nanomaterials for their ability to meet current standards for manufacturing, sterility, and product consistency. This ensures that nanomaterials, once approved, are safe for human use and can be reliably produced at scale for therapeutic applications.

Table 7.3 Comparison of core-shell nanomaterials for biomedical applications.

Material composition	Therapeutic application	Imaging technique	Biocompatibility	Safety profile
Fe_3O_4@SiO_2	Targeted drug delivery, cancer therapy	MRI, Hyperthermia	Non-toxic, biocompatible with minimal immune response	Low risk of toxicity, some concerns about accumulation in liver and spleen
Au@Polymer	Chemotherapy, immunotherapy	CT, fluorescence imaging	Generally biocompatible with surface modifications for enhanced stability	Safe with proper functionalization, potential for mild immune response in certain cases
Ag@SiO_2	Antibacterial treatment, wound healing	CT, MRI, fluorescence	Biocompatible with slight cytotoxicity at high doses	Potential for leaching of silver ions at high concentrations
SiO_2@Polymer	Drug delivery, bioimaging	MRI, fluorescence	Biocompatible, biodegradable	Minimal toxicity; long-term effects still under study
ZnO@Polymer	Photodynamic therapy, cancer treatment	Fluorescence, X-ray	Biocompatible with some cytotoxicity at high concentrations	Risks of cytotoxicity at high doses, requires surface modification
Fe_3O_4@Au	MRI, cancer therapy, targeted drug delivery	MRI, CT, hyperthermia	High biocompatibility, low immunogenicity	Safe when surface-functionalized, concerns over accumulation in organs

7.6 Conclusion

Chapter 6 has provided a comprehensive overview of the transformative potential of core-shell nanostructures in biomedical applications, highlighting their significant roles in advancing therapeutic, diagnostic, and theranostic technologies. As we have explored, core-shell nanomaterials offer a wealth of opportunities due to their multifunctionality, tunability, and biocompatibility, which make them ideal candidates for a range of biomedical applications, from targeted drug delivery to bioimaging and personalized therapies.

It has been shown that core-shell nanostructures can improve drug solubility, bioavailability, and controlled release when used in targeted drug delivery systems, all while minimizing side effects. Surface functionalization with specific ligands enables precise targeting of therapeutic agents to specific cells or tissues, such as cancer cells, ensuring more effective treatment outcomes. The advancement of these systems promises to revolutionize cancer therapy, improving efficacy and reducing toxicity.

In diagnostics, core-shell materials, such as gold- and silica-based nanoparticles, have proven to be powerful agents for biomarker detection, facilitating early detection of illnesses such as neurological problems, cardiovascular ailments, and cancer. The integration of these materials with advanced imaging techniques, such as MRI, fluorescence, and CT, has enhanced the sensitivity and specificity of disease detection, allowing for early-stage diagnosis and timely intervention.

The concept of theranostics, combining therapeutic and diagnostic functions in a single nanostructure, has emerged as a promising avenue for personalized medicine. Dual-functional nanostructures enable real-time monitoring of therapeutic responses, offering the potential for dynamic treatment adjustments. Hybrid materials, such as gold–silver core-shell nanostructures, further exemplify the potential of theranostics, integrating both therapeutic and diagnostic functions in a single platform.

However, the clinical translation of these advanced nanomaterials requires rigorous attention to biocompatibility, safety, and regulatory compliance. Long-term toxicity studies, both in vitro and in vivo, are needed to assure the safe use of these materials in humans. Regulatory agencies like the FDA and EMA are essential in making sure that core-shell materials meet the necessary safety and efficacy standards before they can be used in medical devices or drug delivery systems.

The future of core-shell nanostructures in biomedicine holds immense promise. As research continues to advance, we can expect more refined systems with greater precision, improved biocompatibility, and enhanced therapeutic outcomes. Moreover, with the integration of artificial intelligence and personalized medicine approaches, the development of customized, patient-specific therapies is becoming increasingly feasible.

In conclusion, core-shell nanostructures are at the forefront of biomedical innovation, offering groundbreaking solutions for a variety of medical challenges. The continued evolution of these materials, combined with a focus on safety, efficacy, and regulatory adherence, will pave the way for their successful integration into clinical practice, ultimately improving patient care and outcomes across a range of diseases and conditions.

References

1 Tibbitt MW, Dahlman JE, Langer R. Emerging frontiers in drug delivery. J Am Chem Soc [Internet]. 2016 Jan 27 [cited 2025 Feb 1];138(3): 704–17. Available from: https://pubs.acs.org/doi/full/10.1021/jacs.5b09974

2 Zou H, Luo Z, Yang X, Xie Q, Zhou Y. Toward emerging applications using core–shell nanostructured materials: A review. J Mater Sci [Internet]. 2022 Jun 6 [cited 2025 Feb 1];57(24): 10912–42. Available from: https://link.springer.com/article/10.1007/s10853-022-07328-z

3 Wells CM, Harris M, Choi L, Murali VP, Guerra FD, Jennings JA. Stimuli-responsive drug release from smart polymers. J Funct Biomater [Internet]. 2019 Jul 31 [cited 2025 Feb 1];10(3): 34. Available from: https://www.mdpi.com/2079-4983/10/3/34/htm

4 Wang Y, Gu H, Wang Y, Gu C. Core–shell-type magnetic mesoporous silica nanocomposites for bioimaging and therapeutic agent delivery. Adv Mater [Internet]. 2015 Jan 1 [cited 2025 Feb 1];27(3): 576–85. Available from: https://onlinelibrary.wiley.com/doi/full/10.1002/adma.201401124

5 Khan S, Falahati M, Cho WC, Vahdani Y, Siddique R, Sharifi M, et al. Core–shell inorganic NP@MOF nanostructures for targeted drug delivery and multimodal imaging-guided combination tumor treatment. Adv Colloid Interface Sci. 2023 Nov 1;321: 103007.

6 Chen WH, Luo GF, Zhang XZ, Chen WH, Luo GF, Zhang Z-X. Recent advances in subcellular targeted cancer therapy based on functional materials. Adv Mater [Internet]. 2019 Jan 1 [cited 2025 Feb 1];31(3): 1802725. Available from: https://onlinelibrary.wiley.com/doi/full/10.1002/adma.201802725

7 Wang J, Li Y, Nie G. Multifunctional biomolecule nanostructures for cancer therapy. Nat Rev Mater [Internet]. 2021 May 19 [cited 2025 Feb 1];6(9): 766–83. Available from: https://www.nature.com/articles/s41578-021-00315-x

8 Kumar R, Mondal K, Panda PK, Kaushik A, Abolhassani R, Ahuja R, et al. Core–shell nanostructures: Perspectives towards drug delivery applications. J Mater Chem B [Internet]. 2020 Oct 14 [cited 2025 Feb 1];8(39): 8992–9027. Available from: https://pubs.rsc.org/en/content/articlehtml/2020/tb/d0tb01559h

9 Sharma P, Brown S, Walter G, Santra S, Moudgil B. Nanoparticles for bioimaging. Adv Colloid Interface Sci. 2006 Nov 16; 123–126(SPEC. ISS.):471–85.

10 Spoială A, Ilie CI, Crăciun LN, Ficai D, Ficai A, Andronescu E. Magnetite-silica core/shell nanostructures: From surface functionalization towards biomedical applications—A review. Appl Sci [Internet]. 2021 Nov 22 [cited 2025 Feb 1];11(22): 11075. Available from: https://www.mdpi.com/2076-3417/11/22/11075/htm

11 Schärtl W. Current directions in core–shell nanoparticle design. Nanoscale [Internet]. 2010 Jun 9 [cited 2025 Feb 1];2(6): 829–43. Available from: https://pubs.rsc.org/en/content/articlehtml/2010/nr/c0nr00028k

12 Yetisgin AA, Cetinel S, Zuvin M, Kosar A, Kutlu O. Therapeutic nanoparticles and their targeted delivery applications. Molecules [Internet]. 2020 May 8 [cited 2025 Feb 1];25(9): 2193. Available from: https://www.mdpi.com/1420-3049/25/9/2193/htm

13 Manzari MT, Shamay Y, Kiguchi H, Rosen N, Scaltriti M, Heller DA. Targeted drug delivery strategies for precision medicines. Nat Rev Mater [Internet]. 2021 Feb 2 [cited 2025 Feb 1];6(4): 351–70. Available from: https://www.nature.com/articles/s41578-020-00269-6

14 Mura S, Nicolas J, Couvreur P. Stimuli-responsive nanocarriers for drug delivery. Nat Mater [Internet]. 2013 Oct 23 [cited 2025 Feb 1];12(11): 991–1003. Available from: https://www.nature.com/articles/nmat3776

15 Ghosh Chaudhuri R, Paria S. Core/shell nanoparticles: Classes, properties, synthesis mechanisms, characterization, and applications. Chem Rev [Internet]. 2012 Apr 11 [cited 2025 Feb 1];112(4): 2373–433. Available from: https://pubs.acs.org/doi/full/10.1021/cr100449n

16 Yoo J, Park C, Yi G, Lee D, Koo H. Active targeting strategies using biological ligands for nanoparticle drug delivery systems. Cancers [Internet]. 2019 May 8 [cited 2025 Feb 1];11(5): 640. Available from: https://www.mdpi.com/2072-6694/11/5/640/htm

17 Petkar KC, Chavhan SS, Agatonovik-Kustrin S, Sawant KK. Nanostructured materials in drug and gene delivery: A review of the state of the art. Crit Rev Ther Drug Carrier Syst [Internet]. 2011 [cited 2025 Feb 1];28(2): 101–64. Available from: https://www.dl.begellhouse.com/journals/3667c4ae6e8fd136,3d9c84f85fecce38,3bed07696ba3a974.html

18 Kamaly N, Yameen B, Wu J, Farokhzad OC. Degradable controlled-release polymers and polymeric nanoparticles: Mechanisms of controlling drug release. Chem Rev [Internet]. 2016 Feb 24 [cited 2025 Feb 1];116(4): 2602–63. Available from: https://pubs.acs.org/doi/full/10.1021/acs.chemrev.5b00346

19 Perinelli DR, Cespi M, Bonacucina G, Palmieri GF. PEGylated polylactide (PLA) and poly (lactic-co-glycolic acid) (PLGA) copolymers for the design of drug delivery systems. J Pharm Investig [Internet]. 2019 Apr 22 [cited 2025 Feb 1];49(4): 443–58. Available from: https://link.springer.com/article/10.1007/s40005-019-00442-2

20 Nguyen TT, Mammeri F, Ammar S. Iron oxide and gold based magneto-plasmonic nanostructures for medical applications: A review. Nanomaterials [Internet]. 2018 Mar 7 [cited 2025 Feb 1];8(3): 149. Available from: https://www.mdpi.com/2079-4991/8/3/149/htm

21 Dabbagh A, Abdullah BJJ, Abu Kasim NH, Abdullah H, Hamdi M. A new mechanism of thermal sensitivity for rapid drug release and low systemic toxicity in hyperthermia and thermal ablation temperature ranges. Int J Hyperthermia [Internet]. 2015 Jun 1 [cited 2025 Feb 1];31(4): 375–85. Available from: https://www.tandfonline.com/doi/abs/10.3109/02656736.2015.1006268

22 Gao Q, Zhang J, Gao J, Zhang Z, Zhu H, Wang D. Gold nanoparticles in cancer theranostics. Front Bioeng Biotechnol [Internet]. 2021 Apr 13 [cited 2025 Feb 1];9: 647905. Available from: www.frontiersin.org

23 Rawand MA. Silica core@shell nanoparticles for biomedical applications: A study in pharmaceutical sciences. Åbo Akademi University. 2024 May 31 [cited 2025 Feb 1]. Available from: https://www.doria.fi/handle/10024/188997

24 Cui X, Mathe D, Kovács N, Horváth I, Jauregui-Osoro M, Torres Martin De Rosales R, et al. Synthesis, characterization, and application of core-shell

$Co_{0.16}Fe_{2.84}O_4$@$NaYF_4$(Yb, Er) and Fe_3O_4@$NaYF_4$(Yb, Tm) nanoparticle as trimodal (MRI, PET/SPECT, and optical) imaging agents. Bioconjug Chem [Internet]. 2016 Feb 17 [cited 2025 Feb 1];27(2): 319–28. Available from: https://repositorio.unican.es/xmlui/handle/10902/32912
25 Sneha KR, Sailaja GS. Intrinsically radiopaque biomaterial assortments: A short review on the physical principles, X-ray imageability, and state-of-the-art developments. J Mater Chem B [Internet]. 2021 Oct 27 [cited 2025 Feb 1];9(41): 8569–93. Available from: https://pubs.rsc.org/en/content/articlehtml/2021/tb/d1tb01513c
26 Reiss P, Protière M, Li L. Core/shell semiconductor nanocrystals. Small [Internet]. 2009 Jan 19 [cited 2025 Feb 1];5(2): 154–68. Available from: https://onlinelibrary.wiley.com/doi/full/10.1002/smll.200800841
27 Yu MK, Park J, Jon S. Targeting strategies for multifunctional nanoparticles in cancer imaging and therapy. Theranostics [Internet]. 2012 [cited 2025 Feb 1]; 2(1): 3. Available from: https://pmc.ncbi.nlm.nih.gov/articles/PMC3263514/
28 Tian R, Zhu L, Qin Z, Wang G, Wang J, Zhang H. Glypican-3 (GPC3) targeted Fe_3O_4 core/Au shell nanocomplex for fluorescence/MRI/photoacoustic imaging-guided tumor photothermal therapy. Biomater Sci [Internet]. 2019 Nov 19 [cited 2025 Feb 1];7(12): 5258–69. Available from: https://pubs.rsc.org/en/content/articlehtml/2019/bm/c9bm01248f
29 Kowalska A, Adamska E, Grobelna B. Medical applications of silver and gold nanoparticles and core-shell nanostructures based on silver or gold core: Recent progress and innovations. ChemMedChem [Internet]. 2024 Jun 17 [cited 2025 Feb 1];19(12): e202300672. Available from: https://onlinelibrary.wiley.com/doi/full/10.1002/cmdc.202300672
30 Zhan S, Lou X, Zhou P, Xia F. Sandwich assays based on SPR, SERS, GMR, QCM, microcantilever, SAW, and RRS techniques for protein detection. In: Biosensors Based on Sandwich Assays [Internet]. Springer. 2018 Jan 9 [cited 2025 Feb 1]; 69–91. Available from: https://link.springer.com/chapter/10.1007/978-981-10-7835-4_5
31 Knopp D, Tang D, Niessner R. Review: Bioanalytical applications of biomolecule-functionalized nanometer-sized doped silica particles. Anal Chim Acta. 2009 Aug 4;647(1): 14–30.
32 Tallury P, Payton K, Santra S. Silica-based multimodal/multifunctional nanoparticles for bioimaging and biosensing applications. Nanomedicine [Internet]. 2008 Aug [cited 2025 Feb 1];3(4): 579–92. Available from: https://www.tandfonline.com/doi/abs/10.2217/17435889.3.4.579
33 Masud MK, Na J, Younus M, Hossain MSA, Bando Y, Shiddiky MJA, et al. Superparamagnetic nanoarchitectures for disease-specific biomarker detection. Chem Soc Rev [Internet]. 2019 Dec 9 [cited 2025 Feb 1];48(24): 5717–51. Available from: https://pubs.rsc.org/en/content/articlehtml/2019/cs/c9cs00174c
34 Hou F, Sun S, Abdullah SW, Tang Y, Li X, Guo H. The application of nanoparticles in point-of-care testing (POCT) immunoassays. Anal Methods [Internet]. 2023 May 11 [cited 2025 Feb 1];15(18): 2154–80. Available from: https://pubs.rsc.org/en/content/articlehtml/2023/ay/d3ay00182b

35 Lim K, Kim HK, Le XT, Nguyen NT, Lee ES, Oh KT, et al. Highly red light-emitting erbium- and lutetium-doped core-shell upconverting nanoparticles surface-modified with PEG-folic acid/TCPP for suppressing cervical cancer hela cells. Pharmaceutics [Internet]. 2020 Nov 17 [cited 2025 Feb 1];12(11): 1102. Available from: https://www.mdpi.com/1999-4923/12/11/1102/htm

36 Bharali DJ, Mousa SA. Emerging nanomedicines for early cancer detection and improved treatment: Current perspective and future promise. Pharmacol Ther. 2010 Nov 1;128(2): 324–35.

37 Sabir F, Barani M, Mukhtar M, Rahdar A, Cucchiarini M, Zafar MN, et al. Nanodiagnosis and nanotreatment of cardiovascular diseases: An overview. Chemosensors [Internet]. 2021 Mar 30 [cited 2025 Feb 1];9(4): 67. Available from: https://www.mdpi.com/2227-9040/9/4/67/htm

38 Conklin B, Conley BM, Hou Y, Chen M, Lee KB. Advanced theragnostics for the central nervous system (CNS) and neurological disorders using functional inorganic nanomaterials. Adv Drug Deliv Rev. 2023 Jan 1;192: 114636.

39 Augustine R, Al Mamun A, Hasan A, Salam SA, Chandrasekaran R, Ahmed R, et al. Imaging cancer cells with nanostructures: Prospects of nanotechnology driven non-invasive cancer diagnosis. Adv Colloid Interface Sci. 2021 Aug 1;294: 102457.

40 Sarkar A, Dyawanapelly S. Nanodiagnostics and nanotherapeutics for age-related macular degeneration. J Contr Release. 2021 Jan 10;329: 1262–82.

41 Fitzmorris BC, Cooper JK, Edberg J, Gul S, Guo J, Zhang JZ. Synthesis and structural, optical, and dynamic properties of core/shell/shell CdSe/ZnSe/ZnS quantum dots. J Phys Chem C [Internet]. 2012 Nov 29 [cited 2025 Feb 1];116(47): 25065–73. Available from: https://pubs.acs.org/doi/full/10.1021/jp3092013

42 Das S, Dey MK, Devireddy R, Gartia MR. Biomarkers in cancer detection, diagnosis, and prognosis. Sensors [Internet]. 2023 Dec 20 [cited 2025 Feb 1]; 24(1): 37. Available from: https://www.mdpi.com/1424-8220/24/1/37/htm

43 Xu L, Li Q, Chu CH, Feng Z, Fan H, Cheng L, et al. Advanced biomimetic nanomaterials for non-invasive disease diagnosis. Sec Biomater. 2021 [cited 2025 Feb 1]. Available from: www.frontiersin.org

44 Huang H, Lovell JF. Advanced functional nanomaterials for theranostics. Adv Funct Mater [Internet]. 2017 Jan 1 [cited 2025 Feb 1];27(2): 1603524. Available from: https://onlinelibrary.wiley.com/doi/full/10.1002/adfm.201603524

45 Perez RA, Kim HW. Core–shell designed scaffolds for drug delivery and tissue engineering. Acta Biomater. 2015 Jul 15;21: 2–19.

46 Sajja HK, East MP, Mao H, Wang YA, Nie S, Yang L. Development of multifunctional nanoparticles for targeted drug delivery and noninvasive imaging of therapeutic effect. Curr Drug Discov Technol. 2009 Mar 4;6(1): 43–51.

47 Wang Y, Deng T, Liu X, Fang X, Mo Y, Xie N, et al. Smart nanoplatforms responding to the tumor microenvironment for precise drug delivery in cancer therapy. Int J Nanomed [Internet]. 2024 [cited 2025 Feb 1];19: 6253–77. Available from: https://www.tandfonline.com/action/journalInformation?journalCode=dijn20

48 Joshi R, Missong H, Mishra J, Kaur S, Saini S, Kandimalla R, et al. Nanotheranostics revolutionizing neurodegenerative diseases: From precision diagnosis to targeted therapies. J Drug Deliv Sci Technol. 2023 Nov 1;89: 105067.

49 Shoaib A, Javed S, Tabish M, Khan ME, Zaki M, Alqahtani SS, et al. Applications of nanomedicine-integrated phototherapeutic agents in cancer theranostics:

A comprehensive review of the current state of research. Nanotechnol Rev [Internet]. 2024 Jan 1 [cited 2025 Feb 1];13(1): 20240023. Available from: https://www.degruyter.com/document/doi/10.1515/ntrev-2024-0023/html
50 Goel S, England CG, Chen F, Cai W. Positron emission tomography and nanotechnology: A dynamic duo for cancer theranostics. Adv Drug Deliv Rev. 2017 Apr 1;113: 157–76.
51 Saha RN, Vasanthakumar S, Bende G, Snehalatha M. Nanoparticulate drug delivery systems for cancer chemotherapy. Mol Membr Biol [Internet]. 2010 Oct [cited 2025 Feb 1];27(7): 215–31. Available from: https://www.tandfonline.com/doi/abs/10.3109/09687688.2010.510804
52 Mahmoudi M, Hosseinkhani H, Hosseinkhani M, Boutry S, Simchi A, Shane Journeay W, et al. Magnetic resonance imaging tracking of stem cells in vivo using iron oxide nanoparticles as a tool for the advancement of clinical regenerative medicine. Chem Rev [Internet]. 2011 Feb 9 [cited 2025 Feb 1];111(2): 253–80. Available from: https://pubs.acs.org/doi/full/10.1021/cr1001832
53 Gong J, Song C, Li G, Guo Y, Wang Z, Guo H, et al. Ultrasound-enhanced theranostics of orthotopic breast cancer through a multifunctional core–shell tecto dendrimer-based nanomedicine platform. Biomater Sci [Internet]. 2023 [cited 2025 Feb 1]; 11: 4385. Available from: https://pubs.rsc.org/en/content/articlehtml/2023/bm/d3bm00375b?casa_token=CgEpskGhjp8AAAAA:gtiHsxr8bLcsrE52dtLj4qZpTwVG1Dj6KYoB_dCvZZzGFiyuRrEzKU1Ht6XvAImr7tmYozbobDMZq3A&casa_token=Cay1yXrAkcgAAAAA:xll6wZeEiS-4jQ4kyP65GD6-TBmZliRi6w2-ScFEH7ozNQCKn3XxUCdiKpRieesSQh5U4tNRZH0XYHY
54 Di Pietro P, Strano G, Zuccarello L, Satriano C. Gold and silver nanoparticles for applications in theranostics. Curr Top Med Chem. 2016 Jul 25;16(27): 3069–102.
55 Li W, Chen X. Gold nanoparticles for photoacoustic imaging. Nanomedicine [Internet]. 2015 Jan 1 [cited 2025 Feb 1];10(2): 299–320. Available from: https://www.tandfonline.com/doi/abs/10.2217/nnm.14.169
56 Singh R, Bhateria R. Core–shell nanostructures: A simplest two-component system with enhanced properties and multiple applications. Environ Geochem Health [Internet]. 2020 Nov 8 [cited 2025 Feb 1];43(7): 2459–82. Available from: https://link.springer.com/article/10.1007/s10653-020-00766-1
57 Tabish TA, Dey P, Mosca S, Salimi M, Palombo F, Matousek P, et al. Smart gold nanostructures for light mediated cancer theranostics: Combining optical diagnostics with photothermal therapy. Adv Sci [Internet]. 2020 Aug 1 [cited 2025 Feb 1];7(15): 1903441. Available from: https://onlinelibrary.wiley.com/doi/full/10.1002/advs.201903441
58 Shanmugam V, Selvakumar S, Yeh CS. Near-infrared light-responsive nanomaterials in cancer therapeutics. Chem Soc Rev [Internet]. 2014 Aug 4 [cited 2025 Feb 1];43(17): 6254–87. Available from: https://pubs.rsc.org/en/content/articlehtml/2014/cs/c4cs00011k
59 Hang Y, Wang A, Wu N. Plasmonic silver and gold nanoparticles: Shape- and structure-modulated plasmonic functionality for point-of-caring sensing, bio-imaging and medical therapy. Chem Soc Rev [Internet]. 2024 Mar 18 [cited 2025 Feb 1];53(6): 2932–71. Available from: https://pubs.rsc.org/en/content/articlehtml/2024/cs/d3cs00793f

8

Emerging Applications of Core-shell Nanomaterials

Core-shell nanomaterials have emerged as a critical innovation in the development of advanced technologies due to their unique structural composition and multifunctional properties. This chapter explores the diverse applications of core-shell nanomaterials across several emerging fields, including smart coatings, functional textiles, sensors, optoelectronics, and three-dimensional (3D) printing. By leveraging the distinct characteristics of the core and shell materials, core-shell systems offer enhanced mechanical, thermal, optical, and chemical properties, making them ideal candidates for use in next-generation devices and systems. The chapter delves into the mechanisms behind their performance, such as self-healing, corrosion resistance, signal enhancement, and quantum confinement, highlighting their potential in diverse applications such as environmental monitoring, high-performance actuators, display systems, and customized biomedical devices. With a focus on real-world uses, the chapter discusses how these materials are transforming industries like healthcare, aerospace, and electronics. Finally, it examines the future prospects of core-shell nanomaterials, emphasizing their role in driving technological innovation and solving critical global challenges in sustainability, precision medicine, and advanced manufacturing.

8.1 Introduction

Core-shell nanomaterials are emerging as transformative components in advanced technologies due to their multifunctional properties and unique structural configurations. These nanomaterials, comprising a core material encapsulated by a shell layer, offer tunable physical, chemical, and electronic properties that can be precisely controlled by altering the composition and architecture of the core and shell components [1]. The versatility of core-shell nanostructures has opened up unprecedented possibilities for their application in fields such as smart coatings, sensors, optoelectronics, and additive manufacturing. The growing demand for next-generation materials and devices with enhanced performance, durability, and efficiency has accelerated research and development in this area, making

Core-Shell Nanomaterials: From Fundamentals to Applications, First Edition. Shreya Sharma et al.

core-shell nanomaterials indispensable in various industries. One of the most remarkable features of core-shell nanomaterials is their ability to combine the functionalities of different materials within a single structure. This characteristic allows for the design of materials that exhibit properties beyond the limitations of individual components. For instance, the core can provide structural stability and conductivity, while the shell may offer protective or catalytic functionalities. This synergy makes them highly desirable for applications where multifunctionality is essential. As shown in Figure 8.1, the diverse applications of core-shell nanomaterials span from protective coatings and environmental sensing to advanced quantum technologies.

Smart coatings and functional textiles are among the most promising areas for the application of core-shell nanomaterials. These materials can impart self-healing, anti-corrosive, and superhydrophobic properties to coatings, significantly enhancing their performance and longevity. In textiles, core-shell nanomaterials can be used to develop fabrics with advanced functionalities such as thermal regulation, UV protection, and antibacterial properties. The integration of these materials into coatings and textiles not only improves their mechanical and chemical stability but also adds new functionalities that were previously unattainable [2].

In the realm of sensors and actuators, core-shell nanomaterials have demonstrated exceptional potential in enhancing the sensitivity, selectivity, and response time of sensing devices. Their ability to act as signal amplifiers and transducers makes them ideal candidates for environmental monitoring, healthcare diagnostics, and industrial automation. The shell layer can be functionalized to selectively interact with target analytes, while the core provides a robust platform for signal

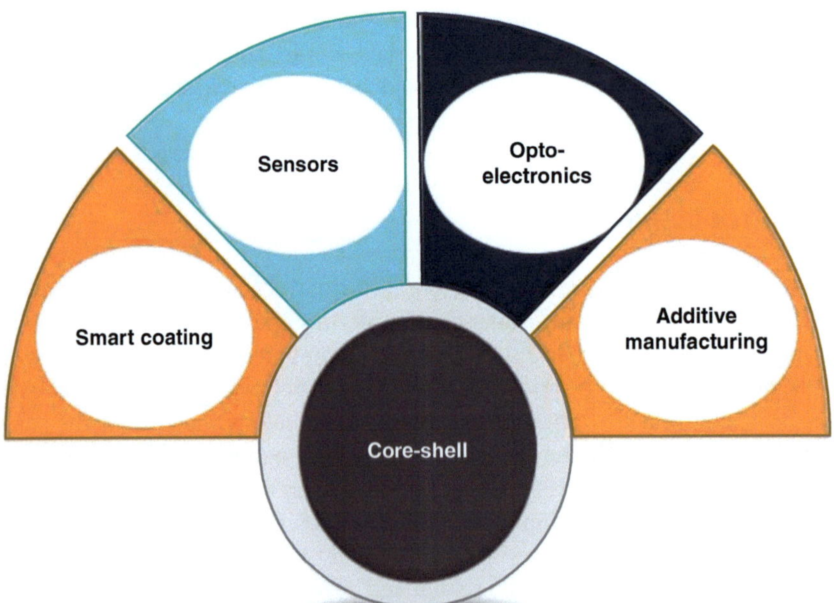

Figure 8.1 Core-shell nanomaterials as enabling components for smart coatings, sensors, optoelectronics, and additive manufacturing.

generation and transduction [3]. This combination results in sensors with superior performance characteristics compared to conventional materials.

Optoelectronics and quantum dots (QDs) represent another exciting frontier for core-shell nanomaterials. These materials have been successfully integrated into light-emitting diodes (LEDs), lasers, and photovoltaic devices, where their unique optical properties contribute to enhanced efficiency and color purity. QD-based core-shell systems, in particular, have revolutionized display technologies by offering unparalleled brightness, color accuracy, and energy efficiency. The ability to fine-tune the emission properties of QDs by manipulating the core-shell structure has paved the way for their widespread adoption in next-generation display and lighting systems [2].

Additive manufacturing, including 3D printing, is yet another domain where core-shell nanomaterials are making a significant impact. The incorporation of these materials into functional inks for 3D printing enables the fabrication of complex structures with tailored properties. Core-shell nanomaterials can enhance the mechanical strength, thermal stability, and electrical conductivity of printed objects, making them suitable for a wide range of applications, from biomedical implants to electronic devices [4]. Their ability to provide multifunctional properties in a single material system aligns perfectly with the goals of advanced manufacturing technologies.

The significance of core-shell nanomaterials in emerging applications lies not only in their multifunctionality but also in their ability to address some of the critical challenges faced by modern technologies. For example, in energy storage and conversion, core-shell nanomaterials have been used to develop high-performance electrodes and catalysts that exhibit superior stability and efficiency [3]. In environmental remediation, these materials have shown great promise in the removal of pollutants and the conversion of waste into valuable resources. Their ability to be tailored for specific applications makes them versatile tools for tackling complex technological challenges. Moreover, the development of core-shell nanomaterials has been driven by advancements in synthesis and characterization techniques. Precise control over the size, shape, and composition of the core and shell components has become possible, allowing researchers to design materials with optimized properties for specific applications. Characterization techniques such as transmission electron microscopy (TEM), X-ray diffraction (XRD), and spectroscopic methods have played a crucial role in understanding the structural and functional aspects of these materials, further accelerating their adoption in various fields.

As industries continue to seek innovative solutions to meet the demands of a rapidly evolving technological landscape, the role of core-shell nanomaterials is expected to become even more prominent. Their ability to seamlessly integrate into existing systems while providing enhanced performance and new functionalities makes them invaluable in the pursuit of sustainable and efficient technologies [5]. The ongoing research in this field is likely to uncover new applications and further improve the properties of core-shell nanomaterials, driving their widespread adoption across different sectors. Core-shell nanomaterials represent a paradigm shift in the development of advanced materials and devices. Their multifunctional properties, tunability, and versatility have positioned them as key enablers of next-generation technologies. From smart coatings and sensors to optoelectronics and additive manufacturing, these materials are poised to play a transformative role in shaping the future of science and technology. As shown in Figure 8.1, the potential applications of core-shell nanomaterials are vast and varied, reflecting their ability to address the diverse needs of modern industries.

8.2 Smart Coatings and Functional Textiles

Smart coatings and functional textiles utilize core-shell nanomaterials to achieve advanced properties such as self-healing, corrosion resistance, and thermal regulation. In coatings, these materials offer enhanced durability, superhydrophobicity, and anti-corrosive features, making them ideal for protective surfaces in harsh environments. Functional textiles, integrated with core-shell nanostructures, provide UV protection, antibacterial effects, and moisture-wicking capabilities [6]. The tunable properties of core-shell systems enable responsive behavior to external stimuli, such as temperature or pH changes, making them valuable for adaptive applications in fashion, healthcare, and industrial fabrics. Their versatility is driving innovation in smart material technologies.

8.2.1 Mechanism and Functionality

The use of core-shell nanomaterials in smart coatings and functional textiles leverages their unique ability to deliver self-healing, anti-corrosion, and superhydrophobic properties. These functionalities arise from the strategic design of nanostructures where the core often provides strength or specific reactivity, while the shell offers protection or enhanced interaction with external stimuli. Self-healing coatings are a remarkable application of core-shell systems. They restore structural integrity upon physical damage, thereby extending the lifespan of materials [7]. Typically, the core-shell nanostructures in such coatings encapsulate healing agents within polymer shells. When a surface is scratched, the shell ruptures, releasing the healing agents, which polymerize or bond to fill the damage. Core-shell microcapsules filled with resins or solvents are common in self-healing coatings for industrial applications such as automotive and aerospace sectors.

Corrosion protection is another crucial area where core-shell nanomaterials excel. Traditional coatings fail over time due to environmental exposure, but core-shell nanostructures provide sustained protection. The core materials, such as metal oxides, offer barrier properties, while the shell can be functionalized to enhance chemical resistance. For instance, Fe_3O_4@SiO_2 nanoparticles are used to prevent moisture penetration and oxidative damage. Additionally, the shell can release corrosion inhibitors when the coating is damaged, offering a self-adaptive defense mechanism [8]. Such smart coatings are highly desirable for infrastructure, marine, and industrial equipment. Superhydrophobic coatings, inspired by the lotus leaf effect, are achieved using core-shell nanostructures with hierarchical roughness and low surface energy materials. The core often consists of silica or metal oxide nanoparticles, while the shell is modified with hydrophobic agents like fluoropolymers. This structure creates surfaces with contact angles above 150°, leading to water-repellent properties. These coatings find applications in self-cleaning surfaces, antifouling marine coatings, and protective layers for electronics.

In textiles, core-shell nanomaterials improve thermal insulation, mechanical strength, and wear resistance. For instance, phase change materials (PCMs) encapsulated in polymeric shells are integrated into fabrics to regulate temperature by absorbing and releasing heat during phase transitions [9]. These smart textiles are particularly valuable for sportswear and military applications where temperature adaptability is crucial. Core-shell systems with antimicrobial functionalities have gained traction in medical and hygiene textiles. Silver nanoparticle cores encapsulated within a biocompatible shell prevent leaching of toxic ions while providing long-lasting antibacterial effects [10]. Moreover, chemical-resistant textiles incorporating core-shell nanostructures are useful in protective gear for industrial workers handling hazardous substances.

The tunability of core-shell nanomaterials enables textiles to respond to various stimuli such as temperature, pH, and light. For example, textiles embedded with thermochromic core-shell nanostructures can change color with temperature variations, enhancing their aesthetic and functional appeal. Similarly, pH-sensitive coatings on textiles can release therapeutic agents for wound care applications. The multifunctionality of core-shell nanomaterials offers transformative advancements in smart coatings and functional textiles. Their ability to provide self-healing, anti-corrosion, superhydrophobic properties, and enhanced mechanical and thermal performance makes them indispensable for next-generation materials [11]. As research continues, the development of environmentally friendly and scalable core-shell systems will further expand their applications in diverse industries.

8.2.2 Applications in Protective and Smart Surfaces

The advent of core-shell nanomaterials has revolutionized protective and smart surface technologies, enabling the development of advanced coatings and textiles with unprecedented functionalities. These applications span critical sectors such as automotive, aerospace, healthcare, and wearable electronics, offering solutions that enhance durability, functionality, and adaptability to dynamic environments. Core-shell nanostructures serve as essential components in these applications due to their ability to integrate multiple properties, such as self-healing, corrosion resistance, and stimuli responsiveness [12].

The automotive and aerospace industries demand materials that can withstand extreme environmental conditions while maintaining structural integrity and aesthetic appeal. Core-shell nanomaterials have proven instrumental in the development of advanced coatings that meet these stringent requirements. One of the primary applications is in self-healing coatings, which significantly extend the lifespan of surfaces by repairing minor scratches and damage autonomously. For instance, core-shell capsules encapsulating healing agents within polymer shells can release their contents upon surface damage, filling cracks and restoring the surface [13]. This self-repair mechanism is invaluable in automotive and aerospace applications, where maintenance costs and downtime must be minimized. In addition, core-shell coatings offer superior anti-corrosion properties. As shown in Figure 8.2, these coatings form a robust barrier that prevents the penetration of moisture and corrosive agents. Metal oxide cores, such as Fe_3O_4, coated with silica or polymer shells, are widely used to protect metallic components from oxidative degradation. These coatings are particularly beneficial for aerospace applications, where exposure to harsh environmental conditions is routine [14]. Superhydrophobic coatings enabled by core-shell nanomaterials further enhance the protective properties of surfaces. These coatings, characterized by their high water contact angles, repel water and contaminants, keeping surfaces clean and free from moisture-related damage. Automotive windshields, aircraft exteriors, and sensor housings are some of the components benefiting from such advancements.

In the realm of functional textiles, core-shell nanomaterials have opened new avenues for innovation in healthcare and wearable electronics. The ability to impart properties such as thermal regulation, antimicrobial activity, and electrical conductivity to fabrics makes these materials highly versatile. Healthcare applications particularly benefit from textiles embedded with core-shell nanostructures. Antimicrobial textiles, incorporating silver nanoparticles encapsulated within biocompatible shells, provide long-lasting protection against pathogens without the risk of toxic ion leaching [15]. These textiles are invaluable in hospital environments, where infection control is a top priority. Temperature-regulating textiles are another significant application. PCMs encapsulated in polymer shells can absorb and release heat during phase transitions, providing dynamic thermal insulation. Such textiles are widely used in sportswear, military uniforms, and therapeutic garments. As shown in Figure 8.2, wearable electronics represent a burgeoning field where core-shell nanomaterials play a crucial role. Conductive

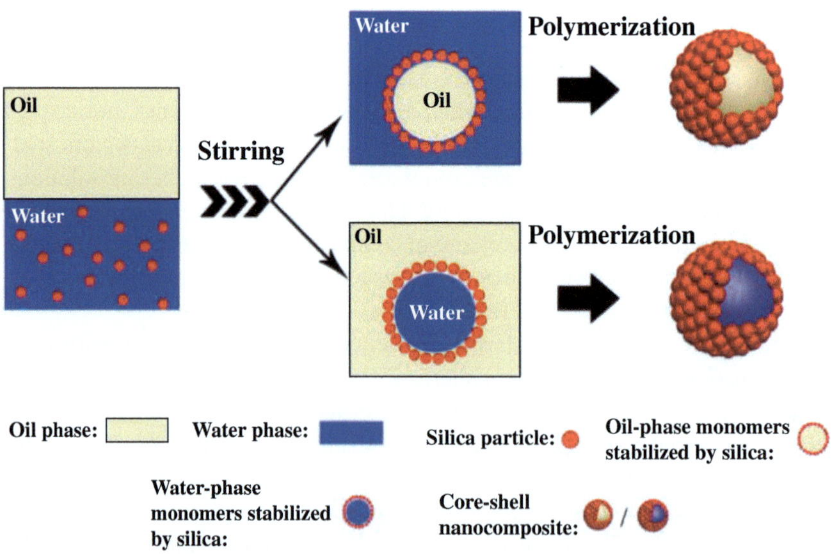

Figure 8.2 Illustration of core-shell nanomaterials enhancing the functional properties of coatings and textiles. [17]/MDPI/CC BY 4.0.

textiles embedded with core-shell nanostructures enable the seamless integration of sensors and circuits into fabrics. These smart textiles can monitor physiological parameters such as heart rate, body temperature, and movement, making them valuable for fitness tracking and medical diagnostics [14]. Moreover, core-shell nanomaterials can impart stimuli-responsive properties to textiles. Fabrics that change color in response to temperature or pH variations are examples of how these materials enhance the functionality and aesthetic appeal of wearable electronics.

The use of core-shell nanomaterials in protective and smart surfaces also aligns with environmental and economic considerations. The extended lifespan of coatings and textiles reduces material waste and maintenance costs. Additionally, the incorporation of self-healing and corrosion-resistant properties minimizes the need for frequent replacements and repairs. The transformative impact of core-shell nanomaterials on protective and smart surfaces is undeniable [16]. From self-healing automotive coatings to antimicrobial and temperature-regulating textiles, these materials are driving innovation across diverse industries. As shown in Figure 8.2, their ability to enhance the functional properties of coatings and textiles underscores their importance in the development of next-generation materials. Continued research and development efforts are expected to unlock even more applications, paving the way for a future where smart surfaces and textiles are integral to everyday life.

8.3 Sensors and Actuators

Core-shell nanomaterials play a transformative role in sensors and actuators by enhancing sensitivity, selectivity, and response times. Their unique structure allows for efficient signal transduction and functional adaptability. In sensors, core-shell systems enable the detection of gases, biomolecules, and environmental pollutants with high precision. For actuators, these materials facilitate responsive motion and mechanical changes upon exposure to external stimuli such as temperature, light, or electric fields [18]. Their multifunctionality, tunability, and nanoscale architecture make them ideal for applications in smart devices, robotics, healthcare diagnostics, and environmental monitoring, driving advancements in next-generation sensing and actuation technologies.

8.3.1 Core-shell Nanomaterials in Sensors

Core-shell nanomaterials have emerged as a powerful class of materials for sensing applications due to their unique structural advantages and multifunctional properties. These materials consist of a core with distinct properties, encapsulated by a shell that can provide additional functionalities or protective attributes. The combination enhances the performance of sensors by improving sensitivity, selectivity, and stability, which are critical parameters in detecting environmental pollutants, gases, and biomolecules. Table 8.1 compares different core-shell nanomaterials and their suitability for gas detection.

Table 8.1 Table comparing different core-shell nanomaterials and their suitability for gas detection, including target gases and detection ranges.

Core-shell composition	Target gas	Detection range (ppm)	Sensor type
$ZnO@SnO_2$	NO_2	0.1–50	Resistive
SnO_2@Graphene	VOCs	0.05–10	Capacitive
$Fe_2O_3@SiO_2$	CO	1–100	Optical
Au@Silica	H_2	10–500	Plasmonic
TiO_2@Polymers	NH_3	5–200	Chemiresistive
CuO@ZnO	O_3	0.1–1	Catalytic
SiO_2@CNTs	CH_4	0.5–100	Resistive
$Ag@TiO_2$	H_2S	0.1–10	Optical

8.3.1.1 Mechanism of Signal Enhancement Using Core-shell Systems

The mechanism by which core-shell nanomaterials enhance signal detection is largely based on their superior electronic, optical, and chemical properties. The core material often acts as the primary signal generator by responding to external stimuli, such as changes in gas concentrations or biomolecule binding events. The shell plays several critical roles, such as:

a. **Surface Modification for Selectivity:** The shell can be functionalized with specific chemical groups or biological receptors, enhancing the selectivity for target molecules or gases. For instance, polymer-coated cores can selectively detect volatile organic compounds (VOCs) in complex environments [19].
b. **Charge Transfer Efficiency:** Core-shell structures facilitate efficient charge transfer between the core and shell, leading to amplified electrical signals in resistive or capacitive gas sensors. For example, ZnO@graphene core-shell systems exhibit enhanced electron mobility and signal transduction efficiency [14].
c. **Stabilization of Core Properties:** The shell often protects the core from environmental degradation, improving sensor stability and longevity. Silica shells, for instance, shield metal nanoparticles from oxidation, ensuring consistent sensor performance over time.
d. **Plasmonic and Catalytic Enhancement:** In optical sensors, core-shell structures such as Au@Ag nanoparticles exhibit enhanced plasmonic effects, increasing the intensity of optical signals. In catalytic gas sensors, core-shell systems can facilitate faster reaction kinetics, improving the detection speed [20].

8.3.1.2 Detection of Environmental Pollutants, Gases, and Biomolecules

Core-shell nanomaterials are widely used in the detection of various environmental pollutants and gases, including carbon monoxide (CO), nitrogen oxides (NOx), and hydrogen (H_2). They also play a crucial role in biosensors for detecting biomolecules such as glucose, proteins, and DNA.

a. **Environmental Monitoring:** Core-shell nanomaterials are employed in gas sensors to monitor harmful pollutants in industrial and urban environments. For example, SnO_2@graphene core-shell systems have been reported to efficiently detect low concentrations of NO_2 [21].
b. **Gas Detection:** Core-shell nanomaterials provide high sensitivity and selectivity for gas detection due to their large surface area and tunable surface properties. Sensors based on Fe_2O_3@SiO_2 and ZnO@SnO_2 core-shell systems have demonstrated excellent detection capabilities for CO and VOCs, respectively [22].
c. **Biosensing Applications:** In healthcare diagnostics, core-shell nanomaterials are used in biosensors to detect specific biomolecules. Gold–silica core-shell nanoparticles, for instance, are widely used in surface-enhanced Raman spectroscopy (SERS) for the detection of proteins and DNA [23].

8.3.1.3 Advantages of Core-shell Nanomaterials in Sensors

- **Enhanced Sensitivity:** The high surface-to-volume ratio of core-shell nanostructures allows for efficient interaction with target analytes.
- **Improved Selectivity:** Functionalized shells provide specificity toward particular gases or biomolecules.
- **Fast Response and Recovery:** The efficient charge transfer and reaction kinetics enabled by core-shell systems result in rapid sensor response times.
- **High Stability:** Protective shells ensure that the core material remains stable under harsh environmental conditions.

8.3.2 Actuator Applications

Core-shell nanomaterials have garnered significant interest in the development of high-performance actuators for applications in robotics, adaptive systems, and smart devices. Actuators, which convert energy into mechanical motion, are fundamental components in systems requiring precise movements and responsiveness [24]. The integration of core-shell nanomaterials in actuators offers enhanced performance, including increased force output, faster response times, and superior mechanical stability.

8.3.2.1 High-performance Actuators in Robotics and Adaptive Systems

Core-shell nanomaterials are particularly suitable for actuator applications in robotics and adaptive systems due to their multifunctional properties. The ability to engineer both the core and shell components allows for customized mechanical, electrical, and thermal responses tailored to specific applications.

- **Robotics:** Core-shell actuators play a crucial role in robotic systems, enabling precise and dynamic movements essential for industrial automation, medical robotics, and humanoid robots [25]. Materials such as carbon nanotube (CNT) core-shell systems have been shown to provide superior actuation performance due to their high tensile strength and flexibility.
- **Adaptive Systems:** In adaptive structures and devices, actuators based on core-shell materials can respond to environmental stimuli such as temperature,

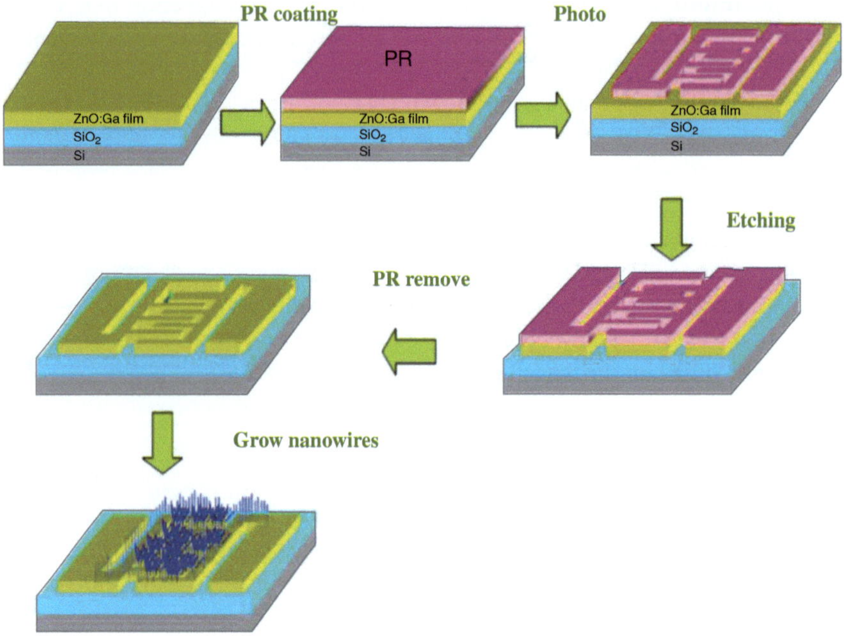

Figure 8.3 Core-shell nanomaterials improving sensor sensitivity and selectivity for environmental monitoring. [29]/MDPI/CC BY 4.0.

pressure, or electrical signals. For example, polymer-coated metallic cores can produce controlled deformation in response to an electric field, making them suitable for shape-memory devices and smart surfaces [26].

As shown in Figure 8.3, core-shell nanomaterials enhance the sensitivity and selectivity of actuators, leading to improved environmental responsiveness and adaptive functionality.

8.3.2.2 Role in Precision Movements and Responsiveness

The role of core-shell nanomaterials in achieving precision movements and responsiveness in actuators can be attributed to their structural and functional advantages.

a. **Enhanced Mechanical Properties:** The core-shell architecture provides a balance between strength and flexibility. The core imparts mechanical strength, while the shell adds flexibility and functionality. For instance, core-shell systems with metal oxide cores and polymer shells can withstand high mechanical stress while maintaining their functional properties [27].
b. **Efficient Energy Conversion:** Core-shell nanomaterials enable efficient energy conversion mechanisms, which are essential for actuators. Piezoelectric core-shell systems, for example, convert electrical energy into mechanical motion with high efficiency [24].
c. **Fast Response Times:** The nanoscale dimensions of core-shell materials facilitate rapid charge transfer and stress distribution, resulting in faster actuation responses.

d. Thermal and Chemical Stability: Core-shell structures often exhibit superior stability under varying environmental conditions, making them suitable for applications in harsh environments [28].

e. Precision Control: The tunability of core-shell materials allows for precise control of actuation parameters such as force, displacement, and speed.

8.3.2.3 Applications in High-performance Actuators

a. Medical Devices: Core-shell actuators are used in biomedical devices for applications such as minimally invasive surgical tools and artificial muscles. The biocompatibility of certain core-shell systems, such as those based on hydrogel coatings, makes them ideal for these applications [28].

b. Wearable Technology: Actuators based on core-shell nanomaterials are increasingly being integrated into wearable devices for applications such as haptic feedback and motion tracking.

c. Automotive and Aerospace: Core-shell actuators are used in automotive and aerospace systems for applications such as vibration damping and adaptive aerodynamics.

d. Smart Textiles: Core-shell actuators embedded in smart textiles enable responsive clothing that can adapt to environmental conditions or provide motion assistance.

Table 8.2 provides a comparative analysis of core-shell nanomaterials used in sensors and actuators, highlighting their specific compositions, functions, and advantages.

Table 8.2 Comparison of core-shell nanomaterials in sensors and actuators.

Core-shell composition	Sensor/ Actuator type	Functionality	Advantages	Applications
CNT@Polymer	Actuator	Precision motion	High flexibility & strength	Robotics
$ZnO@SnO_2$	Gas sensor	Environmental detection	High sensitivity	Air quality monitoring
$Ag@TiO_2$	Actuator	Thermal actuation	Fast response & stability	Adaptive systems
$Fe_2O_3@SiO_2$	Sensor	Optical detection	Superior selectivity	Environmental monitoring
Au@Hydrogel	Actuator	Biomechanical actuation	Biocompatibility	Medical devices
TiO_2@Graphene	Actuator	Electrical actuation	High electrical conductivity	Wearables
SiO_2@ Metal Oxide	Sensor	Chemical sensing	Enhanced selectivity	Industrial sensors

8.4 Optoelectronics and QDs

8.4.1 Role of Core-shell Nanomaterials

Core-shell nanomaterials have become pivotal in the advancement of optoelectronic devices, including LEDs, lasers, and photovoltaic systems. Their unique structural and compositional features enable them to enhance optical properties, energy conversion efficiency, and device stability. By carefully designing the core and shell components, scientists can fine-tune the electronic, optical, and thermal characteristics of these materials, paving the way for more efficient and durable devices in modern technological applications. One of the primary applications of core-shell nanomaterials is in LEDs. Traditional LEDs face limitations in terms of color purity and efficiency, especially for full-spectrum displays. Core-shell QDs, such as CdSe/ZnS or InP/ZnS, have emerged as promising solutions [30]. The core typically provides high photoluminescence efficiency, while the shell acts as a protective barrier, reducing surface defects and enhancing quantum yield. The result is brighter, more stable LEDs with precise color tuning capabilities. Furthermore, core-shell structures help in suppressing non-radiative recombination processes, which are a significant cause of efficiency loss in conventional LED materials. In the field of lasers, core-shell nanomaterials contribute to high-performance and low-threshold laser systems [31]. For example, core-shell semiconductor nanowires and QDs have been integrated into laser cavities to achieve superior optical gain and better emission stability. The shell not only passivates the core but also optimizes the refractive index contrast, enabling strong light confinement. Such advancements are crucial for applications in high-speed communication systems, optical sensing, and holography.

Photovoltaic technology has also greatly benefited from the integration of core-shell nanomaterials. Solar cells require materials that can efficiently absorb sunlight and convert it into electrical energy. Core-shell nanomaterials, such as SiO_2@TiO_2 and ZnO@CdSe, have demonstrated exceptional light-harvesting capabilities. The core usually serves as the primary absorber of sunlight, while the shell facilitates charge separation and transport, reducing recombination losses. Additionally, the shell can protect the core from environmental degradation, significantly enhancing the lifespan and efficiency of solar cells. Hybrid organic–inorganic perovskite solar cells have also adopted core-shell architectures to improve moisture resistance and stability. Beyond enhancing efficiency, core-shell nanomaterials contribute to reducing the environmental impact of optoelectronic devices [32]. By utilizing earth-abundant and non-toxic components in core-shell structures, researchers are addressing the toxicity concerns associated with conventional materials like cadmium-based QDs. This shift toward greener materials is essential for the widespread adoption of nanotechnology in consumer electronics.

Energy efficiency is another critical area where core-shell nanomaterials play a transformative role. By enabling better light absorption, emission, and charge transport, these materials help in reducing the power consumption of optoelectronic devices. For instance, integrating core-shell nanoparticles in organic light emitting diode (OLED) displays has shown significant potential in reducing energy requirements while maintaining high brightness and color purity [33, 34]. The integration of core-shell nanomaterials in LEDs, lasers, and photovoltaics represents a significant leap forward in optoelectronic technology. Their ability to enhance optical properties, improve energy efficiency, and provide long-term stability makes them indispensable for next-generation devices. As research continues to explore novel core-shell compositions and fabrication techniques, the future of optoelectronics looks promising, with smarter, more efficient, and sustainable solutions on the horizon.

8.4.2 QDs-based Core-shell Systems

QDs are semiconductor nanocrystals with unique optical and electronic properties arising from their quantum confinement effects. When incorporated into core-shell structures, QDs exhibit significantly enhanced performance characteristics, making them highly suitable for various optoelectronic applications such as high-resolution displays and advanced photodetectors [35]. Core-shell QD systems have emerged as one of the most promising innovations in nanotechnology, as they combine the advantages of quantum confinement with the functionalization provided by the shell material.

Quantum confinement refers to the phenomenon where the electronic and optical properties of semiconductor nanomaterials are significantly altered due to their reduced size at the nanoscale. When a QD is smaller than the exciton Bohr radius, electrons and holes are confined to the core, resulting in discrete energy levels and the ability to tune the material's optical properties by adjusting the size of the QD [36]. In a core-shell QD system, the core is typically composed of a semiconductor material that exhibits strong quantum confinement effects, while the shell is made of a different material, which enhances the overall stability and optical performance of the QD. The shell in a core-shell structure serves several critical functions. It acts as a passivation layer that reduces surface defects, which can act as non-radiative recombination centers and degrade the QD's efficiency. By minimizing these defects, the shell helps to enhance photoluminescence quantum yield and prolong the emission lifetime of the QDs. Furthermore, the shell can be engineered to modify the electronic properties of the core, allowing for better control of the emission wavelength and enabling the tailoring of QDs for specific applications [36]. For example, core-shell QDs like CdSe/CdS or InP/ZnS are well-known for their high photoluminescence efficiency and stability, as the shell material protects the core from environmental factors such as oxidation and moisture. One of the key benefits of quantum confinement in core-shell QDs is the ability to fine-tune their emission spectra. The size of the core determines the energy gap, and thus the wavelength of light emitted when the QD is excited. By altering the shell thickness and composition, researchers can control the overall optical properties, such as emission wavelength and quantum yield [37]. This size-dependent behavior makes QDs versatile for a wide range of applications, from biological imaging to optoelectronic devices.

The unique properties of core-shell QDs, particularly their size-tunable emission and high quantum efficiency, make them highly suitable for display technologies, particularly in high-resolution and next-generation LED displays. Core-shell QDs are being widely explored in next-generation display systems, such as QLED (Quantum Dot Light Emitting Diode) and OLED technologies [38]. These QDs are utilized in display panels due to their ability to emit pure and saturated colors, offering significant improvements in color accuracy and display brightness. In high-resolution displays, QDs enhance the color gamut, enabling sharper, more vibrant images with more accurate color reproduction. Core-shell QDs like CdSe/ZnS have proven to be highly effective in QLEDs, providing brighter, more efficient displays. This is due to the reduced surface defects and enhanced photoluminescence properties provided by the shell layer [39]. The tunability of core-shell QDs also ensures that they can be tailored to emit light at specific wavelengths, enabling high-quality red, green, and blue pixels in the display, which are critical for achieving high-definition resolutions and color precision. The incorporation of core-shell QDs into displays has led to improvements in energy efficiency, as they require less power to produce brighter and more vivid images.

Core-shell QDs are also becoming an essential part of advanced photodetectors, particularly in applications requiring high sensitivity and resolution, such as in optical communication systems, environmental monitoring, and medical diagnostics. The mechanism of signal enhancement provided by the quantum confinement in core-shell systems allows for more sensitive detection of weak light signals. The high surface-to-volume ratio and well-controlled energy levels in core-shell QDs enable better light absorption, leading to enhanced photodetector performance [40]. In photodetectors, the core-shell QDs can be engineered to absorb light from specific wavelengths, improving their response time and sensitivity. For instance, QDs made from materials like PbS, PbSe, or InAs can be used to detect infrared light, which is useful for a variety of applications, including night vision, security systems, and medical imaging. The improved efficiency and stability of core-shell QDs also result in faster response times and increased longevity of the photodetectors.

Core-shell QD systems provide an innovative solution for enhancing the brightness, efficiency, and performance of high-resolution displays and advanced photodetectors. By leveraging the unique properties of quantum confinement and optimizing the core-shell architecture, these materials can be tailored to meet the specific requirements of a wide range of applications [41]. In display technology, they offer improvements in color accuracy, brightness, and energy efficiency, while in photodetectors, they enhance sensitivity and signal detection capabilities. The ongoing development of core-shell QDs promises to drive significant advances in optoelectronics, making them an integral component of next-generation devices. As shown in Figure 8.4, core-shell QDs enhance brightness and efficiency in display systems, further underscoring their potential in revolutionizing the performance of optoelectronic devices. Their ability to be engineered for specific wavelengths and applications opens up new possibilities for the future of high-performance, energy-efficient technology.

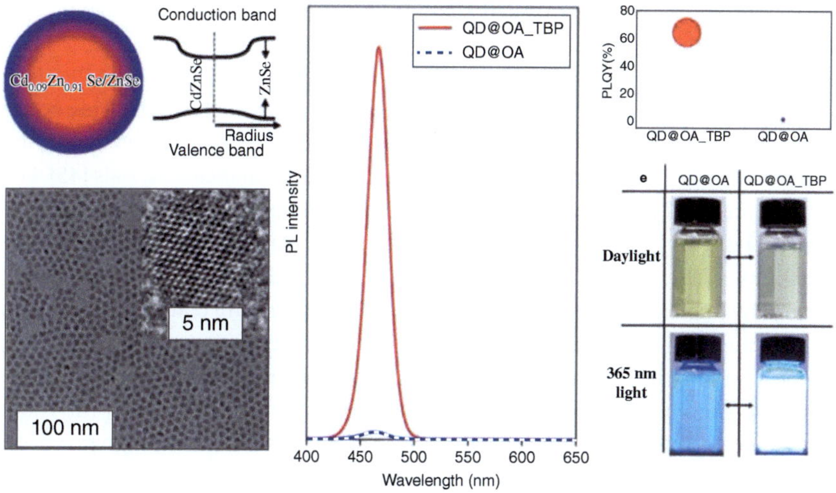

Figure 8.4 Illustration of core-shell quantum dots contributing to brighter and more energy-efficient displays through optimized optical properties, showcasing their structure, energy levels, luminescence characteristics, and photoluminescence quantum yield (PLQY). Reproduced with permission from [42]/Royal Society of Chemistry.

8.5 3D Printing and Additive Manufacturing

8.5.1 Role of Core-shell Nanomaterials in 3D Printing

3D printing, or additive manufacturing, has revolutionized various industries by enabling the production of complex and customized structures with high precision. The integration of core-shell nanomaterials into 3D printing has significantly enhanced the mechanical properties, thermal stability, and precision of printed structures, paving the way for next-generation materials and devices [43]. Core-shell nanomaterials, with their unique multifunctional properties, offer several advantages in additive manufacturing, especially when used as functional inks, enabling the development of high-performance printed objects.

One of the primary benefits of core-shell nanomaterials in 3D printing lies in their ability to improve the mechanical properties of the final printed structure. The combination of different materials in core-shell nanostructures provides a synergistic effect that enhances strength, toughness, and durability. The core component is often chosen for its intrinsic mechanical properties, while the shell is designed to provide additional functionality, such as improved adhesion, reduced friction, or enhanced resistance to wear and tear. For example, core-shell nanoparticles like silica (SiO_2) cores with a polymeric shell or carbon-based core-shell systems can significantly enhance the toughness and impact resistance of 3D-printed materials [44]. These core-shell structures improve the interfacial bonding between particles in the printed layer, resulting in a more uniform and robust material. The enhanced mechanical properties are particularly useful for applications in industries such as automotive, aerospace, and healthcare, where the printed components need to withstand significant mechanical stress and environmental factors.

Moreover, core-shell nanomaterials also play a critical role in improving the thermal stability of 3D-printed parts. The shell layer can be engineered to provide better thermal resistance, helping to mitigate issues like warping or deformation during the printing process, especially when working with high-performance materials [45]. For instance, core-shell nanoparticles with a thermally stable core, such as metal oxide nanoparticles, can be incorporated into the ink formulation to enhance the thermal stability of the printed structure [46]. This ensures that the 3D-printed objects maintain their shape and performance under high-temperature conditions, which is crucial for applications in electronics, aerospace, and energy sectors. Precision in 3D printing is another area where core-shell nanomaterials make a significant impact. The unique structure of core-shell particles allows for better control over the rheological properties of the ink, ensuring that the printed features exhibit high precision and accuracy. The shell material can be tailored to modify the viscosity, surface tension, and flow behavior of the ink, allowing for fine control over the deposition process [47]. This is particularly important in applications such as micro-manufacturing, where precise resolution and smooth surface finish are critical.

Core-shell nanomaterials are increasingly being used as functional inks for 3D printing, enabling the production of multifunctional parts with enhanced properties. These inks are formulated by dispersing core-shell nanoparticles in a suitable solvent, which can then be used for direct ink writing (DIW), inkjet printing, or other additive manufacturing techniques. The functional inks are designed to take advantage of the unique properties of core-shell structures [48], such as high surface area, tunable properties, and the ability to impart specific functionalities like electrical conductivity, magnetism, or optical properties to the printed parts.

In 3D-printing applications, core-shell inks are particularly valuable for producing parts with multifunctionality. For example, core-shell nanomaterials with conductive cores and insulating shells can be used to print electrically conductive patterns on insulating substrates, enabling the fabrication of circuits or sensors [49]. Additionally, core-shell inks that contain magnetic nanoparticles in the core and polymeric shells can be used to print components with controlled magnetic properties, which are crucial for applications in robotics, sensors, and medical devices. The versatility of core-shell nanomaterials as functional inks extends to their ability to be engineered for specific applications. In the field of healthcare, core-shell inks can be used to print scaffolds for tissue engineering, where the core-shell particles can provide structural support, while the shell material can be functionalized to promote cell growth and tissue regeneration [50]. Similarly, in electronics, core-shell inks can be used to fabricate printed sensors, antennas, and even flexible electronics, where the core-shell nanomaterials provide the necessary electrical or optical properties. The incorporation of core-shell materials into 3D printing inks also allows for the development of hybrid materials that combine the properties of different materials in a single printed part. For example, by using core-shell nanoparticles with a conductive core and an insulating shell, a printed part can exhibit both conductive and insulating properties in different regions, making it ideal for applications in printed circuit boards (PCBs) and flexible electronics [51].

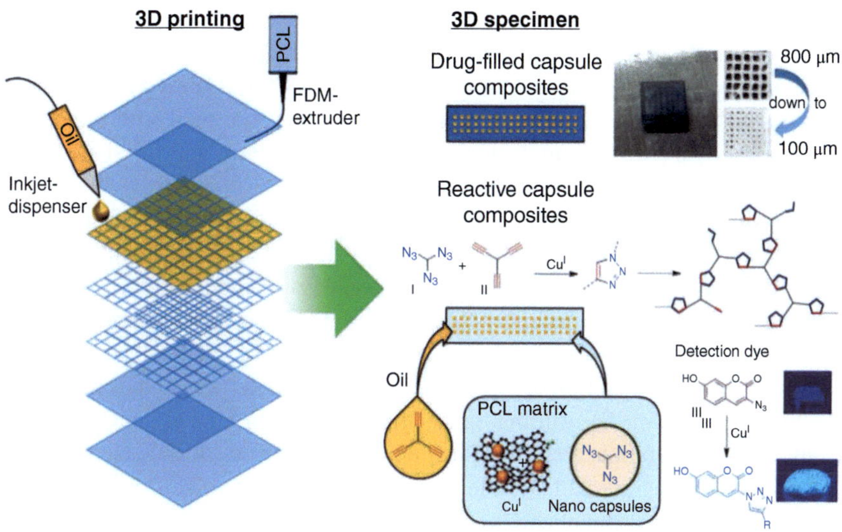

Figure 8.5 Schematic illustration showing the role of core-shell nanomaterials in enhancing the properties and performance of 3D-printed components. [52]/with permission of John Wiley & Sons.

The integration of core-shell nanomaterials into 3D printing has opened up new possibilities for the development of high-performance printed structures with enhanced mechanical properties, thermal stability, and precision. By utilizing core-shell materials as functional inks, additive manufacturing can produce complex and multifunctional parts with tailored properties, meeting the demands of advanced applications in industries such as aerospace, healthcare, electronics, and energy. As shown in Figure 8.5, core-shell nanomaterials play a crucial role in enabling the development of high-performance printed structures, offering new avenues for innovation in additive manufacturing technologies.

8.5.2 Applications in Customized Devices and Prototypes

Core-shell nanomaterials are significantly advancing the capabilities of additive manufacturing, particularly in the production of customized devices and prototypes. Their unique properties, such as tunable mechanical, thermal, and electrical characteristics, allow for the fabrication of a wide range of functional, high-performance components. These capabilities are revolutionizing industries such as biomedical implants, electronics, and architectural components, where precision, personalization, and performance are crucial [53].

In the biomedical field, core-shell nanomaterials are increasingly used for producing customized implants that cater to individual patient needs. Core-shell systems can be designed to mimic the mechanical properties of human tissues, making them ideal for applications such as bone scaffolds, joint replacements, and dental implants. The core material, often a biocompatible metal oxide or polymer, provides structural support, while the shell material can be functionalized to promote tissue growth, prevent bacterial adhesion, or deliver drugs for localized treatment.

For example, core-shell nanoparticles with a biocompatible metal core (such as titanium or gold) and a biopolymer shell can be used in the fabrication of implants that enhance the osseointegration process, thereby improving the success rate of orthopedic surgeries [54]. Additionally, the functionalization of the shell with bioactive molecules can help with wound healing or tissue regeneration, offering significant advantages over traditional materials.

Core-shell nanomaterials also play a vital role in the development of customized electronic devices. The versatility of core-shell structures allows for the integration of different materials that enhance properties like electrical conductivity, flexibility, and optical performance. These materials can be used to create printed circuits, sensors, and even flexible displays, all of which require precise, customized designs. For example, in the production of PCBs, core-shell nanomaterials can be employed in the form of inks for 3D printing, offering improved conductivity and reduced resistance. These materials are crucial for the development of flexible electronics, such as wearable devices and flexible displays, where lightweight, durable, and high-performance components are necessary [55]. Core-shell systems can also be used to create sensors for monitoring health conditions, where the sensors' sensitivity and accuracy are enhanced by the multifunctionality of the core-shell structure.

In the architecture and construction industries, core-shell nanomaterials are being used to design customized components that offer improved structural integrity and performance. For instance, the use of core-shell particles in 3D-printed concrete or composite materials enhances their mechanical properties, making them stronger and more durable. These materials can be tailored to resist environmental stresses like heat, moisture, and corrosion, extending the lifespan of architectural structures [56]. Additionally, core-shell nanomaterials are also being explored for use in 3D-printed building facades and interior design elements. By integrating functional properties such as UV protection, antimicrobial properties, or thermal insulation, these materials contribute to the sustainability and efficiency of buildings. Core-shell systems can also be incorporated into the development of smart building technologies, such as energy-efficient windows or self-healing concrete, which respond dynamically to environmental changes. Table 8.3 shows Comparative analysis of core-shell nanomaterials in additive manufacturing.

Table 8.3 Comparative analysis of core-shell nanomaterials in additive manufacturing.

Material composition	Printing technique	Enhanced properties	Application areas
Metal-core (e.g., TiO_2) and biopolymer shell	Fused Deposition Modeling (FDM)	Biocompatibility, tissue regeneration	Biomedical implants (e.g., joint replacements)
Conductive core (e.g., Ag) and polymer shell	Direct Ink Writing (DIW)	Electrical conductivity, flexibility	Electronics (e.g., flexible displays, sensors)
Ceramic-core (e.g., SiO_2) and metal shell	Stereolithography (SLA)	Durability, corrosion resistance	Architectural components (e.g., building facades)

8.6 Conclusion

Core-shell nanomaterials are transforming the landscape of emerging technologies due to their multifunctionality and tunable properties. These advanced materials are enabling the development of next-generation devices and systems across various sectors, from biomedical applications to electronics, architecture, and additive manufacturing. By combining the distinct advantages of the core and shell materials, core-shell systems offer unique functionalities such as improved mechanical properties, biocompatibility, conductivity, and environmental resistance.

In biomedical fields, core-shell nanomaterials are enhancing the performance of implants, sensors, and drug delivery systems, contributing to more personalized and effective healthcare solutions. In electronics, these materials are paving the way for flexible, high-performance devices, while their integration into 3D-printing technologies is facilitating the creation of customized devices and prototypes with unprecedented precision and functionality. Furthermore, the potential for smart coatings and functional textiles is revolutionizing industries like aerospace, automotive, and healthcare, offering solutions such as self-healing surfaces, anti-corrosion coatings, and wearables with embedded functionalities.

As research and development continue to push the boundaries of core-shell nanomaterials, their applications in fields like sensors, actuators, and optoelectronics are becoming increasingly sophisticated. The continued exploration of their potential will likely lead to the creation of even more innovative materials and devices that could address global challenges in energy, healthcare, and sustainability.

Ultimately, core-shell nanomaterials represent a pivotal breakthrough in materials science, offering a versatile platform for designing and fabricating high-performance components that meet the growing demands of advanced technologies. The convergence of their unique properties with modern fabrication techniques, such as 3D printing and additive manufacturing, holds the promise of shaping the future of multiple industries.

References

1 Ghosh Chaudhuri R, Paria S. Core/shell nanoparticles: Classes, properties, synthesis mechanisms, characterization, and applications. Chem Rev [Internet]. 2012 Apr 11 [cited 2025 Feb 2];112(4): 2373–433. Available from: https://pubs.acs.org/doi/full/10.1021/cr100449n

2 Dastjerdi R, Montazer M. A review on the application of inorganic nano-structured materials in the modification of textiles: Focus on anti-microbial properties. Colloids Surf B Biointerfaces. 2010 Aug 1;79(1): 5–18.

3 Dincer C, Bruch R, Costa-Rama E, Fernández-Abedul MT, Merkoçi A, Manz A, et al. Disposable sensors in diagnostics, food, and environmental monitoring. Adv Mater [Internet]. 2019 Jul 1 [cited 2025 Feb 2];31(30): 1806739. Available from: https://onlinelibrary.wiley.com/doi/full/10.1002/adma.201806739

4 Pinto FM, Dey S, Taft CA, La Porta F. Synthesis of colloidal semiconductor nanocrystals for enhanced optoelectronic technologies. Eng Mater [Internet]. 2024

[cited 2025 Feb 2];Part F3949: 1–47. Available from: https://link.springer.com/chapter/10.1007/978-3-031-75984-0_1

5 Allioui H, Mourdi Y. Exploring the full potentials of IoT for better financial growth and stability: A comprehensive survey. Sensors [Internet]. 2023 Sep 22 [cited 2025 Feb 2];23(19): 8015. Available from: https://www.mdpi.com/1424-8220/23/19/8015/htm

6 Ong PJ, Wang S, Suwardi A, Cao J, Wang F, Yin X, et al. Functional coatings for built environments based on nanotechnology. World Sci Ann Rev Funct Mater [Internet]. 2023 Nov 9 [cited 2025 Feb 2];02: 2330004. Available from: https://www.worldscientific.com/worldscinet/wsarfm

7 He S, Gao Y, Gong X, Wu C, Cen H, He S, et al. Advance of design and application in self-healing anticorrosive coating: A review. J Coat Technol Res [Internet]. 2023 Feb 21 [cited 2025 Feb 2];20(3): 819–41. Available from: https://link.springer.com/article/10.1007/s11998-022-00735-6

8 Jiang R, on Tung S, Tang Z, Li L, Ding L, Xi X, et al. A review of core-shell nanostructured electrocatalysts for oxygen reduction reaction. Energy Stor Mater. 2018 May 1;12: 260–76.

9 Prajapati DG, Kandasubramanian B. A review on polymeric-based phase change material for thermo-regulating fabric application. Polym Rev [Internet]. 2020 Jul 2 [cited 2025 Feb 2];60(3): 389–419. Available from: https://www.tandfonline.com/doi/abs/10.1080/15583724.2019.1677709

10 Lizundia E, Sipponen MH, Greca LG, Balakshin M, Tardy BL, Rojas OJ, et al. Multifunctional lignin-based nanocomposites and nanohybrids. Green Chem [Internet]. 2021 Sep 20 [cited 2025 Feb 2];23(18): 6698–760. Available from: https://pubs.rsc.org/en/content/articlehtml/2021/gc/d1gc01684a

11 Vijayan PP, Puglia D. Biomimetic multifunctional materials: A review. Emerg Mater [Internet]. 2019 Dec 1 [cited 2025 Feb 2];2(4): 391–415. Available from: https://link.springer.com/article/10.1007/s42247-019-00051-7

12 Fu X, Tian L, Fan Y, Ye W, Qiao ZA, Zhao J, et al. Stimuli-responsive self-healing anticorrosion coatings: From single triggering behavior to synergetic multiple protections. Mater Today Chem. 2021 Dec 1;22: 100575.

13 Picone A and Brambilla A. 2025 Atomic-scale insights into metal-oxide/graphene heterostructures. J Phys: Condens Matter 2025;37(25): 253004 https://doi.org/10.1088/1361-648X/addf28

14 Zhu J, Huang X, Song W. Physical and chemical sensors on the basis of laser-induced graphene: Mechanisms, applications, and perspectives. ACS Nano. 2021 Dec 28;15(12): 18708–41.

15 Ielo I, Giacobello F, Castellano A, Sfameni S, Rando G, Plutino MR. Development of antibacterial and antifouling innovative and eco-sustainable sol–gel based materials: From marine areas protection to healthcare applications. Gels [Internet]. 2021 Dec 28 [cited 2025 Feb 2];8(1): 26. Available from: https://www.mdpi.com/2310-2861/8/1/26/htm

16 Harun-Ur-Rashid M, Imran AB, Susan MABH. Green polymer nanocomposites in automotive and packaging industries. Curr Pharm Biotechnol. 2022 May 7;24(1): 145–63.

17 Wang N, Zhou H, Ren J, Gao G, Zhao G, Yang Y, et al. Polymer@SiO_2 core–shell composite particles: Preparation and application. Coatings [Internet].

2023 Feb 1 [cited 2025 Feb 2];13(2): 334. Available from: https://www.mdpi.com/2079-6412/13/2/334/htm

18 Maiti DK, Manna AS, Ghosh S, Ghosh T, Karchaudhuri N, Das S, et al. Smart luminescent materials for emerging sensors: Fundamentals and advances. Chem Asian J [Internet]. 2025 Jan 14 [cited 2025 Feb 2];20(6): e202401328. Available from: https://aces.onlinelibrary.wiley.com/doi/10.1002/asia.202401328

19 Al Mamun A, Yuce MR. Recent progress in nanomaterial enabled chemical sensors for wearable environmental monitoring applications. Adv Funct Mater. 2020 Dec 1;30(51): 2005703.

20 Cao E, Cao Y, Sun M. Surface plasmonic core-shell nanostructures in surface enhanced Raman scattering and photocatalysis. Anal Chem. 2024 Jul 23;96(29): 11623–38.

21 Zou H, Luo Z, Yang X, Xie Q, Zhou Y. Toward emerging applications using core–shell nanostructured materials: A review. J Mater Sci [Internet]. 2022 Jun 6 [cited 2025 Feb 2];57(24): 10912–42. Available from: https://link.springer.com/article/10.1007/s10853-022-07328-z

22 Mirzaei A, Janghorban K, Hashemi B, Neri G. Metal-core@metal oxide-shell nanomaterials for gas-sensing applications: A review. J Nanopart Res [Internet]. 2015 Sep 14 [cited 2025 Feb 2];17(9): 1–36. Available from: https://link.springer.com/article/10.1007/s11051-015-3164-5

23 Li JF, Zhang YJ, Ding SY, Panneerselvam R, Tian ZQ. Core-shell nanoparticle-enhanced Raman spectroscopy. Chem Rev [Internet]. 2017 Apr 12 [cited 2025 Feb 2];117(7): 5002–69. Available from: https://pubs.acs.org/doi/full/10.1021/acs.chemrev.6b00596

24 Pons JL. Emerging actuator technologies: A micromechatronic approach. Google Books [Internet]. John Wiley & Sons. 2005 [cited 2025 Feb 2]. Available from: https://books.google.co.in/books?hl=en&lr=&id=fbmiuULDh7YC&oi=fnd&pg=PR7&dq=Actuators,+which+convert+energy+into+mechanical+motion,+are+fundamental+components+in+systems+requiring+precise+movements+and+responsiveness&ots=PlmhZAG7_Z&sig=mkIrOQrcd8VZ26oe3HeifeZUUU8&redir_esc=y#v=onepage&q&f=false

25 Yin S, Yao DR, Song Y, Heng W, Ma X, Han H, et al. Wearable and implantable soft robots. Chem Rev [Internet]. 2024 [cited 2025 Feb 2];124: 11585–11636. Available from: https://pubs.acs.org/doi/full/10.1021/acs.chemrev.4c00513

26 Sun L, Wang TX, Chen HM, Salvekar AV, Naveen BS, Xu Q, et al. A brief review of the shape memory phenomena in polymers and their typical sensor applications. Polymers [Internet]. 2019 Jun 15 [cited 2025 Feb 2];11(6): 1049. Available from: https://www.mdpi.com/2073-4360/11/6/1049/htm

27 Ma X, Li Y, Hussain I, Shen R, Yang G, Zhang K. Core–shell structured nanoenergetic materials: Preparation and fundamental properties. Adv Mater [Internet]. 2020 Jul 1 [cited 2025 Feb 2];32(30): 2001291. Available from: https://onlinelibrary.wiley.com/doi/full/10.1002/adma.202001291

28 AlTowireb SM, Goumri-Said S. Core-Shell structures for the enhancement of energy harvesting in piezoelectric nanogenerators: A review. Sustain Energy Technol Assess. 2023 Feb 1;55: 102982.

29 Chen IC, Lin SS, Lin TJ, Hsu CL, Hsueh TJ, Shieh TY. The assessment for sensitivity of a NO_2 gas sensor with $ZnGa_2O_4$/ZnO core-shell nanowires—A novel approach. Sensors [Internet]. 2010 Mar 30 [cited 2025 Feb 2];10(4): 3057–72. Available from: https://www.mdpi.com/1424-8220/10/4/3057/htm

30 Sahiner N, Butun S, Ilgin P. Hydrogel particles with core shell morphology for versatile applications: Environmental, biomedical and catalysis. Colloids Surf A Physicochem Eng Asp. 2011 Aug 5;386 (1–3): 16–24.

31 Gupta SK, Kumar P, Singh DP. Quantum dots for modern display devices. In: Graphene, Nanotubes and Quantum Dots-Based Nanotechnology: Fundamentals and Applications. Woodhead Publishing. 2022 Jan 1; 899–932.

32 Zhang C, Chen J, Kong L, Wang L, Wang S, Chen W, et al. Core/shell metal halide perovskite nanocrystals for optoelectronic applications. Adv Funct Mater [Internet]. 2021 May 1 [cited 2025 Feb 2];31(19): 2100438. Available from: https://onlinelibrary.wiley.com/doi/full/10.1002/adfm.202100438

33 Saleem MI, Batool A, Hur J. Cutting-edge developments in metal halide perovskites core/shell heterocrystals: From photodetectors to biomedical applications. Small [Internet]. 2025 Jan 1 [cited 2025 Feb 2];21(2): 2407032. Available from: https://onlinelibrary.wiley.com/doi/full/10.1002/smll.202407032

34 Zhang DD, Xu JL, Sun HB, Zhang DD, Xu JL, Sun B-H. Toward high efficiency organic light-emitting diodes: Role of nanoparticles. Adv Opt Mater [Internet]. 2021 Mar 1 [cited 2025 Feb 2];9(6): 2001710. Available from: https://onlinelibrary.wiley.com/doi/full/10.1002/adom.202001710

35 Zou Y, Sun Z, Wang Q, Ju Y, Sun N, Yue Q, et al. Core-shell magnetic particles: Tailored synthesis and applications. Chem Rev [Internet]. 2025 Jan 22 [cited 2025 Feb 2];125(2): 972–1048. Available from: https://pubs.acs.org/doi/full/10.1021/acs.chemrev.4c00710

36 Vasudevan D, Gaddam RR, Trinchi A, Cole I. Core–shell quantum dots: Properties and applications. J Alloys Compd. 2015 Jul 5;636: 395–404.

37 Singh Selopal G, Zhao H, Wang ZM, Rosei F, Selopal GS, Wang ZM, et al. Core/shell quantum dots solar cells. Adv Funct Mater [Internet]. 2020 Mar 1 [cited 2025 Feb 2];30(13): 1908762. Available from: https://onlinelibrary.wiley.com/doi/full/10.1002/adfm.201908762

38 Kim J, Roh J, Park M, Lee C. Recent advances and challenges of colloidal quantum dot light-emitting diodes for display applications. Adv Mater [Internet]. 2024 May 1 [cited 2025 Feb 2];36(20): 2212220. Available from: https://onlinelibrary.wiley.com/doi/full/10.1002/adma.202212220

39 Li S, Xu X, Lin Q, Sun J, Zhang H, Shen H, et al. Bright and stable yellow quantum dot light-emitting diodes through core–shell nanostructure engineering. Small [Internet]. 2024 Jun 1 [cited 2025 Feb 2];20(24): 2306859. Available from: https://onlinelibrary.wiley.com/doi/full/10.1002/smll.202306859

40 Ding M, Guo Z, Chen X, Ma X, Zhou L. Surface/interface engineering for constructing advanced nanostructured photodetectors with improved performance: A brief review. Nanomaterials [Internet]. 2020 Feb 19 [cited 2025 Feb 2];10(2): 362. Available from: https://www.mdpi.com/2079-4991/10/2/362/htm

41 Chen G, Ågren H, Ohulchanskyy TY, Prasad PN. Light upconverting core–shell nanostructures: Nanophotonic control for emerging applications. Chem Soc Rev [Internet]. 2015 Mar 10 [cited 2025 Feb 2];44(6): 1680–713. Available from: https://pubs.rsc.org/en/content/articlehtml/2015/cs/c4cs00170b

42 Zhao B, Chen L, Liu W, Wu L, Lu Z, Cao W. High efficiency blue light-emitting devices based on quantum dots with core-shell structure design and surface modification. RSC Adv [Internet]. 2021 Apr 14 [cited 2025 Feb 2];11(23): 14047–52. Available from: https://pubs.rsc.org/en/content/articlehtml/2021/ra/d0ra10173g

43 Shahbazi M, Jäger H, Ettelaie R, Chen J, Kashi PA, Mohammadi A. Dispersion strategies of nanomaterials in polymeric inks for efficient 3D printing of soft and smart 3D structures: A systematic review. Adv Colloid Interface Sci. 2024 Nov 1;333: 103285.

44 Galogahi FM, Zhu Y, An H, Nguyen NT. Core-shell microparticles: Generation approaches and applications. J Sci Adv Mater Devices. 2020 Dec 1;5(4): 417–35.

45 Zhang Y, Zhang F, Yan Z, Ma Q, Li X, Huang Y, et al. Printing, folding and assembly methods for forming 3D mesostructures in advanced materials. Nat Rev Mater [Internet]. 2017 Mar 29 [cited 2025 Feb 2];2(4): 1–17. Available from: https://www.nature.com/articles/natrevmats201719

46 Li H, Zhang H, Chang A, Ma X, Rong J, Yang L. A novel core-shell structure NTC ceramic with high stability fabricating by an in-situ ink-jet printing method. J Eur Ceram Soc. 2021 Jul 1;41(7): 4167–74.

47 Abdolmaleki H, Kidmose P, Agarwala S, Abdolmaleki H, Kidmose P, Agarwala S. Droplet-based techniques for printing of functional inks for flexible physical sensors. Adv Mater [Internet]. 2021 May 1 [cited 2025 Feb 2];33(20): 2006792. Available from: https://onlinelibrary.wiley.com/doi/full/10.1002/adma.202006792

48 Saadi MASR, Maguire A, Pottackal NT, Thakur MSH, Ikram MM, Hart AJ, et al. Direct ink writing: A 3D printing technology for diverse materials. Adv Mater [Internet]. 2022 Jul 1 [cited 2025 Feb 2];34(28): 2108855. Available from: https://onlinelibrary.wiley.com/doi/full/10.1002/adma.202108855

49 Pajor-Świerzy A, Szczepanowicz K, Kamyshny A, Magdassi S. Metallic core-shell nanoparticles for conductive coatings and printing. Adv Colloid Interface Sci. 2022 Jan 1;299: 102578.

50 Milojević M, Vihar B, Banović L, Miško M, Gradišnik L, Zidarič T, et al. Core/shell printing scaffolds for tissue engineering of tubular structures. J Vis Exp [Internet]. 2019 Sep 27 [cited 2025 Feb 2];2019(151): e59951. Available from: https://www.jove.com/t/59951/coreshell-printing-scaffolds-for-tissue-engineering-tubular

51 Tong C. Conductive materials for printed flexible electronics. Springer Ser Mater Sci [Internet]. 2022 [cited 2025 Feb 2];317: 119–57. Available from: https://link.springer.com/chapter/10.1007/978-3-030-79804-8_3

52 Rupp H, Binder WH. 3D Printing of core–shell capsule composites for post-reactive and damage sensing applications. Adv Mater Technol [Internet]. 2020 Nov 1 [cited 2025 Feb 2];5(11): 2000509. Available from: https://onlinelibrary.wiley.com/doi/full/10.1002/admt.202000509

53 Karim MR, Siddiqui IH, Assaifan AK, Aijaz MO, Alnaser IA. Review on cutting-edge innovations in carbon nanomaterials reinforced silicone rubber composites

for flexible electronics and healthcare devices. ACS Appl Polym Mater [Internet]. 2024 Mar 19 [cited 2025 Feb 2];6(23): 14235–59. Available from: https://pubs.acs.org/doi/abs/10.1021/acsapm.4c02511?casa_token=AHRlMsA-TLgAAAAA:_5yoCq4tTt5QQIfakTQnCm1D7c04bwjKu138KiWhlZD4NusQfjMRtf7ywYN8VjOJ7BOsT6HzBS_TnWOh&casa_token=bSVVDmMCvAEAAAAA:ukkuE-FMHURm6et5PqRYkSUH-UTAv_ZjD00A0CktjOMdILR5SQiVT9vP0Vn_s2jNVexS4BxRhk2CLjRr

54 Wang N, Fuh JYH, Dheen ST, Senthil Kumar A. Functions and applications of metallic and metallic oxide nanoparticles in orthopedic implants and scaffolds. J Biomed Mater Res B Appl Biomater [Internet]. 2021 Feb 1 [cited 2025 Feb 2];109(2): 160–79. Available from: https://onlinelibrary.wiley.com/doi/full/10.1002/jbm.b.34688

55 Fratto E. Printed assembly and nanosoldering of core/shell nanowires and graphene oxide for environmental sensing and e-textile applications. University of Massachusetts Lowell. 2022.

56 Mahmood A, Perveen F, Chen S, Akram T, Irfan A. Polymer composites in 3D/4D printing: Materials, advances, and prospects. Molecules [Internet]. 2024 Jan 9 [cited 2025 Feb 2];29(2): 319. Available from: https://www.mdpi.com/1420-3049/29/2/319/htm

9

Challenges and Opportunities in Core-shell Nanotechnology

Core-shell nanomaterials have emerged as a revolutionary class of materials with remarkable properties that enable a wide range of applications across diverse industries, including energy, healthcare, environmental sustainability, and electronics. These materials, consisting of a core and a shell with distinct compositions and functionalities, offer enhanced performance through synergistic interactions between the two components. This chapter explores the current state of core-shell nanotechnology, focusing on their synthesis, multifunctionality, and integration into advanced applications such as smart coatings, sensors, energy storage, and biomedical therapies. Key challenges, including issues related to synthesis complexity, scalability, environmental impact, and regulatory barriers, are discussed, alongside emerging trends and future opportunities in research. Furthermore, the potential of artificial intelligence (AI) and machine learning (ML) to accelerate the design and optimization of core-shell nanomaterials is highlighted, emphasizing the importance of interdisciplinary collaboration in advancing this field. The chapter concludes by exploring the transformative impact of core-shell nanomaterials in addressing global challenges related to energy, the environment, and medicine, underscoring their potential to shape the future of technology and sustainability.

9.1 Introduction

Core-shell nanotechnology is at the forefront of materials science innovation, offering unique opportunities for enhancing the properties and functionalities of nanomaterials. At its core, core-shell nanotechnology involves the design and synthesis of nanoparticles composed of two distinct materials: a core material and a shell material that surrounds it. These materials often have complementary properties that enable the creation of hybrid nanostructures with multifunctional capabilities, making them ideal for a wide range of applications [1]. The core-shell approach allows researchers to control the properties of materials at the nanoscale level, manipulating their size, shape, and composition to achieve enhanced performance in various technological sectors such as medicine, energy, electronics, and environmental

Core-Shell Nanomaterials: From Fundamentals to Applications, First Edition. Shreya Sharma et al.

science. The core-shell structure is often designed with specific properties tailored to meet the demands of particular applications. For example, the core material can serve as the main structural element, while the shell can be designed to improve stability, functionality, or reactivity. This unique structural design allows for significant advantages over traditional bulk materials, such as increased surface area, improved catalytic activity, and enhanced optical, magnetic, or electrical properties [2]. The versatility of core-shell nanomaterials enables their use in multiple fields, including drug delivery, diagnostics, bioimaging, energy storage, and environmental monitoring.

However, despite the vast potential of core-shell nanotechnology, numerous challenges hinder its widespread adoption and commercialization. These challenges span across synthesis methods, scalability, environmental and toxicological concerns, and the regulatory landscape. The development of core-shell materials involves complex synthesis techniques that require precise control over size, shape, and composition. Achieving uniformity in the fabrication process is critical, as variations in the core-shell structure can significantly impact the performance of the material [3]. Additionally, the scalability of these materials from the laboratory to industrial production remains a significant challenge, as many synthesis techniques are still not economically viable for large-scale manufacturing.

Environmental and safety concerns surrounding the use of core-shell nanomaterials have also raised questions regarding their long-term effects on human health and the environment. The potential for nanomaterials to leach toxic substances or accumulate in biological systems remains a critical area of research. As the field advances, it is essential to develop appropriate guidelines for assessing the biocompatibility, toxicity, and environmental impact of core-shell nanomaterials. Moreover, there are significant economic and regulatory barriers to overcome. The approval processes for nanomaterial-based products by regulatory agencies, such as the Food and Drug Administration (FDA) and European Medicines Agency (EMA), are often complex and require comprehensive testing to ensure the safety and efficacy of these materials [4]. Despite these challenges, the potential of core-shell nanotechnology remains vast. Addressing these obstacles is crucial for realizing the full potential of these materials in a variety of applications. The research community has made significant strides in overcoming some of these challenges, and new approaches are emerging to improve synthesis techniques, enhance material performance, and address safety concerns. Innovations in AI and ML have already begun to play a pivotal role in accelerating the discovery and optimization of core-shell nanomaterials [5]. These technologies can help researchers design more efficient and targeted materials, streamline synthesis processes, and predict the behavior of nanomaterials in various environments.

The aim of the chapter is to provide an in-depth analysis of the current challenges and opportunities in core-shell nanotechnology. It will explore the complexity of synthesis, scalability issues, and the economic and regulatory barriers that limit the widespread adoption of core-shell nanomaterials. Additionally, the chapter will examine emerging trends in research, such as the integration of AI and ML in nanomaterial design and the potential breakthroughs in energy, environment, and medicine. The chapter will also highlight the importance of interdisciplinary collaboration in advancing core-shell nanotechnology, as the integration of expertise from various fields is crucial to overcoming existing challenges and realizing the full potential of these advanced materials. In the subsequent sections, the chapter will first address the synthesis challenges associated with core-shell nanomaterials, examining the methods used to fabricate these materials and the issues related to uniformity and reproducibility. It will then discuss the scalability and industrialization challenges, including the economic feasibility and technical barriers to large-scale production [6]. Following this, the chapter will explore environmental and toxicological concerns, discussing the potential risks associated with the use of nanomaterials in various applications. The regulatory aspects of core-shell nanomaterials will be reviewed, focusing on the guidelines provided by regulatory agencies and the approval processes for nanomaterial-based products.

Emerging trends in core-shell nanotechnology will also be covered, with a particular emphasis on the role of AI and ML in accelerating the development of advanced nanomaterials. The integration of these technologies is expected to revolutionize the field, enabling more efficient design processes and facilitating the discovery of novel materials with enhanced properties. Finally, the chapter will conclude by discussing the potential breakthroughs in energy, environmental, and medical applications, highlighting the transformative impact core-shell nanomaterials could have in addressing global challenges. By the end of the chapter, readers will have a comprehensive understanding of the current landscape of core-shell nanotechnology, the challenges that must be overcome, and the opportunities that exist for advancing the field. The information presented here will be valuable for researchers, industry professionals, and policymakers working in the field of nanotechnology, as well as those interested in the broader implications of these technologies for society.

9.2 Synthesis Complexity and Control Issues

Achieving uniformity in core-shell structures is one of the major challenges in the synthesis of core-shell nanomaterials. The unique properties of these materials stem largely from the combination of two different materials in a single nanoparticle. To maximize the performance of core-shell systems, it is crucial that the core and shell layers are homogeneous in terms of size, composition, and thickness. Variations in these parameters can lead to inconsistencies in material properties, undermining their effectiveness in applications ranging from drug delivery to sensors and electronics [7]. A primary issue lies in controlling the size of the core and ensuring that the shell forms a uniform layer around it. Small deviations in the thickness of the shell or the size of the core can significantly impact the surface area, catalytic activity, and stability of the core-shell nanomaterial.

The problem of size control in core-shell structures is particularly evident when attempting to scale up synthesis processes. In laboratory-scale production, it is possible to manipulate the experimental conditions with precision to obtain nanoparticles with desired characteristics. However, replicating these conditions on an industrial scale introduces variability, leading to non-uniform nanoparticles. This lack of uniformity is especially problematic when these nanomaterials are used in applications that require high consistency, such as in medical treatments or in sensors for environmental monitoring [8]. Variations in size can affect not only the structural integrity but also the functional performance of the materials, making it difficult to predict and control their behavior in real-world applications. Another issue related to core-shell nanostructures is the stability of the interface between the core and the shell materials. The interaction between the core and the shell can significantly influence the overall stability and performance of the nanomaterial [9]. If the interface is not well-defined or if the materials used for the core and the shell do not bind well together, the nanoparticle may experience degradation over time or lose its functional properties. This issue is exacerbated when the shell is very thin, as the core material may leach into the surrounding environment, potentially reducing the efficacy of the material or causing toxicity in biological applications.

To address these synthesis challenges, researchers have developed a variety of novel techniques and precision synthesis methods. One such technique is the use of controlled precipitation or solution-based processes, where careful adjustments of temperature, pH, and precursor concentration allow for the formation of core-shell nanostructures with well-controlled size and shell thickness. Another promising approach is the use of templating methods, where a sacrificial template is used to guide the deposition of the shell material, ensuring uniform coverage and stability [10]. High-resolution imaging and characterization tools, such as transmission electron microscopy (TEM) and scanning electron microscopy (SEM), have been invaluable in evaluating the uniformity and quality of core-shell nanomaterials, providing real-time insights into the synthesis process. Additionally, the development of surface chemistry techniques has played a significant role in improving the stability and uniformity of core-shell structures. By tuning the surface chemistry of both the core and the shell materials, researchers can achieve better adhesion between the two components, ensuring a more stable interface. Another technique gaining attention is atomic layer deposition (ALD), which allows for the deposition of ultrathin and highly uniform layers around the core material, leading to greater control over the shell thickness and the interface between the materials [11].

Despite these advancements, significant challenges remain in achieving consistent, scalable, and cost-effective synthesis of core-shell nanomaterials. As the demand for these materials grows, the ability to produce high-quality core-shell structures with precise control over their size, shell thickness, and interface stability will be essential [12]. Future research is likely to focus on improving the scalability of these synthesis techniques, making them viable for large-scale production, while also refining methods to address the inherent variability that can arise in core-shell nanostructures. Figure 9.1 illustrates various synthesis techniques for core-shell nanomaterials and how they can result in different outcomes in terms of uniformity and performance. As shown, different methods, such as sol–gel processes, coprecipitation, and chemical vapor deposition (CVD), can each yield core-shell nanoparticles with varying levels of consistency in size, shell thickness, and overall structural integrity.

9.3 Scalability and Industrialization Challenges

Transitioning from laboratory-scale to large-scale production of core-shell nanomaterials presents a number of significant challenges. While small-scale synthesis allows for tight control over experimental conditions, this level of precision becomes increasingly difficult to maintain when scaling up the production process. In laboratory settings, it is relatively easy to control the temperature, concentration

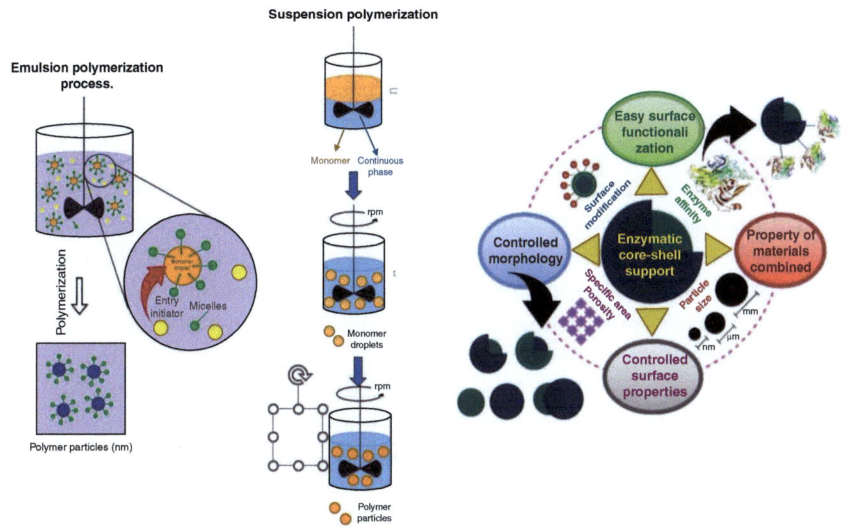

Figure 9.1 Comparison of synthesis methods and the control of core-shell nanostructures. This illustration demonstrates the differences in outcomes resulting from various synthesis techniques, highlighting the challenges in achieving uniformity in core-shell structures. Reproduced from [13] copyright © 2022 Wiley-VCH GmbH.

of precursors, reaction time, and other parameters to obtain well-defined core-shell nanostructures. However, when attempting to transition these processes to industrial-scale production, these parameters can become difficult to regulate consistently across large batches [14]. This leads to reproducibility issues, where each batch of nanomaterials may have slight variations in size, composition, or properties, making it difficult to guarantee uniformity in the final product.

One of the key challenges in large-scale production is the issue of reproducibility. In small-scale synthesis, it is possible to adjust parameters like precursor concentrations, solvent types, and temperature with great precision. However, when scaling up, the equipment size, flow dynamics, and energy input all change, potentially altering the reaction conditions in ways that are difficult to predict [15]. This variation can lead to inconsistencies between different batches, meaning the final core-shell nanomaterials may have differing characteristics. This lack of reproducibility is problematic for industries that require high-quality, uniform nanomaterials for use in sensitive applications such as drug delivery, diagnostics, or electronics. In these applications, even small deviations in size or functionality can significantly impact performance and safety. In addition to reproducibility issues, there are challenges related to consistency when producing core-shell nanomaterials in large batches. In laboratory-scale synthesis, processes are often carried out in small volumes, allowing for tight control of the reaction environment [16]. However, in large-scale production, the scale-up can lead to inhomogeneity in the reaction mixture, where different parts of the batch may undergo slightly different reaction conditions. This leads to variability in the size, structure, and properties of the core-shell nanomaterials. For example, there could be differences in shell thickness or core composition that result in inconsistent material properties, such as mechanical strength or drug release profiles, which could affect the final application.

To overcome these scalability challenges, several strategies have been proposed to enhance the scalability of core-shell nanomaterial production. One promising solution is the use of continuous flow processes, which have become increasingly popular in nanomaterial synthesis [17]. Unlike batch processing, which involves the production of nanomaterials in a single container over a set period of time, continuous flow systems allow for the constant injection of reagents into a flowing stream, creating a more controlled and steady environment. This method can reduce variability between batches and improve the uniformity of the resulting nanomaterials. Continuous flow reactors also provide better heat and mass transfer, which can further improve the quality and consistency of the final product. Automated systems also offer potential solutions for large-scale production. Automation helps ensure that reaction conditions are tightly controlled across large batches, reducing human error and the variability that comes with manual operation [18]. Automated systems can also provide real-time monitoring and feedback, adjusting parameters such as temperature, pressure, and reactant concentration in real time to maintain optimal conditions for synthesis. This ability to fine-tune reaction parameters on the fly can help ensure greater consistency and reproducibility in the final nanomaterial products.

Another strategy for scaling up core-shell nanomaterial production involves optimizing the precursor materials and reaction conditions to improve the efficiency of the synthesis process. By reducing the complexity of the synthesis steps or simplifying the raw material requirements, manufacturers can achieve higher yields and reduce the overall cost of production. In particular, the development of more cost-effective, environmentally friendly precursor materials is crucial to making large-scale synthesis more feasible for industrial applications [19]. Figure 9.2 illustrates the scalability process of core-shell nanomaterials, providing a flowchart that highlights the key steps in transitioning from laboratory-scale synthesis to industrial-scale production. As shown, the transition involves not only improving the precision and consistency of the synthesis methods but also addressing the technical challenges that arise when moving from small-scale to large-scale operations. It also emphasizes the potential solutions, such as continuous flow reactors and automated systems, that could mitigate the challenges of scalability and reproducibility.

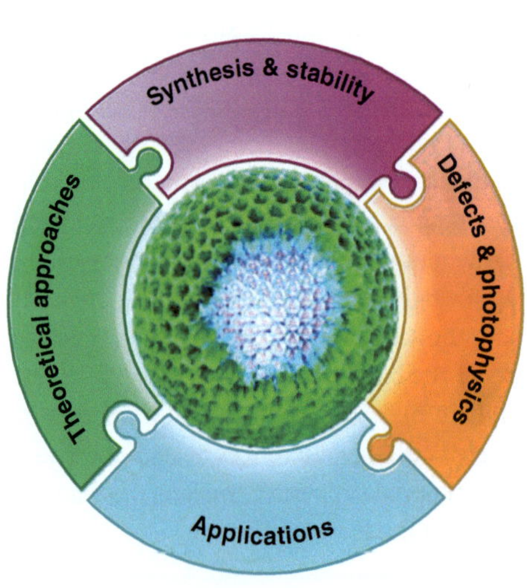

Figure 9.2 Challenges and potential solutions for the large-scale production of core-shell nanomaterials. This flowchart outlines the key steps involved in scaling up the synthesis process and the strategies for ensuring uniformity, consistency, and efficiency in industrial production. [20]/American Chemical Society/CC BY 4.0.

9.4 Environmental and Toxicological Concerns

Environmental and toxicological concerns surrounding core-shell nanomaterials have garnered significant attention due to their potential risks at various stages, from production to disposal. The unique properties of these nanomaterials—such as their high surface area, reactivity, and ability to interact with biological systems—can lead to unforeseen consequences for the environment and living organisms. As core-shell nanomaterials are increasingly integrated into advanced technologies, understanding their environmental and health impacts becomes crucial to ensuring their safe use [21]. One major concern is the potential environmental impact during the production, use, and disposal of core-shell nanomaterials. The synthesis processes involved in creating these materials often require the use of hazardous chemicals, solvents, or energy-intensive procedures, which could lead to the release of harmful substances into the environment. During production, by-products and waste materials may contain residual nanoparticles, which, if not properly managed, could contaminate water, soil, or air [22]. Additionally, the use of core-shell nanomaterials in various applications—such as in drug delivery systems, coatings, or sensors—could result in the release of nanomaterials into the environment through product wear and tear or improper disposal. The long-term persistence and mobility of these particles in the environment raise concerns about their accumulation in ecosystems and potential toxicity to non-target organisms.

Once core-shell nanomaterials are released into the environment, their behavior and fate depend on several factors, including their size, surface properties, and chemical composition [23]. Nanoparticles are known to have high reactivity, which may lead to the formation of reactive oxygen species (ROS) or other by-products that could pose environmental risks. For example, metal-containing core-shell nanomaterials may undergo oxidation, releasing metal ions that could accumulate in aquatic environments and harm aquatic life. The high surface-to-volume ratio of these materials also makes them more bioavailable, increasing the likelihood of bioaccumulation and posing risks to the food chain. In addition to environmental risks, the potential toxicological effects of core-shell nanomaterials on living organisms need to be carefully considered [24]. Core-shell materials, due to their small size and unique physicochemical properties, can easily penetrate biological membranes and enter cells, tissues, or organs. This can lead to unintended interactions with biological systems, resulting in toxicity. Toxicological studies have shown that nanoparticles can cause cellular damage, inflammation, oxidative stress, and even disrupt the functioning of vital organs, depending on their composition and exposure levels. For example, metal-based core-shell nanomaterials, such as gold or silver nanoparticles, have been shown to induce cytotoxicity in certain cell lines, while silica-based core-shell materials may cause lung inflammation when inhaled [25]. These findings underscore the importance of conducting thorough toxicity assessments before the widespread adoption of core-shell nanomaterials in various industries.

To address these concerns, extensive risk assessment studies are needed to evaluate the potential hazards posed by core-shell nanomaterials to both human health and the environment. These studies should involve testing the biocompatibility and toxicity of core-shell materials through in vitro and in vivo experiments. Understanding the release mechanisms of nanoparticles, their interaction with biological systems, and their long-term effects is crucial for identifying potential risks. Regulatory frameworks and guidelines, such as those established by the U.S. FDA and the EMA, must be adapted to include nanomaterials to ensure their safe use [26]. This involves setting acceptable exposure limits, determining safe disposal methods, and implementing proper labeling to inform users about the potential risks associated with nanomaterials. Furthermore, public awareness and transparent communication about the environmental and health risks of core-shell nanomaterials are essential for promoting responsible innovation. Regulatory bodies must work with researchers, industries, and environmental groups to develop standards and best practices for the safe design, use, and disposal of these materials [27]. Additionally, developing more sustainable and environmentally friendly synthesis methods for core-shell nanomaterials—such as green chemistry approaches—can help reduce their environmental footprint and ensure that they contribute positively to technological advancements without causing harm. Figure 9.3 provides a visual representation of the lifecycle of core-shell nanomaterials, highlighting the potential environmental and toxicological risks at various stages, from production to disposal. As shown in the diagram, these materials can pose different types of risks depending on how they are synthesized, used, and disposed of. The lifecycle analysis emphasizes the need for careful monitoring and assessment at each stage to minimize the environmental and health impacts associated with core-shell nanomaterials.

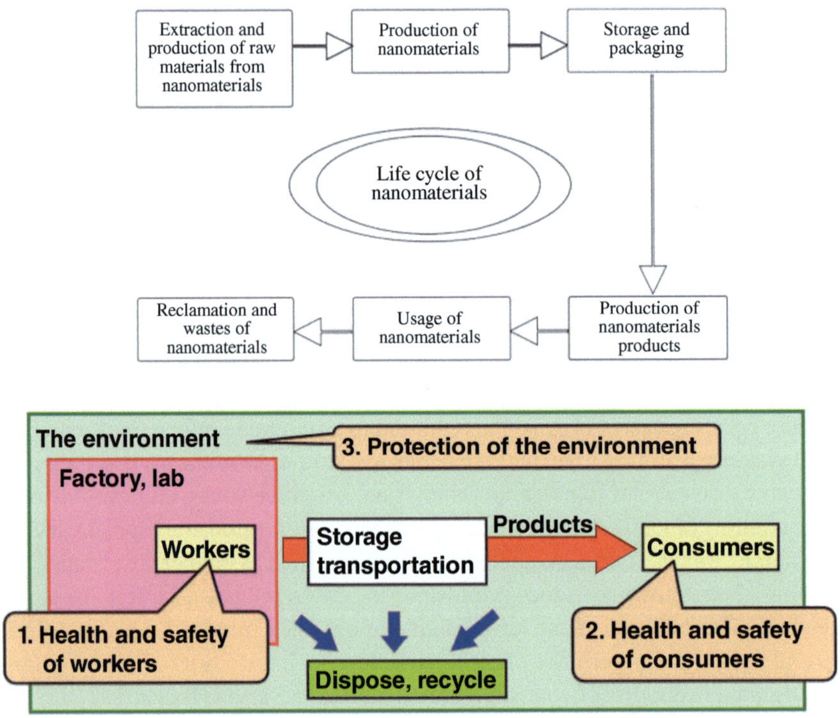

Figure 9.3 Environmental and toxicological considerations for core-shell nanomaterials from production to disposal. The diagram illustrates the various stages of a core-shell nanomaterial's lifecycle ([28]/with permission of Technology Center), and the potential risks associated with each stage ([29]/with permission of Fuji Technology Press), highlighting the importance of risk management throughout their use and disposal.

9.5 Economic and Regulatory Barriers

9.5.1 Cost-related Challenges in the Commercialization of Core-shell Nanomaterials

- **High Production Costs:** The synthesis of core-shell nanomaterials often involves complex and resource-intensive methods, such as CVD, sol–gel processes, and high-temperature treatments. These methods can increase manufacturing costs significantly.
- **Material Costs:** Specialized precursors, high-purity materials, and advanced equipment are required, driving up the cost of production for core-shell nanomaterials.
- **Scalability:** The transition from small-scale laboratory production to large-scale manufacturing is challenging and costly, requiring significant investment in infrastructure and technology to ensure high throughput and quality control [30].

9.5.2 Economic Feasibility in Manufacturing and Integrating These Materials into Existing Products

- **Integration Costs:** Incorporating core-shell nanomaterials into existing product lines (e.g., sensors, medical devices, or coatings) can be costly, requiring modifications in manufacturing processes and equipment.
- **Market Competition:** The high cost of core-shell nanomaterials can make it difficult for companies to compete with conventional materials, which may offer similar properties at a fraction of the cost.
- **Cost-benefit Analysis:** For widespread adoption, companies must evaluate whether the performance improvements provided by core-shell nanomaterials justify the higher production costs, particularly when cost-effective alternatives exist [31].

9.5.3 Regulatory Hurdles and Approval Processes for Nanomaterial-based Products

- **Lack of Standardized Regulations:** There is no universally accepted set of guidelines for the regulation of nanomaterials, leading to uncertainty in the approval process. Regulatory bodies like the FDA and EMA are still in the process of developing comprehensive guidelines for the approval of nanomaterial-based products [32].
- **Safety and Toxicity Assessments:** Nanomaterials, including core-shell structures, may require extensive testing for biocompatibility, toxicity, and environmental impact. These tests are time-consuming and costly, slowing down product approval.
- **Labeling and Documentation:** Regulatory requirements for labeling and documentation can be stringent, requiring manufacturers to provide detailed information about the composition, synthesis methods, and potential risks of nanomaterials in their products [33].

Table 9.1 illustrates the economic and regulatory analysis of core-shell nanomaterials in various sectors.

Table 9.1 Economic and regulatory analysis of core-shell nanomaterials in various sectors. The table compares manufacturing costs, synthesis techniques, regulatory approval processes, approval timelines, and the industries in which different core-shell nanomaterials are used.

Core-shell nanomaterial	Manufacturing cost	Manufacturing technique	Regulatory approval process	Approval timeline	Industries of use
Fe_3O_4@SiO_2 (magnetic nanoparticles)	High	Coprecipitation, sol–gel	Extensive toxicity and stability testing required	1–2 years	Medical diagnostics, drug delivery
Au@Polymer (gold-polymer)	Moderate	Chemical reduction, polymerization	Extensive safety and environmental impact assessment	1–2 years	Sensors, catalysis, medical devices
SiO_2@Ag (silver-silica)	Moderate	Sol–gel, chemical deposition	Requires toxicological evaluation and environmental analysis	1–2 years	Electronics, coatings, antibacterial agents
ZnO@TiO_2 (zinc oxide-titanium dioxide)	Low	Hydrothermal synthesis	Minimal testing required for industrial applications	6 months–1 year	Solar cells, photocatalysis, sensors
C@CdS (carbon-cadmium sulfide)	High	Microwave-assisted synthesis	Extensive testing for cadmium toxicity and environmental impact	2–3 years	Optoelectronics, sensors

9.6 Emerging Trends in Core-shell Research

Emerging trends in core-shell nanotechnology are reshaping numerous industries by providing solutions to complex challenges in fields ranging from energy storage and medicine to electronics and environmental sustainability. Core-shell nanomaterials have witnessed significant advancements, driven by innovations in synthesis techniques, functionalization, and novel applications that leverage their multifunctional properties [34]. Researchers continue to explore new ways of improving the performance, stability, and scalability of these materials while opening up new possibilities for high-impact applications.

One of the key emerging trends in core-shell nanotechnology is the development of novel materials and functionalities. Traditional core-shell structures typically consist of a core material surrounded by a shell that can be tuned for various properties, such as optical, electrical, or chemical characteristics [35]. However, new approaches are enabling the design of more complex architectures with multiple layers or hybrid compositions that enhance the overall performance of the nanomaterials. For instance, multicomponent core-shell systems that incorporate several elements within the core and shell are being developed to improve specific properties like photostability, electrical conductivity, and catalytic activity. These novel materials show promise in areas such as photovoltaics, energy storage, and catalysis [36]. Another significant research trend is the integration of core-shell materials with advanced functionalization techniques. By attaching bioactive molecules, targeting ligands, or responsive groups to the surface of core-shell nanomaterials, researchers can create tailored materials with enhanced biocompatibility, selectivity, and responsiveness [37]. This trend is particularly prominent in the fields of drug delivery and biomedicine, where core-shell structures are being used to design materials that can respond to environmental stimuli such as pH, temperature, or light. This has opened doors for more efficient and targeted drug delivery systems, which can offer controlled release and reduced side effects in cancer therapy and other treatments. Similarly, core-shell systems functionalized with bioactive molecules have shown potential in biosensing and diagnostics, where they can detect specific biomarkers at low concentrations, facilitating early-stage disease detection.

In addition to these innovations, new synthesis techniques are at the forefront of advancing core-shell nanotechnology. While traditional methods such as coprecipitation, sol–gel processes, and CVD have been widely used for producing core-shell nanomaterials, researchers are now exploring green chemistry approaches, microwave-assisted synthesis, and continuous flow processes to enhance scalability, reduce costs, and improve the uniformity and quality of the materials [38]. Microwave-assisted methods, for example, allow for faster and more efficient synthesis of nanomaterials with better control over size and structure. This opens the door to mass production, which is a crucial step toward the commercialization of core-shell-based technologies. Moreover, researchers are investigating additive manufacturing (3D printing) to precisely fabricate core-shell structures with complex geometries, enabling new applications in customized biomedical implants, electronic components, and sensors.

Emerging applications of core-shell nanomaterials are also broadening the scope of their impact. In energy applications, core-shell nanomaterials are being increasingly explored for use in batteries, supercapacitors, and solar cells. For example, core-shell nanostructures are being designed to enhance the performance of lithium-ion batteries by improving the capacity, cycle stability, and rate performance of the anode materials [39]. In solar cells, core-shell materials with tailored optical properties are used to optimize light absorption and charge transport, leading to more efficient energy conversion. Additionally, supercapacitors benefit from the increased surface area and enhanced electrochemical properties of core-shell systems, providing higher energy densities and faster charge–discharge cycles.

In the field of medicine, core-shell nanomaterials are poised to revolutionize several aspects of healthcare. The combination of drug delivery and imaging capabilities in a single nanostructure is one of the most promising areas of research. These theranostic nanomaterials allow for the simultaneous monitoring of treatment efficacy while delivering targeted therapies, especially in the treatment of cancer [40]. Moreover, core-shell systems that integrate multiple functionalities, such as pH-sensitive drug release or light-triggered therapy, are opening up new therapeutic options. The biocompatibility and biosafety of core-shell materials are continually being improved, making them more viable for clinical applications. Electronics is another sector where core-shell nanomaterials are making significant strides. For instance, quantum dots (QDs) and other core-shell structures are being used in displays and lighting technologies to enhance brightness, efficiency, and resolution [41]. Core-shell systems with tunable optical properties are also being investigated for use in advanced photodetectors and laser devices. As the demand for smaller, faster, and more energy-efficient devices grows, core-shell nanomaterials are expected to play an increasingly critical role in the next generation of nanoelectronics.

Finally, environmental sustainability is another major area of focus in core-shell nanomaterial research. These materials are being designed for applications in water purification, pollution control, and carbon capture, where their high surface area and reactivity can help remove toxic substances from the environment. Core-shell structures with photocatalytic properties, for example, can be used to break down organic pollutants or reduce harmful gases like carbon dioxide, providing an efficient solution to environmental challenges.

9.7 Integration with AI and ML

The integration of AI and ML into the design, synthesis, and optimization of core-shell nanomaterials is revolutionizing the field of nanotechnology. AI and ML techniques are providing powerful tools to overcome some of the inherent challenges in working with these materials, such as controlling their size, shape, and functionality. By leveraging the computational power and pattern recognition capabilities of AI/ML, researchers are able to predict and design core-shell nanostructures with a high degree of precision, improving both the efficiency and scalability of their development [42]. AI and ML play a pivotal role in the design and synthesis of core-shell nanomaterials by enabling predictive modeling and data-driven approaches. For example, ML algorithms can analyze large datasets of material properties and synthesis parameters to identify relationships that would be difficult to discern manually. These models can predict how changes in synthesis conditions (e.g., temperature, pH, precursor concentration) will affect the structure and performance of the nanomaterials. As a result, AI/ML-driven approaches can help optimize the synthesis processes, reducing trial-and-error experimentation and leading to faster and more reliable outcomes [43]. Moreover, AI can assist in automating the synthesis of core-shell nanomaterials, enhancing reproducibility and consistency in large-scale production.

AI/ML applications extend beyond synthesis to the prediction of material properties and performance outcomes. ML models can be trained to predict various characteristics of core-shell materials, such as their optical, electrical, mechanical, and chemical properties. For instance, by analyzing data from previous experiments, AI can predict how a core-shell structure will behave under certain conditions, such as its response to external stimuli like light, temperature, or pH [44]. These predictions are invaluable for accelerating the development of new materials with desired properties, especially when exploring novel material combinations or complex synthesis pathways. This can drastically reduce the time required to discover functional nanomaterials and bring them to practical applications.

Moreover, AI and ML are becoming increasingly essential for optimizing the performance of core-shell nanomaterials in real-world applications. For example, in drug delivery, AI models can analyze how different nanomaterial formulations will interact with biological systems, predicting factors such as cellular uptake, biodegradability, and targeted release [45]. Similarly, in energy storage applications, ML can optimize the electrochemical performance of core-shell systems, improving battery efficiency, stability, and cycle life. By continuously learning from experimental data, AI can fine-tune the properties and performance of core-shell nanomaterials for specific applications, making them more effective and reliable.

Looking to the future, AI-driven innovations in nanotechnology hold tremendous promise. AI and ML can enable the creation of adaptive nanomaterials that can autonomously respond to their environment, enhancing their functionality in fields like medicine, environmental monitoring, and materials science. For example, AI can help design self-healing materials, where the system autonomously detects and repairs damage to the core-shell structure. Additionally, AI can be employed in multifunctional nanomaterials that combine sensing, drug delivery, and imaging capabilities, making them highly effective in theranostics and personalized medicine [46]. As computational power continues to grow and more advanced algorithms are developed, the synergy between AI, ML, and core-shell nanomaterials is likely to lead to groundbreaking discoveries and innovations across multiple industries.

The integration of AI and ML with core-shell nanotechnology is opening up new possibilities for faster, more efficient material design and optimization, making it easier to develop nanomaterials with highly tailored properties. The combination of AI-driven design, predictive modeling, and performance optimization holds great potential for advancing a wide range of applications, including drug delivery, energy storage, and environmental monitoring, while also paving the way for next-generation technologies [47]. The future of AI in nanotechnology looks exceptionally promising, offering unprecedented opportunities to enhance the functionality and impact of core-shell nanomaterials.

9.8 Interdisciplinary Approaches for Advanced Applications

Interdisciplinary approaches have become a cornerstone of advanced applications in core-shell nanotechnology, where the convergence of expertise from multiple scientific and engineering disciplines enhances the functionalities of nanomaterials. The collaboration between materials scientists, engineers, chemists, biologists, and other fields allows for a holistic approach to the design, synthesis, and application of core-shell nanomaterials, addressing complex challenges in areas such as healthcare, energy, and environmental protection [48]. These collaborations are driving innovation, enabling the development of novel materials with tailored properties and performance characteristics that would not be possible within a single discipline.

The impact of interdisciplinary approaches on enhancing core-shell nanomaterial functionalities is profound. Materials scientists provide deep insights into the structural and mechanical properties of nanomaterials, while engineers focus on the scalability, manufacturing, and integration of these materials into devices. Chemists play a pivotal role in designing novel synthesis methods and functionalization strategies, while biologists contribute by understanding how these materials interact with living systems, facilitating applications in drug delivery, bioimaging, and diagnostics [49]. The synergy between these fields fosters the creation of advanced core-shell systems that meet specific application requirements, such as controlled drug release, targeted therapy, and high-performance sensors. One of the key benefits of interdisciplinary collaboration is the ability to address complex problems from multiple perspectives. For example, in the development of theranostic nanomaterials for cancer treatment, chemists and biologists work together to design nanostructures that can carry both therapeutic agents and imaging probes [50]. Engineers help optimize the manufacturing process for large-scale production, while materials scientists focus on ensuring that the nanomaterials maintain stability and performance under real-world conditions. This teamwork results in more effective, multifunctional nanomaterials that can deliver treatments and monitor their progress in real time.

Several case studies highlight the success of interdisciplinary collaborations in core-shell nanotechnology. For instance, a collaboration between materials scientists and biologists led to the development of core-shell nanoparticles for targeted drug delivery, where the core material carried the drug and the shell was functionalized with targeting ligands to improve the delivery to specific cells [51]. Another example is the work done in the field of bioimaging, where chemists and biologists developed QD-based core-shell systems that allow for high-resolution imaging of tumors, enabling better diagnosis and monitoring of cancer treatments [52]. These case studies exemplify the power of collaboration between diverse scientific disciplines in pushing the boundaries of nanotechnology. Table 9.2 illustrates a few

Table 9.2 Overview of successful interdisciplinary projects in core-shell nanomaterial research.

Project title	Collaboration fields	Focus area	Achievements	Application areas
Theranostic nanoparticles for cancer	Materials Science, Biology, Chemistry	Targeted drug delivery and bioimaging	Developed core-shell nanoparticles that deliver drugs while enabling tumor imaging	Cancer therapy, diagnostic imaging
QD-based bioimaging	Chemistry, Biology, Engineering	High-resolution imaging	Created QD-based core-shell nanomaterials for improved tumor imaging	Cancer detection, molecular imaging
Self-healing coatings for aerospace	Materials Science, Engineering	Smart coatings	Developed self-healing, anti-corrosion coatings for aerospace applications	Aerospace, automotive coatings
Core-shell nanomaterials for energy storage	Materials Science, Chemistry, Engineering	Energy storage and batteries	Designed core-shell structures for high-capacity, long-lifetime batteries	Energy storage, electric vehicles
Biocompatible core-shell nanoparticles for drug delivery	Chemistry, Biology, Engineering	Drug delivery systems	Engineered core-shell nanoparticles with improved biocompatibility and drug release control	Drug delivery, medical devices

notable interdisciplinary research projects in core-shell nanotechnology, highlighting the collaboration across multiple disciplines to tackle specific challenges and create materials with diverse applications, ranging from **drug delivery** and **bioimaging** to **energy storage** and **smart coatings**. These examples underline the importance of pooling expertise from various fields to advance core-shell nanomaterial research and translate it into real-world innovations.

9.9 Potential Breakthroughs in Energy, Environment, and Medicine

Core-shell nanomaterials are poised to play a transformative role in addressing some of the most pressing global challenges in energy, the environment, and medicine. Their unique structural characteristics, including enhanced surface area, tunable properties, and the ability to encapsulate and deliver diverse functional elements, make them ideal candidates for a wide range of applications across these critical sectors.

9.9.1 Energy Applications: Next-generation Energy Storage, Conversion, and Harvesting

In the energy sector, core-shell nanomaterials hold significant promise for revolutionizing energy storage, conversion, and harvesting technologies. The core-shell structure allows for the encapsulation of active materials within protective shells, enhancing their stability, cycle life, and performance. One of the most exciting applications is in energy storage, where core-shell nanomaterials are being used in next-generation batteries and supercapacitors. For instance, core-shell nanostructures in lithium-ion batteries can increase the capacity and improve the lifespan by preventing the degradation of the active materials, allowing for more efficient storage and longer-lasting energy solutions [53]. In supercapacitors, core-shell materials help to improve charge/discharge rates and enhance energy density. Moreover, core-shell materials can be integrated into solar cells to enhance energy conversion efficiency. The tunable optical properties of these materials, including the ability to tailor light absorption and electron transport, make them highly effective in photovoltaic devices. Core-shell systems are also being explored for energy harvesting, where they can be used to capture and convert ambient energy sources, such as heat or light, into usable electrical energy.

9.9.2 Environmental Applications: Waste Treatment, Water Purification, and Pollution Control

Core-shell nanomaterials offer promising solutions for various environmental applications, particularly in areas like waste treatment, water purification, and pollution control. The high surface area and functionality of the materials enable them to interact effectively with pollutants, contaminants, and waste products. In waste treatment, core-shell nanostructures can be engineered to capture and neutralize toxic compounds or to break down hazardous chemicals into less harmful substances. This could be especially useful in treating industrial waste and hazardous materials that are difficult to degrade by traditional methods. For water purification, core-shell nanomaterials can be used to remove heavy metals, pathogens, and organic contaminants, ensuring cleaner and safer water supplies [54]. Their ability to adsorb or degrade pollutants efficiently can help meet global water purification needs, especially in regions with limited access to clean water. Core-shell nanomaterials are also valuable in pollution control, where they can serve as highly effective catalysts in removing greenhouse gases, particulate matter, and other air pollutants. They can be incorporated into filters or catalysis systems to improve air quality, especially in urban environments or areas heavily impacted by industrial emissions.

9.9.3 Medical Applications: Targeted Drug Delivery, Theranostics, and Tissue Engineering

In medicine, core-shell nanomaterials hold the potential to revolutionize drug delivery, theranostics, and tissue engineering. The ability to encapsulate therapeutic agents within the core and functionalize the shell with targeting ligands enables precise targeted drug delivery. This approach minimizes side effects, increases therapeutic efficacy, and allows for controlled release of drugs over time, especially in the treatment of cancer, autoimmune diseases, and chronic conditions. Theranostics, the combination of therapy and diagnostics in a single nanomaterial platform, is another area where core-shell nanomaterials are making strides [55]. For example, nanoparticles designed with core-shell structures can simultaneously deliver a therapeutic agent while providing real-time imaging for monitoring the treatment's progress. These multifunctional materials offer great promise in personalized medicine, where they can be tailored to the specific needs of the patient. In tissue engineering, core-shell materials can be used to create scaffolds that mimic the natural extracellular matrix, providing structural support for cell growth and tissue regeneration. The core-shell design can also allow for the gradual release of growth factors or other bioactive molecules that promote tissue healing and regeneration, making them suitable for applications in regenerative medicine.

9.9.4 Future Opportunities for Core-shell Nanomaterials

The potential breakthroughs of core-shell nanomaterials extend far beyond their current applications. With continued research and development, these materials are expected to play an integral role in addressing global challenges, including energy sustainability, environmental conservation, and medical advancements. The future of core-shell nanomaterials lies in the integration of novel materials, advanced functionalization techniques, and multifunctional properties that can address complex, real-world problems in a wide variety of fields [56]. As global challenges such as climate change, resource depletion, and the aging population continue to escalate, core-shell nanomaterials have the potential to provide solutions that are more efficient, sustainable, and cost-effective. In the energy sector, they can contribute to cleaner and more efficient energy systems. In the environment, they offer innovative methods for tackling pollution and conserving water resources. And in medicine, they provide new opportunities for precision therapies and improving healthcare outcomes.

Figure 9.4 illustrates the wide range of potential applications for core-shell nanomaterials across energy, environmental, and medical sectors. It highlights the material's diverse capabilities, from enhancing energy efficiency and storage to providing solutions for environmental remediation and advancing medical therapies. The diagram underscores the significant impact that core-shell nanotechnology could have in addressing some of the most critical global challenges in the coming years.

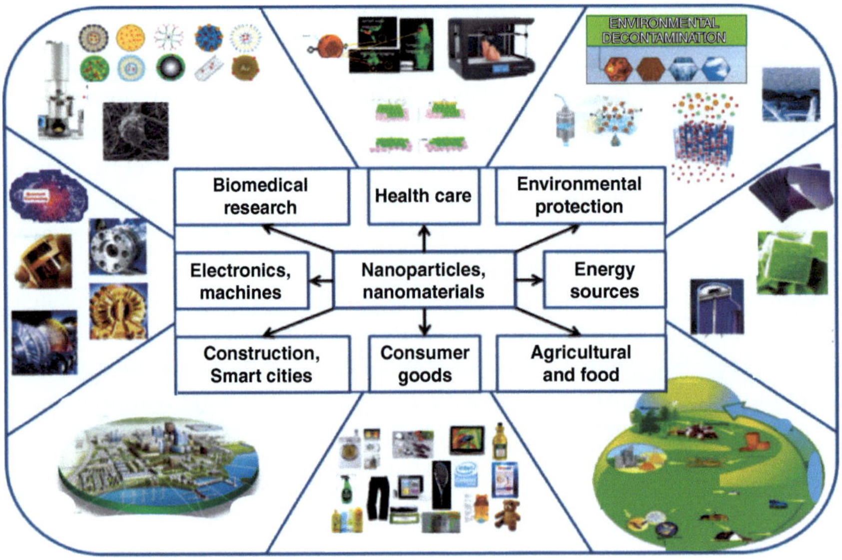

Figure 9.4 Potential breakthroughs and future applications of core-shell nanomaterials in global challenges. Reproduced with permission from [57]/Sciendo.

9.10 Conclusion

Core-shell nanomaterials represent a promising frontier in nanotechnology, with the potential to address some of the most pressing challenges in energy, the environment, and medicine. Their unique structural characteristics, such as enhanced surface area, tunable properties, and the ability to integrate multiple functionalities into a single system, make them ideal candidates for a wide range of applications. As explored in the chapter, core-shell nanomaterials are already making significant strides in fields like energy storage, water purification, and targeted drug delivery, and their impact is expected to grow as research and development in this area continue.

The future of core-shell nanomaterials is bright, with ongoing advancements in synthesis techniques, scalability, and functionalization opening new possibilities for even more innovative applications. By overcoming challenges related to uniformity, toxicity, and economic feasibility, these materials can be further integrated into industrial processes and consumer products, offering solutions that are not only technologically advanced but also sustainable and economically viable. Additionally, the integration of AI, ML, and interdisciplinary collaborations will accelerate the development of next-generation core-shell systems that can meet the evolving needs of industries and societies worldwide.

As we look to the future, the potential breakthroughs in energy, environmental conservation, and medical advancements offered by core-shell nanomaterials hold the promise of creating a more sustainable and healthier world. The continued exploration of their multifunctional properties, coupled with innovations in synthesis and application techniques, will be crucial in unlocking their full potential and translating these scientific advancements into practical solutions for global challenges.

References

1 Mir SH, Nagahara LA, Thundat T, Mokarian-Tabari P, Furukawa H, Khosla A. Review—organic-inorganic hybrid functional materials: An integrated platform for applied technologies. J Electrochem Soc. 2018;165(8): B3137–56.
2 Li W, Liu J, Zhao D. Mesoporous materials for energy conversion and storage devices. Nat Rev Mater [Internet]. 2016 May 4 [cited 2025 Feb 1];1(6): 1–17. Available from: https://www.nature.com/articles/natrevmats201623
3 Ma X, Li Y, Hussain I, Shen R, Yang G, Zhang K. Core–shell structured nanoenergetic materials: Preparation and fundamental properties. Adv Mater [Internet]. 2020 Jul 1 [cited 2025 Feb 1];32(30): 2001291. Available from: https://onlinelibrary.wiley.com/doi/full/10.1002/adma.202001291
4 Sucheta, Britya NV, Jha M, Yadav R, Chahal K, Nagori K, et al. Clinical and regulatory consideration for nanobiomaterials. In: Biomaterial-Inspired Nanomedicines for Targeted Therapies [Internet]. Springer. 2024 [cited 2025 Feb 1]; 451–76. Available from: https://link.springer.com/chapter/10.1007/978-981-97-3925-7_17
5 Hassan SADH, Almaliki MNS, Hussein ZA, Albehadili HM, Banoon SR, Al-Abboodi A, et al. Development of nanotechnology by artificial intelligence: A comprehensive review. J Nanostruct [Internet]. 2023 Oct 1 [cited 2025 Feb 1];13(4): 915–32. Available from: https://jns.kashanu.ac.ir/article_114406.html
6 Wang J. Barriers of scaling-up fuel cells: Cost, durability and reliability. Energy. 2015 Feb 1;80: 509–21.
7 Pal K, Asthana N, Aljabali AA, Bhardwaj SK, Kralj S, Penkova A, et al. A critical review on multifunctional smart materials 'nanographene' emerging avenue: Nano-imaging and biosensor applications. Crit Rev Solid State Mater [Internet]. 2022 [cited 2025 Feb 1];47(5): 691–707. Available from: https://www.tandfonline.com/doi/abs/10.1080/10408436.2021.1935717
8 Roy S, Kumar R, Acooli A, Roy S, Chatterjee A, Chattaraj S, et al. Transforming nanomaterial synthesis through advanced microfluidic approaches: A review on accessing unrestricted possibilities. J Compos Sci [Internet]. 2024 Sep 25 [cited 2025 Feb 1];8(10): 386. Available from: https://www.mdpi.com/2504-477X/8/10/386/htm
9 Feng H-p, Tang L, Zeng G-m, Zhou Y, Deng Y-c, Ren X, et al. Core-shell nanomaterials: Applications in energy storage and conversion. Adv Colloid Interface Sci. 2019 May 1;267: 26–46.
10 Pavlenko V, Khosravi HS, Żółtowska S, Haruna AB, Zahid M, Mansurov Z, et al. A comprehensive review of template-assisted porous carbons: Modern preparation methods and advanced applications. Mater Sci Eng R Rep. 2022 Jun 1;149: 100682.
11 Hu Y, Lu J, Feng H. Surface modification and functionalization of powder materials by atomic layer deposition: A review. RSC Adv. 2021 [cited 2025 Feb 1];11: 11918–42. Available from: https://pubs.rsc.org/en/content/articlehtml/2021/ra/d1ra00326g
12 Salvatore KL, Wong SS. Exploring strategies toward synthetic precision control within core-shell nanowires. Acc Chem Res [Internet]. 2021 Jun 1 [cited 2025 Feb 1];54(11): 2565–78. Available from: https://pubs.acs.org/doi/full/10.1021/acs.accounts.1c00041

13 Gurgel D, Vieira YA, Henriques RO, Machado R, Oechsler BF, Junior AF, et al. A comprehensive review on core-shell polymeric particles for enzyme immobilization. ChemistrySelect [Internet]. 2022 Dec 13 [cited 2025 Feb 1];7(46): e202202285. Available from: https://onlinelibrary.wiley.com/doi/full/10.1002/slct.202202285

14 Clomburg JM, Crumbley AM, Gonzalez R. Industrial biomanufacturing: The future of chemical production. Science [Internet]. 2017 Jan 6 [cited 2025 Feb 1]; 355(6320). Available from: https://www.science.org/doi/10.1126/science.aag0804

15 Zhu LT, Chen XZ, Ouyang B, Yan WC, Lei H, Chen Z, et al. Review of machine learning for hydrodynamics, transport, and reactions in multiphase flows and reactors. Ind Eng Chem Res [Internet]. 2022 Jul 20 [cited 2025 Feb 1];61(28): 9901–49. Available from: https://pubs.acs.org/doi/full/10.1021/acs.iecr.2c01036

16 Cambié D, Bottecchia C, Straathof NJW, Hessel V, Noël T. Applications of continuous-flow photochemistry in organic synthesis, material science, and water treatment. Chem Rev [Internet]. 2016 Sep 14 [cited 2025 Feb 1];116(17): 10276–341. Available from: https://pubs.acs.org/doi/full/10.1021/acs.chemrev.5b00707

17 Viswanadham N, Narahari Y. Performance modeling of automated systems. Google Books [Internet]. PHI Learning Pvt. Ltd. 2015.

18 Dhiman S, Yadav A, Debnath N, Das S. Application of core/shell nanoparticles in smart farming: A paradigm shift for making the agriculture sector more sustainable. J Agric Food Chem [Internet]. 2021 Mar 24 [cited 2025 Feb 1];69(11): 3267–83. Available from: https://pubs.acs.org/doi/full/10.1021/acs.jafc.0c05403

19 Oliveira MLS, Izquierdo M, Querol X, Lieberman RN, Saikia BK, Silva LFO. Nanoparticles from construction wastes: A problem to health and the environment. J Clean Prod. 2019 May 10;219: 236–43.

20 Ahmed GH, Yin J, Bakr OM, Mohammed OF. Successes and challenges of core/shell lead halide perovskite nanocrystals. ACS Energy Lett [Internet]. 2021 Apr 9 [cited 2025 Feb 1];6(4): 1340–57. Available from: https://pubs.acs.org/doi/full/10.1021/acsenergylett.1c00076

21 Lead JR, Batley GE, Alvarez PJJ, Croteau MN, Handy RD, McLaughlin MJ, et al. Nanomaterials in the environment: Behavior, fate, bioavailability, and effects—An updated review. Environ Toxicol Chem [Internet]. 2018 Aug 1 [cited 2025 Feb 1];37(8): 2029–63. Available from: https://onlinelibrary.wiley.com/doi/full/10.1002/etc.4147

22 More S, Bampidis V, Benford D, Bragard C, Halldorsson T, Hernández-Jerez A, et al. Guidance on risk assessment of nanomaterials to be applied in the food and feed chain: Human and animal health. EFSA J [Internet]. 2021 Aug 1 [cited 2025 Feb 1];19(8): e06768. Available from: https://onlinelibrary.wiley.com/doi/full/10.2903/j.efsa.2021.6768

23 Chan Y, Ng SW, Soon L. Polymeric and inorganic nanoparticles targeting chronic respiratory diseases. In: Advanced Drug Delivery Strategies for Targeting Chronic Inflammatory Lung Diseases [Internet]. Springer. 2022 Jan 1 [cited 2025 Feb 1]; 375–433. Available from: https://link.springer.com/chapter/10.1007/978-981-16-4392-7_18

24 Habumugisha T, Zhang Z, Uwizewe C, Yan C, Ndayishimiye JC, Rehman A, et al. Toxicological review of micro- and nano-plastics in aquatic environments: Risks to ecosystems, food web dynamics and human health. Ecotoxicol Environ Saf. 2024 Jun 15;278: 116426.

25 Turan NB, Engin GO, Bilgili MS. Environmental nanotechnology: implications and applications. Google Books [Internet]. Elsevier. 2022.

26 Lis Y, Roberts MH, Kamble S, Guo JJ, Raisch DW. Comparisons of food and drug administration and European medicines agency risk management implementation for recent pharmaceutical approvals: Report of the international society for pharmacoeconomics and outcomes research risk benefit management working group. Value Health. 2012 Dec 1;15(8): 1108–18.

27 Amoabediny H, Naderi A, Malakootikhah J, Koohi M, Mortazavi A, Naderi M, et al. Guidelines for safe handling, use and disposal of nanoparticles. J Phys Conf Ser [Internet]. 2009 May 1 [cited 2025 Feb 1];170(1): 012037. Available from: https://iopscience.iop.org/article/10.1088/1742-6596/170/1/012037

28 Vambol S, Vambol V, Suchikova Y, Deyneko N. Analysis of the ways to provide ecological safety for the products of nanotechnologies throughout their life cycle. East Eur J Enterp Technol. 2017 [cited 2025 Feb 1];1(10): 27–36. Available from: http://www.irbis-nbuv.gov.ua/cgi-bin/irbis_nbuv/cgiirbis_64.exe?C21COM=2&I21DBN=UJRN&P21DBN=UJRN&IMAGE_FILE_DOWNLOAD=1&Image_file_name=PDF/Vejpte_2017_1(10)__5.pdf

29 Yoshizawa G, Suzuki T, Takemura M. Approach to environmental, health and safety issues of nanotechnology in Japan. J Disaster Res [Internet]. 2011 Oct;6(5): 506–13. Available from: https://www.researchgate.net/publication/291871945

30 Bramsiepe C, Sievers S, Seifert T, Stefanidis GD, Vlachos DG, Schnitzer H, et al. Low-cost small scale processing technologies for production applications in various environments—Mass produced factories. Chem Eng Process: Process Intensif. 2012 Jan 1;51: 32–52.

31 Chen L, Zhang Y, Chen Z, Dong Y, Jiang Y, Hua J, et al. Biomaterials technology and policies in the building sector: A review. Environ Chem Lett. [Internet]. 2024 Jan 29 [cited 2025 Feb 1];22(2): 715–50. Available from: https://link.springer.com/article/10.1007/s10311-023-01689-w

32 Ali F, Neha K, Parveen S. Current regulatory landscape of nanomaterials and nanomedicines: A global perspective. J Drug Deliv Sci Technol. 2023 Feb 1;80: 104118.

33 Aulic S, Laurini E, Marson D, Skoko N, Fermeglia M, Pricl S. Regulatory, safety, and toxicological concerns of nanomaterials with their manufacturing issues. In: Nano-Pharmacokinetics and Theranostics: Advancing Cancer Therapy. Elsevier. 2021 Jan 1; 93–115.

34 Jones CF, Resina L, Ferreira FC, Sanjuan-Alberte P, Esteves T. Conductive core-shell nanoparticles: Synthesis and applications. J Phys Chem C [Internet]. 2024 Jul 11 [cited 2025 Feb 1];128(27): 11083–100. Available from: https://pubs.acs.org/doi/full/10.1021/acs.jpcc.4c02012

35 Ghosh Chaudhuri R, Paria S. Core/shell nanoparticles: Classes, properties, synthesis mechanisms, characterization, and applications. Chem Rev [Internet]. 2012 Apr 11 [cited 2025 Feb 1];112(4): 2373–433. Available from: https://pubs.acs.org/doi/full/10.1021/cr100449n

36 Lu H, Tournet J, Dastafkan K, Liu Y, Ng YH, Karuturi SK, et al. Noble-metal-free multicomponent nanointegration for sustainable energy conversion. Chem Rev [Internet]. 2021 Sep 8 [cited 2025 Feb 1];121(17): 10271–366. Available from: https://pubs.acs.org/doi/full/10.1021/acs.chemrev.0c01328

37 Zou Y, Sun Z, Wang Q, Ju Y, Sun N, Yue Q, et al. Core-shell magnetic particles: Tailored synthesis and applications. Chem Rev [Internet]. 2025 Jan 22 [cited 2025 Feb 1];125(2): 972–1048. Available from: https://pubs.acs.org/doi/full/10.1021/acs.chemrev.4c00710

38 Mandal K, Ghose S, Mandal M, Majumder D, Talukdar S, Chakraborty I, et al. Notes on useful materials and synthesis through various chemical solution techniques. In: Chemical Solution Synthesis for Materials Design and Thin Film Device Applications. Elsevier. 2021 Jan 1; 29–78.

39 Shen L, Li H, Uchaker E, Zhang X, Cao G. General strategy for designing core-shell nanostructured materials for high-power lithium ion batteries. Nano Lett [Internet]. 2012 Nov 14 [cited 2025 Feb 1];12(11): 5673–8. Available from: https://pubs.acs.org/doi/full/10.1021/nl302854j

40 Lim EK, Kim T, Paik S, Haam S, Huh YM, Lee K. Nanomaterials for theranostics: Recent advances and future challenges. Chem Rev [Internet]. 2015 Jan 14 [cited 2025 Feb 1];115(1): 327–94. Available from: https://pubs.acs.org/doi/full/10.1021/cr300213b

41 Fan J, Han C, Yang G, Song B, Xu R, Xiang C, et al. Recent progress of quantum dots light-emitting diodes: Materials, device structures, and display applications. Adv Mater [Internet]. 2024 Sep 1 [cited 2025 Feb 1];36(37): 2312948. Available from: https://onlinelibrary.wiley.com/doi/full/10.1002/adma.202312948

42 Ebrahimi F, Kumari A, Dellinger K. Integration of nanoengineering with artificial intelligence and machine learning in Surface-Enhanced Raman Spectroscopy (SERS) for the development of advanced biosensing platforms. Adv Sensor Res [Internet]. 2024 Dec 20 [cited 2025 Feb 1];4(2): 2400155. Available from: https://onlinelibrary.wiley.com/doi/full/10.1002/adsr.202400155

43 Tang Y, Zhang Y, Zhang D, Liu Y, Nussinov R, Zheng J. Exploring pathological link between antimicrobial and amyloid peptides. Chem Soc Rev [Internet]. 2024 Aug 27 [cited 2025 Feb 1];53(17): 8713–63. Available from: https://pubs.rsc.org/en/content/articlehtml/2024/cs/d3cs00878a

44 Alshawwa SZ, Kassem AA, Farid RM, Mostafa SK, Labib GS. Nanocarrier drug delivery systems: Characterization, limitations, future perspectives and implementation of artificial intelligence. Pharmaceutics [Internet]. 2022 Apr 18 [cited 2025 Feb 1];14(4): 883. Available from: https://www.mdpi.com/1999-4923/14/4/883/htm

45 Massaro A. Intelligent materials and nanomaterials improving physical properties and control oriented on electronic implementations. Electronics [Internet]. 2023 Sep 6 [cited 2025 Feb 1];12(18): 3772. Available from: https://www.mdpi.com/2079-9292/12/18/3772/htm

46 Voola PK, Gangu K, Pandian PKG, Goel P Dr, Jain A Prof Dr. Ai-driven predictive models in healthcare: Reducing time-to-market for clinical applications. SSRN Elec J [Internet]. 2021 Nov 20 [cited 2025 Feb 1]; Available from: https://papers.ssrn.com/abstract=4984971

47 Ramesh M, Janani R, Deepa C, Rajeshkumar L. Nanotechnology-enabled biosensors: A review of fundamentals, design principles, materials, and applications. Biosensors [Internet]. 2022 Dec 27 [cited 2025 Feb 1];13(1): 40. Available from: https://www.mdpi.com/2079-6374/13/1/40/htm

48 Jha RK, Jha PK, Chaudhury K, Rana SVS, Guha SK. An emerging interface between life science and nanotechnology: Present status and prospects of reproductive healthcare aided by nano-biotechnology. Nano Rev [Internet]. 2014 Jan [cited 2025 Feb 1];5(1): 22762. Available from: https://www.tandfonline.com/doi/abs/10.3402/nano.v5.22762

49 Yao J, Yang M, Duan Y. Chemistry, biology, and medicine of fluorescent nanomaterials and related systems: New insights into biosensing, bioimaging, genomics, diagnostics, and therapy. Chem Rev [Internet]. 2014 Jun 25 [cited 2025 Feb 1];114(12): 6130–78. Available from: https://pubs.acs.org/doi/full/10.1021/cr200359p

50 Bardhan R, Lal S, Joshi A, Halas NJ. Theranostic nanoshells: From probe design to imaging and treatment of cancer. Acc Chem Res [Internet]. 2011 Oct 18 [cited 2025 Feb 1];44(10): 936–46. Available from: https://pubs.acs.org/doi/full/10.1021/ar200023x

51 Anani T, Rahmati S, Sultana N, David AE. MRI-traceable theranostic nanoparticles for targeted cancer treatment. Theranostics [Internet]. 2021 [cited 2025 Feb 1];11(2): 579. Available from: https://pmc.ncbi.nlm.nih.gov/articles/PMC7738852/

52 Mondal J, Lamba R, Yukta Y, Yadav R, Kumar R, Pani B, et al. Advancements in semiconductor quantum dots: Expanding frontiers in optoelectronics, analytical sensing, biomedicine, and catalysis. J Mater Chem C Mater [Internet]. 2024 Jul 18 [cited 2025 Feb 1];12(28): 10330–89. Available from: https://pubs.rsc.org/en/content/articlehtml/2024/tc/d4tc01396d

53 Jiang L, Yuan X-Z, Liang J, Zhang J, Wang H, Zeng G-m. Nanostructured core-shell electrode materials for electrochemical capacitors. J Power Sources. 2016 Nov 1;331: 408–25.

54 Saravanan A, Senthil Kumar P, Jeevanantham S, Karishma S, Tajsabreen B, Yaashikaa PR, et al. Effective water/wastewater treatment methodologies for toxic pollutants removal: Processes and applications towards sustainable development. Chemosphere. 2021 Oct 1;280: 130595.

55 Mukherjee A, Paul M, Mukherjee S. Recent progress in the theranostics application of nanomedicine in lung cancer. Cancers [Internet]. 2019 Apr 29 [cited 2025 Feb 1];11(5): 597. Available from: https://www.mdpi.com/2072-6694/11/5/597/htm

56 Abdel-Karim R. Nanotechnology-enabled biosensors: A review of fundamentals, materials, applications, challenges, and future scope. Biomed Mater Devices [Internet]. 2024 Jan 17 [cited 2025 Feb 1];2(2): 759–77. Available from: https://link.springer.com/article/10.1007/s44174-023-00147-z

57 Singer A, Markoutsa E, Limayem A, Mohapatra S, Mohapatra SS. Nanobiotechnology medical applications: Overcoming challenges through innovation. Eurobiotech J. 2018 Jul 1;2(3): 146–60.

Index

Page numbers in *italics* refer to figures, and page numbers in **bold** refer to tables

n